ENVIRONMENT AND CLIMATE CHANGE

ENVIRONMENT AND CLIMATE CHANGE

Editor

Havlíček Červenka

scitus academics

Environment and Climate Change

Edited by **Havlíček Červenka**

This Book is Printed on Acid-Free Paper. ♾

ISBN: 9781681179117

Library of Congress Control Number: 2017948939

SCITUS Academics LLC,

616, Corporate Way, Suite 2, 4766,

Valley Cottage, NY 10989

www.scitusacademics.com

Printed in United States of America.

Preface

Climate change and variability are concerns of human being. Climate Change is increasingly one of the most severe national security threats which will have significant impacts on natural resources, ecosystem and biodiversity. The triple whammy problems of rapid increase in population, climate change and pressure on resources are major constraints to adequate agricultural production. Among the three, climate change is the most pressing challenges that the world faces today. At the same time, it is likely to trigger food insecurity, human migration, economic, and social depression, environmental and political crisis, thereby affecting national development. It is important to note that these global averages may hide regional variations. Projected reductions in agricultural yields due to climate change by 2050 are larger for some crops than those estimated for the past half century, but smaller than projected increases to 2050 due to rising demand and intrinsic productivity growth. Productive agriculture is essential to feed a growing population and sustain modern civilization. Climate affects agriculture, a fact well known to every farmer. Year-to-year variations in harvest are largely due to variations in temperature and precipitation that can make the difference between bountiful "bumper" crops and economic ruin.

This volume brings together state-of-the-art applied studies related to agriculture, development, and the environment. This book reviews the state of agricultural climate change mitigation globally, with a focus on identifying the feasibility, opportunities and challenges for achieving mitigation among farmers. It analyzes the global consequences to crop yields, production, and risk of hunger linking climate and socioeconomic scenarios. This volume also explores how large-scale programs for smallholder adaptation to climate change might link effectively with community-led adaptation initiatives. It identifies key success factors and barriers for considering local priorities, capacities and lessons in large-scale adaptation programs. It highlights the key roles of extension services and farmers' organizations as mechanisms for linking between national-level and community-level adaptation, and a range of other success factors which include participative and locally driven vulnerability assessments, tailoring of adaptation technologies to local contexts, mapping local institutions and working in partnership across institutions.

Table of Contents

CHAPTER 1

Agriculture, Climate Change and Food Security

Nehemia Kiprutto, Laban K. Rotich, Geoffrey K. Riungu

School of Tourism, Hospitality and Events Management, Moi University, Eldoret, Kenya

ABSTRACT

Agriculture is one of the main economic activities in the world, with its significance magnified especially among the world's poor. This paper seeks to explore how climate change would impact on the agricultural sector and consequently food security. Agricultural activities in most developing countries entirely depend on rainfall patterns, on the contrary, climate change has emerging as an environmental challenge with adverse impacts expected on food security. This is mainly through incidences of changes in water availability, floods and drought. For example, changes in temperature and precipitation including drought affects crop and livestock yield, hydrologic balances, input supplies and other components of agricultural systems. Equally important, environmental degradation is one of the drivers of climate change that further undermines sustainability of agriculture. Climate change is real and already taking place. According to the inter-governmental panel on climate change (IPCC), warming temperatures are projected to cause more frequent and more intense extreme weather events such as heavy rain storms, flooding and tropical storms and in some cases drought in many parts of the country. Agriculture is one of the most climate sensitive sectors, with projections that 800 million people are currently food insecure. Additionally, it has the potential to undermine advances in poverty reduction and sustainable development. Thus, fundamental changes in agricultural systems are needed because climate change poses new and serious challenges for farmers hence food insecurity. Most important, agricultural adaptation and mitigation pathways need to be developed so as to avoid further threat on food security.

Keywords: Agriculture, Climate Change, Extreme Events, Food Security

1. INTRODUCTION

Agriculture constitutes the backbone of most African economies. It is the largest contributor to GDP; the biggest source of foreign exchange, accounting for about 40% of the continent's foreign currency earnings; and the main generator of savings and tax revenues. In addition, about two-thirds of manufacturing value-added is based on agricultural raw materials. Agriculture remains crucial for pro-poor economic growth in most African countries, as rural areas support 70% - 80% of the total population [1] . On the contrary, climate change is considered as posing the greatest threat to agriculture and food security in the 21st century, especially in many poor, agriculture-based countries of sub-Saharan Africa (SSA) with their low capacity to effectively cope [2] [3] . Most important, African agriculture is already under stress as a result of population increase, industrialization and urbanization, competition over resource use, degradation of resources, and insufficient public spending for rural infrastructure and services, yet the impact of climate change is likely to exacerbate these stresses even further [1] . Therefore, the global agricultural industry will have to meet the demands of food from rising populations in the face of degraded soil and water stress as a result from climate change. Special attention goes to the fact that, agricultural productivity needs to be increased in order to meet the demand of the growing population in the present and coming decades, this means more water is already a challenge because of its scarcity and competition for it.

Also, the United Nations Development Programme [4] warns that the progress in human development achieved over the last decade may be slowed down or even reversed by climate change, as new threats emerge to water and food security, agricultural production and access, and nutrition and public health [1] .

Equally important, the impacts of climate change include; sea level rise, droughts, heat waves, floods and rainfall variation which could by 2080, push another 600 million people into malnutrition and increase the number of people facing water scarcity by 1.8 billion [4] . A number of countries in Sub-Saharan Africa (SSA) are already experiencing considerable water stress as a result of insufficient and unreliable rainfall, changing rainfall patterns or flooding. Similarly, the impacts of climate change including predicted increases in extremes are likely to add to this stress, leading to additional pressure on water availability, accessibility, supply and demand [1] (Table 1).

2. MATERIALS AND METHODS

This paper builds on recent reports of climate change effects on agriculture and food security [1] [5] . Majority of the information relating directly and indirectly to agriculture, climate change and food security were sourced from published articles using the following keywords; agriculture, climate change and food security. In addition, the various reports were carefully reviewed and were found authentic. Also, majority of the findings from the published articles expressed deep concerns about climate change impacts on agriculture, posing serious threats to food security especially in developing countries. Moreover, future predictions indicated food insecurity in developing countries including Kenya coupled with extreme weather events such as droughts, floods and rising temperatures. However, the articles strongly underscored mitigation pathways and adaptation strategies as solutions that would play a major and critical role in reducing climate change impacts on agriculture including food security. All the articles were published between the years (1996-2009).

The possible causes of food insecurity coupled with climate change effects on agriculture were summarized in a tabular form. Problem analysis was performed using a fishbone chart approach to identify climate change impacts on agricultural production and food security. The fishbone chart approach explained through the fishbone diagram (Figure 1), also known as cause and effect diagram refers to an orderly chart that categorizes the many potential causes in a problem in an orderly way. It was invented by Kaoru Ishikawa in 1943, further; spacecraft engineers have been using these diagrams for many years to diagnose mechanical problems [6] . The table provides a comprehensive view of future food security situations comprised with agricultural yield output in the face of climate change negative impacts. This method strives to highlight climate change influence on agricultural production as well as food security, identified from the reviewed reports, and aided in the development of a framework that provides the basis for future detailed investigations and experiments.

3. AGRICULTURE AND CLIMATE CHANGE

Although the different climate change models are not clear with respect to rainfall and periods of drought, temperature projections are generally more reliable. Increased evaporation and evapotranspiration with associated soil moisture deficits will impact rain fed agriculture [7] . Similarly, fertilizer application (agricultural soils), livestock and manure management, rice cultivation, savanna clearing and burning of agricultural residues are ma jor sources of emissions from the agricultural sector. Firstly, nitrous oxide (N_2O) that is apparently the largest source of GHG emission from agriculture naturally produced in the soil through nitrification and denitrification. Also, nitrogen fertilizer application, livestock manure and sewage sludge are some of the agricultural activities that result in emission of nitrous oxide (N_2O) that in turn contribute to climate change. In addition, rice production contributes to emissions from anaerobic decomposition of organic matter. Similarly, natural digestive processes in ruminants such as cattle and sheep accounts for majority of methane production. Moreover, manure management includes the handling, storage and treatment of manure, and accounts for 7 percent of agricultural emissions. As a result, methane emissions from enteric fermentation are projected to increase by 32 percent by 2020, with China, Brazil, India, the U.S. and Pakistan being the likely top sources [8] .

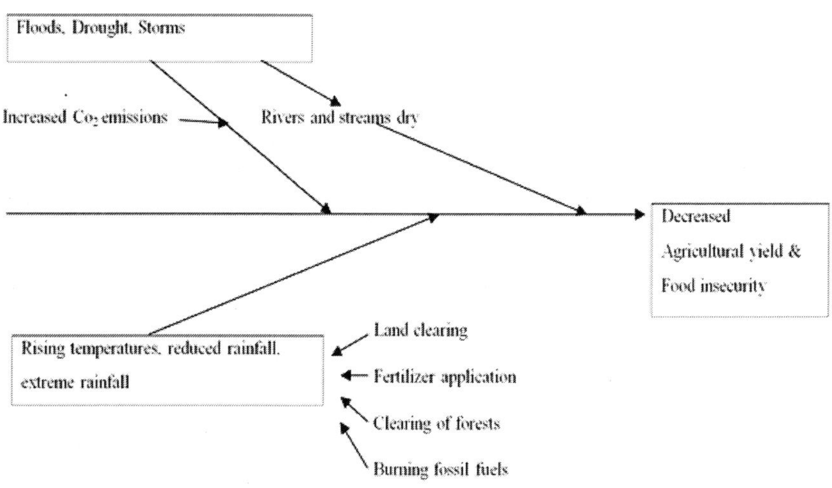

Figure 1. Cause effect model for climate change impacts on agriculture and food security. Source: Author, 2014.

Table 1. Articles on impacts of climate change on agriculture and food insecurity.

Source (Main Content) Article & Findings
World Food Programme (WFP) August, 2009: Kenya in need of access to emergency food assistance—Kenya is food insecure
Awuor, Cynthia. 2009: Increasing drought in Kenya, in understanding climate change adaptation—Kenya experiencing drought
Government of Kenya (Gok), April, 2010: National climate change response strategy (NCCRS)—Strategies to curb climate change
Food and Agricultural Organization (FAO) 2010: Analysis of climate change and variability risks in the Smallholder sector–Climate change impacting smallholder farmers
International Food Policy Research Institute (IFPRI) October, 2010: Climate variability and climate change: Impacts on Kenyan agriculture—Agriculture being impacted by climate change
International Food Policy Research Institute (IFPRI) 2007: Temperature on the rise—rising temperatures
http://www.ifpri.org/climate-change December, 2014: Reducing hunger, poverty and environmental degradation in the highlands, curbing human activities that contribute to climate change
Government of Kenya (GOK) 2002: A framework for combating desertification in Kenya in the context of the United Nations convention to combat desertification—Curbing desertification in Kenya
Climate Knowledge & Development Network (CDKN) 4th June, 2012: Agriculture, food security and compatible development case studies—Climate change affecting Agriculture
International Food Policy Research Institute (IFPRI) December, 2010: Food security and climate change, challenges to 2050 and beyond
www.fao.org/docrep/013/11683e.pdf August, 2014: The state of food insecurity in the world—Kenya is food insecure
http://www.igidr.ac.in/pdf/publication/WP-2011-014.pdf December 2014: Climate change, rural livelihoods and agriculture (focus on food security)
http://www.ifpri.org/climate-change December 2014: Securing a place for agriculture at the international climate change—negotiations 2020 panel discussion—Agriculture under threat
ftp://ftp.fao.org/docrep/fao/009/j9271e/j9271e.pdf November, 2014: Adaptation to climate change in agriculture, forestry and fisheries: perspective, framework and priorities—Adapting agriculture to climate changes
http://www.ipcc.ch/pdf/assessment-report/ar4/syr/ar4_syr.pdf December, 2014: IPCC, Climate change report—Climate change is happening
http://www.ipcc.ch/publications_and_data/ar4/wg1/en/ch10.html November, 2014: Global climate change scenarios
http://www.economics-ejournal.org/economics/journalarticles/2013-34 December, 2014: US food security and climate: Agricultural futures—Climate change threatening food security
http://ec.europa.eu/research/research-eu December, 2014: New challenges for agricultural research: Climate change, food security, rural development, agricultural knowledge systems—Climate change posing as a threat to advancement in agriculture
www.ccafs.cgiar.org/commission November, 2014: Food security, farming, and climate change to 2050—Climate change posing as a threat to advancement in agriculture

Apparently, demand for beef and dairy products is expected to rise globally, with sharp increases in consumption and production forecast for the developing world. By 2020, over 60 percent of meat and milk consumption is expected to take place in the developing world, and the production of beef, meat, poultry, pork, and milk is likely to at least double from 1993 levels [9] . Emissions from agriculture are expected to rise due to increased demand for agricultural production, improved nutrition, and the changing dietary preferences of growing populations that favor larger shares of meat and dairy products [9] .

Recent estimates show that for each 1˚C rise in average temperature dry land farm profits in Africa will drop by nearly 10% [10] . Furthermore, increased evaporation of open water storage would be expected to reduce water availability for irrigation purposes and generation of hydropower. Keep in mind, the Fourth Assessment Report of the Intergovernmental

Panel on Climate Change [11] predicts decreasing rainfall in northern and southern Africa, increasing rainfall over the Ethiopian/East African Highlands and a considerable increase in frequency of floods and drought.

Mitigation and a reduction of the carbon dioxide gas in the atmosphere could help in a great way in addressing climate change effects on agriculture yield. In addition, both mitigation and adoption strategies will require support from governments and policy makers. Mitigation basically means measures undertaken with the aim of reducing the amount of emissions so as to enhance absorption capacity of greenhouse gases. Also, carbon storage through crop management "best practices" for example no-till agriculture or reduce the amount of carbon emission into the atmosphere through, burning, tillage, and soil erosion. There are numerous "best" management practices in agriculture that raise SOC, including reducing the amount of bare fallow, restoring degraded soils, improving pastures and grazing land, irrigation, crop and forage rotation, and no-tillage practices [12] . Most important, production of liquid fuels from grains and oilseeds would on a greater way help in mitigation and also cutting down continued dependence and reliance on fossil fuel. Equally important, improving the efficiency of fertilizer application or turning to organic production can decrease the amount of nutrient load and N_2O emissions.

In general, improved management farm practices for example livestock and manure management, fertilizer management and improved rice cultivation would in a greater way help in reducing green gas emissions into the atmosphere that contribute to climate change. Manure management practices include use of solid covers for storage tanks and lagoons, separating solids from slurry.

Estimations predict that cereal imports will increase in developing countries by 10 to 40 percent by 2080. Also, Africa will become the region with the highest population of food insecure, accounting for up to 75 percent of the world total by 2080 [5] .

Adaptation to climate change is another avenue that can be initiated in order to combat climate change effects. Adaptation to climate change is an adjustment made to a human, ecological or physical system in response to a perceived vulnerability [13] . Adaptation responses can be categorized by the level of ownership of the adaptation measure or strategy. Individual level or autonomous adaptationsare considered to be those that take place in reaction to climatic stimuli (after manifestation of initial impact), that is, as a matter of course and without the directed intervention of any public agency [12] . Adaptation agenda is very large, actions should be undertaken focusing on adaptation capacity, strategies and policy at the national and international levels so as to move adaptation forward. Policy driven or planned adaptationis often interpreted as being the result of a deliberate policy decision on the part of a publicagency, based on an awareness that conditions are about to change or have changed, and that action is required tominimize losses or benefit from opportunities [14] .

4. FOOD SECURITY AND CLIMATE CHANGE

Food security concerns have been on the international agenda since 1948. Inhabitants of the planet earth have the right from hunger and most important, right to a standard of living including food. On the contrary, over the years, diminishing world food supplies coupled with large-scale food shortages have been experienced globally. This has resulted to situations of hunger that calls for increased food production.

Food security is defined as a 'situation when all people, at all times, have physical, social and economic access to sufficient, safe, and nutritious food that meets their dietary needs and food preferences for an active and healthy life' [12] . All dimensions of food security are likely to be affected by climate change. Importantly, food security will depend not only on climate and socio-economic impacts on food production, but also (and critically so) on economic growth, changes to trade flows, stocks, and food aid policy [1] .

Food insecurity affects communities around the world because of the reason that poverty prevents assured access to food supplies. Food insecurity also contributes to degradation and depletion of natural resources, migration to urban areas and cross borders, and political and economic instability. Most important, urbanization through rapid population growth, economic recession and structural adjustments policies that have reduced government spending and decreased employment opportunities influence all aspects of food production and consumption. Similarly, high dependence on commercially processed food and market system to acquire food supplies are aspects of food security associated with urban areas. Equally important, it is claimed that the world's resources are adequate to produce enough food for its population. However, current farming practices, including land clearing and inefficient use of fertilizers and organic residues, make agriculture a significant contributor to greenhouse gas emissions on the planet [15] .

As global demand for food, fodder and bio energy crops grows, many agricultural systems are depleting soil fertility, biodiversity and water resources. For instance, in many regions there are large gaps between potential and actual crop yields. In addition, inadequate access to productive resources in terms of access to land, security tenure, availability of water and other inputs are major challenges in achieving sustained agricultural productivity. Every year, an estimated 12 million hectares of agricultural land, which could potentially produce 20 million tonnes of grain, are lost to land degradation, adding to the billions of hectares that are already degraded [15] . Consequently, global food prices have risen dramatically in the last few years and are forecast to rise further and become more volatile. Generally, climate change will affect agricultural production, specifically areas suffering from food insecurity are expected to experience more negative impacts. To reduce the effect of climate change on food supplies, livelihoods and economies, we must greatly increase adaptive capacity in agriculture both to long-term climatic trends and to increasing variability as an urgent priority [15] .

Majority of African countries are already experiencing considerable water stress as a result of insufficient and unreliable rainfall, changing rainfall patterns or flooding. Climate change impacts including predicted increases in extremes are likely to add to this stress, leading to additional pressure on water availability, accessibility, supply and demand. For Africa, it is estimated that 25% of the population (approximately 200 million people) currently experience water stress, with more countries expected to face high risks in the future, hence food and water insecurity [1] . Thus, food security remains one of the greatest challenges as a result of water stress. In addition, food security problem is also compounded by poor performance of the agriculture sector and low-value crop production due to re-current droughts and floods hence poverty in many parts of the country. Kenya, particularly its agricultural sector is prone to climate change impacts because of heavy reliance on rain-fed agriculture. Climate change will therefore pose a major food security problem as normal weather patterns are affected. Droughts have severely affected food production over the past two decades. Furthermore, adverse weather conditions and patterns have resulted into reduced planting and crop failure leading to food shortages as well as undermining access to food [5] .

Climate change is also likely to have a significant negative impact on agricultural production, prompting output reductions that will greatly affect parts of the developing world. Adaptation, including crop choice and timing, has the ability to partially compensate for production declines in all regions. Finally, climate change adaptation and mitigation have to proceed simultaneously. Since adaptation becomes costlier and less effective as the magnitude of climate changes increases, mitigation of climate change remains essential. The greater the level of mitigation that can be achieved at affordable costs, the smaller the burden placed on adaptation. Policies focused on mitigating GHG emissions, if carefully designed, can help generate a new development strategy; one that encourages the creation of new value in pro-poor investments by increasing the profitability of environmentally sustainable practices. To achieve this goal, it will be necessary to streamline the measurement and enforcement of offsets, financial flows, and carbon credits for investors. It will also be important to enhance global financial facilities and to reform their governance, namely to simplify rules and to increase the funding flows for mitigation in developing countries [5] .

5. CONCLUSIONS

Findings from the literature review indicate that climate change poses serious challenges to farmers because of gradual increases in temperature, increased variability in annual rainfall patterns and greater occurrences of extreme events such as drought and floods, hence food insecurity. Indeed, climate change poses great threats to agriculture, which is sensitive to changes in climatic conditions, with outcomes affecting food security, livelihoods and economic prosperity. Therefore, in order to address food security in the agricultural sector, we must address continued food gap, low global food reserves, rising food prices, rising agricultural input prices and needs for increased biofuel production. This means fundamental changes in the agricultural sector so as to adequately address climate change effects. Changes in agricultural systems and their management are needed to enable to adapt to climate change effects, including mitigation pathways. In addition, science based solutions need to be developed and the best delivery mechanisms to be initiated.

The rural poor together with vulnerable farmers in developing countries are in dire need of science based solutions so as to accelerate successful adaptation of agricultural systems to climate change. Such solutions for example practices, technologies and policies will play a major role in enhancing the ability of farmers and other resource users to better manage land, the environment and the most importantly food security.

Although there are viable mitigation technologies in the agricultural sector, especially in developing countries, some key constraints need to be overcome. First, rules of access which still do not credit developing countries for reducing emissions by avoiding deforestation or improving soil carbon sequestration must be changed. Second, operational rules, with their high transaction costs for developing countries and small farmers and foresters in particular, must be streamlined [5] .

CITE THIS PAPER

Nehemia Kiprutto, Laban K. Rotich, Geoffrey K. Riungu (2015) Agriculture, Climate Change and Food Security. *Open Access Library Journal*,**02**,1-7.

REFERENCES

1. Ludi, E. (2009) Climate Change, Water and Food Security, Background Note. Overseas Development Institute, London.
2. Smith, P., Martino, D., Cai, Z., Gwary, D., Janzen, H., Kumar, P., McCarl, B., Ogle, S., O'Mara, F., Rice, C., Scholes, B. and Sirotenko, O. (2001) Agriculture. In: Metz, B., Davidson, O.R., Bosch, P.R., Dave, R. and Meyer, L.A., Eds., Climate Change 2007: Mitigation. Contribution of Working Group III to the Fourth Assessment Report of the Intergovernmental Panel on Climate Change, Cambridge University Press, Cambridge, and New York.
3. Rosegrant, M.W., Ewing, M., Yohe, G., Burton, I., Huq, S. and Valmonte-Santos, R. (2008) Climate Change and Agriculture, Threats and Opportunities. German Technical Cooperation (GTZ).
4. UNDP (2008) Fighting Climate Change, Human Solidarity in a Divided World. United Nation Development Programme, New York.
5. FAO (2002) The State of Food Insecurity in the World 2001. Food and Agricultural Organization, Rome.
6. Skjei, S. (2005) Using a Fishbone Diagram to Troubleshoot a Satellite Link. The Orbiter, April, 2005.
7. Adger, W.N., Agrawala, S., Mirza, M.M.Q., Conde, C., O'Brien, K., Pulhin, J., Pulwarty, R., Smit, B. and Takahashi, K. (2007) Assessment of Adaptation Practices, Options, Constraints and Capacity. In: Parry, M.L., Canziani, O.F., Palutikof, J.P., van der Linden, P.J. and Hanson, C.E., Eds., Climate Change 2007: Impacts, Adaptation and Vulnerability. Contribution of Working Group II to the Fourth Assessment Report of the Intergovernmental Panel on Climate Change, Cambridge University Press, Cambridge, 717-743.
8. IPCC (2007) Summary for Policy Makers. In: Pachauri, R.K. and Reisinger, A., Eds., Climate Change 2007: Synthesis Report. Contribution of Working Groups I, II and III to the Fourth Assessment Report of the Intergovernmental Panel on Climate Change, Intergovernmental Panel of Climate Change, Geneva.
9. Bates, B.C., Kundzewicz, Z.W., Wu, S.H. and Palutikof, J.P., Eds. (2008) Climate Change and Water. Technical Paper of the Intergovernmental Panel on Climate Change. IPCC Secretariat, Geneva.
10. Nellemann, C., MacDevette, M., Manders, T., Eickhout, B., Svihus, B., Prins, A. and Kaltenborn, B., Eds. (2009) The Environmental Food Crisis. The Environment's Role in Averting Future Food Crises. A UNEP Rapid Response Assessment. UNDP, Arendal.
11. UNDP (2006) Global Mitigation of Non-CO2 Greenhouse Gases. Office of Atmospheric Programs, United States Environmental Protection Agency, Washington DC.
12. Beddington, J., Asaduzzaman, M., Fernandez, A., Clark, M., Guillou, M., Jahn, M., Erda, L., Mamo, T., Van Bo, N., Nobre, C.A., Scholes, R., Sharma, R. and Wakhungu, J. (2011) Achieving Food Security in the Face of Climate Change: Summary for Policy Makers from the Commission on Sustainable Agriculture and Climate Change. CGIAR Research Program on Climate Change, Agriculture and Food Security (CCAFS), Copenhagen.

13. Delgado, C., Rosegrant, M., Steinfeld, H., Ehui, S. and Courbois, C. (1999) Livestock to 2020: The Next Food Revolution. Food, Agriculture, and Environment. Discussion Paper No.28, International Food Policy Research Institute, Washington D.C.
14. Pittock, B. and Jones, R.N. (2000) Adaptation to What and Why? Environmental Monitoring and Assessment, 61, 9-35.
15. Shah, M., Fischer, G. and van Velthuizen, H. (2008) Food Security and Sustainable Agriculture. The Challenges of Climate Change in Sub-Saharan Africa. International Institute for Applied Systems Analysis, Laxenburg.

CHAPTER 2

A Review on Relationship between Climate Change and Agriculture

Yohannes H*

Center for Environmental Science, College of Natural Science, Addis Ababa University, Addis Ababa, Ethiopia

ABSTRACT

This paper reviews various articles and documents on relationship between climate change and agriculture. The two-way relationship of climate change and agriculture is of great significance in particular to developing countries due to their large dependence on agricultural practice for livelihoods and their lack of infrastructure for adaptation when compared to developed countries. Agricultural activities are affected by climate change affects due to their direct dependence on climatic factors. In high latitude areas with low temperature, increased temperature due to climate change could allow for longer growing season. Agriculture affects climate through emissions of greenhouse gases (GHGs) such as carbon dioxide, methane and nitrous oxide. These emissions come directly from use of fossil fuels, tillage practices, fertilized agricultural soils and livestock manure in large proportion. Conversely, agriculture could be a solution for climate change by the widespread adoption of mitigation and adaptation actions. This happens with the help of best management practices such as organic farming, agroforestry practice and manure management etc.

Keywords: Adaptation; Agriculture; Climate change; Greenhouse gas emissions; Mitigation

INTRODUCTION

Globally, climate change (CC) is the most serious environmental threat that adversely affects agricultural productivity [1]. According to inter-governmental panel on climate change [2] (IPCC) report, climate change refers to any change in climate over time, due to natural variability or as a result of human activity. This climate change mainly caused by greenhouse gases (GHGs) accumulation in the atmosphere, which results in increased greenhouse effect. Climate change and agriculture are interrelated processes, both of which take place on a global scale and their relationship is of particular importance as the imbalance between world population and world food production increases. Based on some projections, changes in temperature, rainfall and severe weather events are expected to reduce crop yield in many regions of the developing world, particularly sub-Saharan Africa and parts of Asia [3]. The impact and consequences of climate change for agriculture tend to be more severe for countries with higher initial temperatures, areas with marginal or already degraded lands and lower levels of development with little adaptation capacity [4].

On the other hand, various studies indicate that current agricultural activities are a significant source of GHGs that aggravate climate disruption [5]. The practice of agriculture is very different between developing and developed counties, which results in variation of agricultural contribution to climate change. In developing countries, GHG emission from agriculture sector is much more because of large number of cattle and inadequate manure management, improper use of agro-chemicals and mismanagement of the land. In turn, CC impact becomes more serious in developing countries due to their dependence is on agriculture. In Ethiopia, agriculture supports the livelihoods of the majority of people by providing 80% of employment and it contributes 43% of the (gross domestic product) GDP [6]. This could be a typical example of the impact of climate change on the vulnerable rural communities in the developing world.

On the other side, agricultural sector has large potential to mitigate and adapt climate change. According to IPCC [2], mitigation is an intervention to reduce the emissions sources or enhance the GHG sinks, whereas adaptation is the adjustment in natural or human systems in response to actual or expected climatic change or their effects, to reduce harm or exploit beneficial opportunities. Sustainable and organic agricultural systems can help reduce agricultural GHG emissions through energy conservation, lower levels of carbon-based inputs, lower use of synthetic fertilizer and other features that minimize GHG emissions and sequester carbon in the soil. During mitigation and adaptation of climate change through different agricultural activities, there might be many challenges or barriers such as financial, policy and implementation barrier etc. In general, Agricultural activity could be a source of GHGs as well as a sink, notably through the storage of carbon in the soil organic matter and in biomass and influenced by CC [7].

Climate Change and Agriculture

Impact of climate change on agriculture

Agriculture is an economic activity that is highly dependent upon weather and climate in order to produce the food and fiber necessary to sustain human life. Not surprisingly, agriculture is deemed to be an economic activity that is expected to be vulnerable to climate variability and change. It involves natural processes that frequently require fixed proportions of nutrients, temperatures, precipitation, and other conditions [8]. Agriculture represents less than 2 percent of GDP in high-income countries, and 2.9 percent for the world as a whole. It is more important for low-income countries, contributing large GDP [9]. This is particularly so for countries like Ethiopia; contribute about 43% GDP [6].

Climate change affects agriculture in a number of ways; including through changes in average temperatures; rainfall and climate extremes with an important impact on soil erosion (i.e. floods, drought, etc): changes in pests and diseases, changes in atmospheric carbon dioxide, changes in the nutritional quality of some foods, changes in growing season, and changes in sea level [10]. Crop yields show a strong correlation with temperature change and with the duration of heat or cold waves, and differ based on plant maturity stages during extreme weather events [11]. Modified precipitation patterns will enhance water scarcity and associated drought stress for crops and alter irrigation water supplies. They also reduce the predictability for farmers' planning [12]. In an indirect way, a change in temperature and moisture levels may lead to a change in the absorption rate of fertilizers and other minerals, which determine yield output. In short, the rise in temperature along with the reduction in rainfall reduces agricultural productivity if both are beyond the threshold that is suitable for crop production [13]. According to Ignaciuk and Mason-D'Croz [14], climate change currently decreases the yield of maize, rice, wheat, potatoes and vegetables and continue to reduce seriously by 2050 globally.

Climate change is also likely to affect the livestock sector both by affecting the quantity and quality of feed and by affecting the frequency and severity of extreme climate events. There is a limited body of literature that deals with climate change impacts on livestock, but livestock sector may be particularly vulnerable to the effects of climate change [11].

Climate change regional impacts are likely to be substantial and variable, with some regions benefiting from an altered climate and other regions adversely affected. Generally, food production is likely to decline in most critical regions (e.g. subtropical and tropical areas), whereas agriculture in developed countries may actually benefit where technology is more available and if appropriate adaptive adjustments are employed [15]. In relation, crop productivity is projected to increase slightly at mid to high latitudes for local mean temperature increases of up to 1-3°C depending on the crop, and then decrease beyond that in some regions. At lower latitudes, especially seasonally dry and tropical regions, crop productivity is projected to decrease for even small local temperature increases (1-2°C), which would increase risk of hunger [16]. Warmer weather was expected to bring longer growing seasons in northern areas, and plants everywhere were expected to benefit from carbon fertilization [8] (Figure 1).

Agriculture is central to the survival of millions of people in many countries of sub-Saharan Africa (SSA). It is the number one provider of employment and livelihood in developing countries [17]. The impacts of climate change on agriculture have significant consequences on livelihoods, food production, and the overall economy of countries, particularly those with agriculture-based economies in the developing world because agriculture contributing 29 percent of developing countries' GDP and 65 percent of developing countries' populations, [18]. As Lobell et al.[19] study in 12 food-insecure regions of the world reported that climate change could significantly impact agricultural production and food security up to 2030 particularly for sub-Saharan Africa and South Asia due to both changes in mean temperatures and rainfall as well as increased variability associated with both. The study from Nigeria done by Okoloye et al. [20], shows that in recent years the climatic elements are extremely changed and flood and drought frequency are increasing. According to Guiteras [21], climate change is likely to impose significant costs on the Indian economy by affecting crop yield.

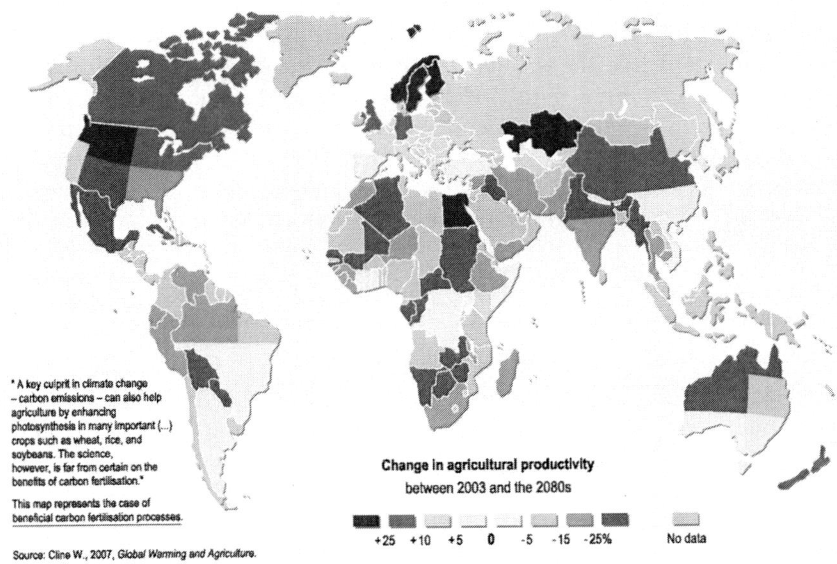

Figure 1: Projected impact of climate change on agricultural yield in different latitude.

From African countries, Ethiopia is one of the most at risk from climate change impacts on agricultural productivity and food security [22]. Gebreegziabher et al. [23] reported that its low adaptive capacity, geographical location and topography make the country highly vulnerable to the adverse impacts of climate change. In addition, dependency of most population on climate sensitive sectors for livelihood i.e. worsens Ethiopia's vulnerability to the impacts of climate change [24]. Over the last decades, the temperature in Ethiopia increased at about 0.2°C per decade, on the other hand, precipitation remained fairly stable over the last 50 years when averaged over the country. However, the spatial and temporal variability of precipitation is high [25]. MacDonald and Simon [26] also reported that farmers living in Ethiopia's semi-arid and arid lowlands that have less diversified assets and are heavily reliant on rain-fed agriculture are, along with their livestock, particularly vulnerable to climate change. It is estimated that the 1984-85 climate induced drought reduced Ethiopia's agricultural production by 21 percent, which led to a 9.7 percent fall in the GDP [27]. Bezabih et al. [28] also point out that climate variability and change in Ethiopia has significant impact in different crop yield. Particularly, Bayrau et al. [29], study shows that change in climate will likely have an overall significant impact in reducing the productivity of sugarcane and cotton in Ethiopia.

According to Afar National Regional State (ANRS), 2010 [30] report, all pastoral regions in Ethiopia are highly prone to the adverse impacts of climate change, while the problem is more prevalent in the North Eastern lowlands of the country mainly Afar region. Many studied done in different part of Ethiopia reported that climate change has tremendous impact in livestock resources [31,32]. Likewise, According to Gebreegziabher, [33] 2014 study in Nile basin of Ethiopia, Changing rainfall and temperature patterns due to climate change have different effects on crops and livestock.

The contribution of agricultural activities to climate change

According to Thornton and Lipper [34], agriculture contributes 30-40% of anthropogenic GHG emissions. Three-quarters of agricultural GHG emissions occur in developing countries, and this share may rise above 80% by 2050 [35]. According to FAO [36] report in developing countries there is a significant increase in GHGs emission from 2001-2011 (14%), the increase occurred, due to an expansion of total agricultural outputs. In Ethiopia particularly, a GHG emission from agriculture accounts about 80% [37]. As the global population and the demand for food continue to grow, total GHG emissions from the agricultural sector are projected to increase over time [38].

Agriculture creates both direct and indirect emissions. Direct emissions come from fertilized agricultural soils and livestock manure. While indirect emissions come from runoff and leaching of fertilizers, emission from land-use changes, use of fossil fuels for mechanization, transport and agro-chemical and fertilizer production [35,39]. The most significant indirect emissions are changes in natural vegetation and traditional land use, including deforestation and soil degradation. Yibekal et al. [40] found that deforestation (for agricultural expansion and fuel wood) is the main cause of climate change in Ethiopia. Intensive tillage is also one of traditional land use practices which involve continuously disturb the land. This practice increases CO_2 emissions by causing decomposition of SOM and soil erosion [41].

According to the World Bank [42], agriculture contributes about half of the global emissions of two of the most potent non-CO_2 GHG: nitrous oxide (N_2O) and methane (CH_4). These non-carbon GHGs have powerful greenhouse effects and have greater longevity than CO_2(Table 1).

Globally, agriculture contributes to 58 percent of total N_2O emission [35,43]. It creates 4.5 million tons of nitrous oxide annually [44]. Various management practices in the agricultural land can lead to production and emission of nitrous oxide, range from fertilizer application to methods of irrigation, tillage and cattle and feedlots. The use of synthetic fertilizer for agriculture is a major source of nitrous oxide emissions. Apart from this, large quantities of natural gas are used to make synthetic fertilizers because it is the main ingredient. The production process also takes a lot of energy so their impact on climate change is actually larger when we factor this in. Industrialized farming practices have worsened this loss and the result has been increased emissions. Continuous cropping may result in using of large chemical fertilizer [42]. According to Amdu [45], Study from Ethiopia reported that farmers widely used huge amount of chemical fertilizer to boost their product and considered as one of the main activities to adapt CC. And the study also reported that the farmers consider the negative effect of agro-chemicals only for their crops. This indicates that the farmers only care about their crop and select suitable condition to apply not environmental side effect of fertilizers. Because of this all, agriculture is the most important human source of nitrous oxide emissions. In Ethiopia, GHG emissions in 2010 from fertilizer use and N2O emission from crop residues reintroduced into the soil are approximately 10 Mt CO2 equivalent emissions and 3 Mt CO2 equivalent emissions, respectively [46].

As earlier mentioned, methane is another non-CO2 GHG which have large contribution in agriculture GHGs emission. Livestock, especially cattle, produce methane as part of their digestion. This process is called enteric fermentation. The sector is responsible for 47 percent of the world's methane emissions. The way in which manure is managed contributes largely. When the manure of livestock is not used as a fertilizer or left in fields during grazing. Many of these systems create conditions that are favorable for nitrous oxide producing bacteria [39].

Ethiopia has the largest livestock population in Africa and the tenth largest in the world [46]. Therefore, it is obvious the impact these large number of livestock for climate change if there is improper management. This reflects the fact that livestock farming goes together with high methane emissions. As earlier mentioned, Ethiopia's GHG emissions are dominated by agriculture, which contributes 80% of the total GHG emissions. On the other hand, According to [47] report, about 50% of Ethiopian GHG emissions is attributed to agriculture sector. According to FDRE, GHG emissions from livestock are estimated to 65 Mt CO2 equivalent emissions in 2010. Ethiopia's GHG emissions are closely linked to basic needs of the population: Food production (through livestock farming) and heating.

Major environmental conventions on climate change and agriculture

Since the first world climate conference takes place in 1979 in Geneva, There are different documents that mention about agriculture in various environmental conventions and agreements, even if the attention for this sector is not enough despite its important. Agriculture and food security under a changing climate have come up the international agenda in recent years [48] (FAO, 2014b). Therefore, this review includes some of environmental conventions are presented as follow:

Article 4 (c) of UNFCCC [49], describe that all parties should Promote and cooperate in the development, application and diffusion, including transfer of technologies, practices and processes that control, reduce or prevent anthropogenic emissions of GHG in all relevant sectors, including agriculture. Also article 4 (e) of UNFCCC [50,51] states about the need of cooperation in preparing for adaptation to the impacts of climate change; develop and elaborate appropriate and integrated plans for coastal zone management, water resources and agriculture, and for the protection and rehabilitation of areas, particularly in Africa, affected by drought and desertification, as well as floods.

The linkage between climate change and agriculture is clearly addressed directly in Article 2 of Kyoto protocol which states: promotion of sustainable forms of agriculture in light of climate change considerations. In Article 3.4 of Kyoto protocol also mentioned about the carbon sink partly be accomplished efficiently by organic agriculture. Article 10 of Kyoto protocol also stated how to mitigate climate change and measures to facilitate adequate adaptation to climate change by concerning different sectors including agriculture [52] (FAO, 2008).

The Bali roadmap, IPCC's 4th assessment (2007) report, indicates that actions aimed at safeguarding food security and rural livelihoods under climate change in the coming decades must focus on interactions between adaptation and mitigation strategies in the agricultural and forestry sectors in order to address climate, environmental, social and economic concerns expressed within both the UNFCCC and the millinium development goals (MDGs). It clearly describes the potential future effects of climate change on agriculture and identified current knowledge gaps to be addressed [49].

Article 2 of UNFCCC, COP14 (in Poznan, Poland), (2008); addresses mitigation and adaptation in the agricultural sector and makes specific reference to food security when it confirms that climate change mitigation should be achieved within a time-frame sufficient to ensure, among other things, that food production is not threatened, ecosystems can adapt naturally, and economic development is pursued in a sustainable manner [53].

In copenhagen accord (2009), there was discussion about how agriculture is seriously affected by climate change. In addition it mentioned that the role of agriculture to mitigate and adapt climate change. It was a good progress but no specific decisions were taken in Copenhagen for agriculture sector with regards to climate change [54].

The cancun agreements in (2010) allow for consideration of agriculture as a driver of deforestation and thus can be considered under adaptation actions. Agriculture already figured prominently in national adaptation programmes of action (NAPAs) formulated by least-developed countries. NAPAs are now to inform the new national adaptation plans, which, in accordance with the Cancun Agreements, are to be prepared by developing countries. Also, following the fifteenth session of the COP, a number of developing countries indicated their intention to undertake nationally appropriate mitigation actions (NAMAs) related to agriculture.

In 2014, the IPCC 5th assessment report Working Group II published an update of the potential future effects of climate change on agriculture, identifying knowledge gaps and increased urgency for adaptation in the sector [48]. This is a part of Lima international climate change conference agreement.

In 2015 Paris (COP 21) Draft article 1 it is mentioned that recognizing the fundamental priority of safeguarding food security and ending hunger, and the particular vulnerabilities of food production systems to the adverse impacts of climate change. Article 92 explain that to pursue a transformation towards sustainable development that fosters climate resilient and low greenhouse gas emission societies and economies, and that does not threaten food production and distribution [50]. These all conventions and agreements indicate that, there is an increase to concern about agriculture under climate change situation.

Agriculture as a solution for climate change

The agricultural sector holds significant climate change mitigation potential through reductions of GHG emissions and enhancement of agricultural sequestration [7]. In addition, it also has significant role to adapt climate change. Adaptation alone is not enough to offset the effects of climate change, and thus still need to be supplemented by concerted mitigation efforts [8]. Mostly, when we implement adaptation measure, we enhance mitigation capacity of particular area such as practicing different land use managements (soil and water conservation measure, manure and fertilizer management) in the agricultural field will help us to sequester substantial amount of carbon in the field and reduce emission of methane and nitrous oxide which are the main GHG emission means. Therefore, the management activities are interrelated and help us to adapt and mitigate climate change. Agricultural activities are relatively affordable form of mitigation option, for which many technical options are already readily available [55].

Table 1: An overview of agricultural greenhouse gases with the trends as currently predicted (adapted from IPCC, 2009).

	CO_2	CH_4	N_2O
Atmosphere life time (yr)	120	14.5	120
Direct GWP (Global Warming Potential	1	24.5	320
Pre-industrial concentration	280 ppmv	0.8 ppmv	288 ppbv
Current annual increase (%)	0.5	0.9	0.25
Major Agricultural Source	Deforestation	Ruminants	Synthetic N fertilizers
		Wetland rice	Animal Excreta
		Biomass burning	Biological N fixation
Percentage of global source stemming from agriculture	30	40	25
Predicted change 1990-2020	-	+	+

Global adoption of organic agriculture (OA) has the potential to sequester up to the equivalent of 32% of all current man-made GHG emissions [56]. OA is a production system that sustains the health of soils, ecosystems and people. In OA, soil fertility is maintained mainly through farm internal inputs (organic manures, legume production, wide crop rotations etc); energy-demanding synthetic fertilizers and plant protection agents are rejected; and there is less or no use of fossil fuel [16]. In relation, improved cropland management (lower use of synthetic fertilizers, reduced tillage etc), Reducing industrial livestock production and improving feeding and grazing land management, Restoration of organic soils and degraded lands to increase soil carbon sinks, Improved water and rice management, Land-use change and agro-forestry, Increasing efficiency in fertilizer production and behavioural changes of food consumers (reducing the meat content) could also be main climate change mitigation measures in agriculture sector [57].

As earlier mentioned, the agricultural sector has also a potential to adapt to climate change in many areas. Climate change adaptation is a continuous process requiring location-specific response. Adaptation should enable agricultural systems to be more resilient to the consequences of climate change [58]. Farming systems and farmers will differ enormously in their capacities to respond to climate change. Differentiated adaptation strategies and enhanced climate risk management support to agriculture and farming households are critical to counter the impacts of climate change [17]. These adaptation measures could include in particular the choice and change of species and varieties, the adaptation of the field works to the calendar (more flexibility), the adaptation of plant production practices (i.e. fertilization, plant protection, irrigation, etc.) or the adoption of plant production practices that increase the soil organic matter content or the soil coverage by plants, manure management and agroforestry practices. Some of them discussed below how these practices serve as adaptation means:

Change crop variety

It involves switching from one crop variety to another in response to climatic stresses and changes. Study done by Komba and Muchapondwa [59], in Tanzania explained that Tanzania's farmers try to adapt climate change by using drought resistance crops. Introducing Avena species (Ingedo) species in Ethiopia as fodder crop and through time it replace the dominant stable crop i.e., barley in the highland and serve as one means to adapt CC [45].

Change in cropping pattern

Application of changes in how crops are cycled within a season. Farmers in the drought-prone semi-arid areas of Brazil have realized that several varieties of a single crop species can occupy a common land area, incorporating several bean varieties, maize and sorghum, among others, to increase harvest potential arid climate stresses [60]. Likewise, In Ethiopia farmers try to adapt CC by planting diversifies crops (homestead maize and other crops).

Change in cropping calendar

It is another common adaptation to climate change at the farm level, which largely involves altering the timing of farm activities to suit climatic variations or changes. In Philippines, farmers adapt to the early onset of rainy season through early cultivation of upland farms, which results in high agricultural production for the season and higher household income from farm activities [61]. Tanzania's farmers also used planting date changing practice to adapt CC [62]. In addition, according to Rhodes et al. [63], most West Africa counties such as Burkina Faso, Niger and Senegal already develop and implement a mathematical model for different crops to plant under changed climate by shifting planting date to adapt CC. Changing planting dates of crops helps greatly for the farmers live in East Gojam (Choke mountain in Ethiopia) and East Hararghe for CC adaptation respectively.

Farm management practices

Change in current farm management practices such as OA practice focus on maintaining diverse farming systems (i.e. planting different crop species) also helps diversify potential sources of income for farmers, making the farming household more resilient to adverse impacts of climate on agricultural production [63]. According to Rhodes et al. [62], also crop residue management practice is considered one of the best climate smart actions. In addition, Smallholder farmers in sub-humid Southwestern Cameroon have been adapting to variations in rainfall through different soil and water conservation practices. Ditch and check dam constructions are the major soil and water conservation (SWC) practice that helps to adapt CC in Ethiopia. Similarly, study in Ethiopia revealed that terracing and different water harvesting practices widely used to adapt the changed climate.

Livestock management

It related to livestock and their waste management. Particularly to adapt climate change farmers from central Africa implement different adaptation strategies such as Breeding locally adapted livestock species, diversifying livestock types, proper resource management practices and alternative feed production technologies (use of agricultural byproductsor household and industrial waste products are needed to produce feed) [64]. In Ethiopia (East Gojjam), farmers has made different local adjustment of avoid the impact of CC such as planting trees that serve as a shed for their livestock, locate their livestock near to the clinic (due to pest and disease infestation increased) and keep their animals in home and feed them through cut and carry system in times of feed and water scarcity.

In addition, during livestock waste management, manure should be collected as quickly as possible to limit GHG emission and to reach high energy potential from manure. In South Asia and China, some livestock producers are coming up with innovative ways of storing and removing animal waste and produce biogas and the device break down bio-solids in underground fermentation units which produce gas for cooking, lighting and energy for on-farm activities. Solids broken down in fermentation units can also be used as fertilizers in crop production, or sold for cash [65]. In this way, Livestock waste management can serve as both climate change mitigation and adaptation means.

Agroforestry

Given the large contribution of land use conversion and the forestry sector to GHG emissions, agroforestry presents an opportunity to counter the adverse impacts of climate change through the joint action of adaptation and mitigation [66]. Trees on farms enhance the coping capacity of small farmers to climate risks through crop and income diversification, soil and water conservation and efficient nutrient cycling and conservation [61]. Different researches on agroforestry in West Africa related to climate change have focused on its carbon sequestration potential and effect on soil fertility [67]. Similarly, most central Africa countries studies reported that significant crop yield increases for maize, sorghum, millet, cotton and groundnut when grown in proximity to Faidherbia albida tree [64].

Barriers to adapt and mitigate climate change in agriculture sector

According to Smith et al. [68] study, the following condition considered as the main barrier to mitigate CC; it need long time (15-60 years) to obtain maximum storage that is carbon sequestration in soils or terrestrial biomass, depending on management practice, management history and the system being modified. Costs that are related to measurement and monitoring and transaction cost (incentive-based system) also another barrier: When there is lack of clear ownership about the land results in less incentive to implement different land management activities become another problem. On the other hand, According to IPCC Fourth Assessment Report, There are often technological and financial limits that prevent the scale of adaptation that we would need. The other barriers mentioned by FAO 2011, are funding and implementation barriers. Nhemachena and Hassan [69], identify the important determinants of adaptation in South Africa, Zambia and Zimbabwe as access to credit and extension and also awareness. Ishaya and Abaje [70] also found that lack of awareness and knowledge about climate change and adaptation strategies, lack of capital and improved seeds, and lack of water for irrigation played an important role in hindering adaptation in Nigeria. Similarly, Okolo et al. [71] study finding revealed that policy barrier such as lack of local context and little attention to local situation, limited capacity to implement the policy, lack of coordination etc. hinder farmers' to practice different adaptation activities. Bryan et al. [72] study in Ethiopia and South Africa access to credit is the main barrier that affects CC adaptation by farmers. Another study done by Amdu, in Ethiopia reported that lack of knowledge, improper policy and lack of land use policy implementation, labor and water shortages, and information are the main barriers to CC adaptation. According to Rhodes et al., review on west Africa region, lack of scientific research that deliver information this result in limited knowledge to understand adaptation strategies and policy gap are the main barrier to implement effective adaptation measures.

CONCLUSION

From this extensive review, it is concluded that globally, climate change has relationship with agriculture in one or another way. This relationship becomes strong in developing countries because their livelihood depends on agricultural activities and this activities mostly depend on climatic condition, For instance in Ethiopia, almost all farm activity is rain fed. In relation, the impact of climate change is very serious in developing counties due to their limited adaptive capacity and lack of technology and also they are the main emitter of noncarbon GHGs from their cattle and farm management mainly from use of synesthetic fertilizers. Those are the main direct emitters. There are also indirect emitters such as land use change; from runoff and leaching of fertilizers; use of fossil fuels for mechanization; transport and agro-chemical and fertilizer production.

On the other hand, by the help of the right farming practice agriculture could be the main solution for climate change by mitigation and adaptation response. Within the current and projected situation of climate change globally, only climate change mitigation is not enough so long term solution is important by combining climate change adaptation in agriculture sector. Such practices could be organic agriculture, manure management, agroforestry practice etc.

Know a days, the significant relation of climate change and agriculture sector become well known. In recent years, even if the attention is not enough, the significant relation of climate change and agriculture sector becomes acknowledged. To happen this, International conventions and agreements has significant role in addition to research role.

RECOMMENDATIONS

Based on this review, the following recommendations are forwarded:

- Agriculture has significant contrition for climate change disruption but the attention given for this situation is low. So, there should be more research and awareness creation on this is important.

- It is obvious that climate change has significant negative role particularly for developing countries farmers. So adequate mitigation and adaptation measure implementation should be widely practices.
- International agreements and convention about climate change and agriculture contribute to distribute information widely over globe still need more focus on this nexus as they worry about CO2 emission and practical support starting from local farmers is very vital.

REFERENCES

1. Enete AA, Amusa AT (2010) Challenges of Agricultural Adaptation to Climate Change in Nigeria: a Synthesis from the Literature. Field Actions Science Reports.
2. IPCC (Intergovernmental Panel for Climate Change). (2007a) Climate Change 2007: The Fourth Assessment Report. Impacts, Adaptation and Vulnerability. [https://www.ipcc.ch/publications_and_data/publications_ipcc_fourth_assessment_report_wg2_report_impacts_adaptation_and_vulnerability.htm]
3. Gornall J, Betts R, Burke E (2010) Implications of climate change for agricultural productivity in the early twenty-first century. Philos Trans R Soc Lond B Biol Sci. 365:2973-2989.
4. Keane J, Page S, Kergna A, Kennan J (2009) Climate Change and Developing Country Agriculture: An Overview of Expected Impacts, Adaptation and Mitigation Challenges, and Funding Requirements
5. Parvatha PR (2014) Climate Resilient Agriculture for Ensuring Food Security. Springer. pp:1-15.
6. IMF (International Monetary Fund) (2012) Ethiopian Economic Outlook'Ethiopian Economy. EFD research brief series pp.12-03.
7. [http://www.africaneconomicoutlook.org/en/] Acceessed on: Feb 25, 2016.
8. Vuren DPV, Ochola WO, Riha S, Gampietro M, Ginze H, et al. (2009) Outlook on agricultural change and its drivers. In: McIntyre BD, Herren HR, Wakhungu J, Watson RT (ed.). Agriculture at a Crossroads, Island Press, Washington, DC.
9. Ackerman F, Stanton EA (2013) Climate Impacts on Agriculture: A Challenge to Complacency? Global Development and Environment Institute Working Paper No. 13-01.
10. World Bank (2008a) World Bank data on agricultural value added as a share of GDP in 2008
11. Hoffmann U (2013) Section B: Agriculture: a key driver and a major victim of global warming, in: Lead Article, in: Chapter 1, in Hoffmann 2013, pp: 3-5.
12. OECD (Organization for Economic Co-operation and Development) (2014) Climate change, water and agriculture: Towards resilient agricultural and water systems. [http://dx.doi.org.10.1787/9789264209138-en].
13. Tirado R, Cotter J (2010) Ecological Farming: Drought-resistant Agriculture. Greenpeace Research Laboratories, University of Exeter, United Kingdom.
14. Cabral L, Ludi E, Peskett L, Stevens C (2007) Climate Change and Agriculture: Agricultural Trade, Markets and Investment. Overseas Development Institute, London.
15. Ignaciuk A, Mason-D'Croz D (2014) Modelling Adaptation to Climate Change in Agriculture. OECD Food, Agriculture and Fisheries Papers70: 58.
16. OECD (2015) Agriculture and Climate Change. Trade and Agriculture Directorate. [http://www.oecd.org/tad/sustainable-agriculture/agriculture-and-climate-change.htm].

17. IPCC (2007b) Contribution of Working Group II to the Fourth Assessment Report of the Intergovernmental Panel on Climate Change-Summary for Policymakers. [http://www.ipcc.ch/pdf/assessment-report/ar4/wg2/ar4-wg2-spm.pdf] Retrieved on Dec 10, 2015.
18. Campbell B, Mann W, Meléndez-Ortiz R, Streck C, Tennigkeit T, et al. (2011) Agriculture and Climate Change. A Scoping Report.
19. Padgham J (2009) Agricultural development under climate change: Opportunities and challenges for adaptation. World Bank, Washington DC.
20. Lobell DB, Burke MB, Tebaldi C, Mastrandrea MD, Falcon WP, et al. (2008) Prioritizing climate change adaptation needs for food security in 2030. Journal of Science 319: 607-610.
21. Okoloye CU, Aisiokuebo NI, Ukeje JE, Anuforom AC, Nnodu ID (2013) Rainfall Variability and the Recent Climate Extremes in Nigeria. Journal of Meteorology& Climate Science 11:49-57.
22. Guiteras R (2009) The Impact of Climate Change on Indian Agriculture. University of Maryland.
23. Evangelista P, Nicholas, Young N, Burnett J (2013) How will climate change spatially affect agriculture production in Ethiopia? Case studies of important cereal crops. J Climatic Change 119: 855-873.
24. Gebreegziabher Z, Jesper S, Mekonnen A, Alemu A (2011) Climate Change and the Ethiopian Economy: A Computable General Equilibrium Analysis. http://www.rff.org/files/sharepoint/WorkImages/Download/EfD-DP-11-09.pdf]
25. United Nations Environment Programme (2011) Framework for UNDP Ethiopia's Climate Change, Environment and Disaster Risk Management Portfolio.
26. IPCC (2007c) Climate Change 2007: Synthesis Report. Contribution of Working Groups I, II and III to the Fourth Assessment Report of the Intergovernmental Panel on Climate Change.[https://www.ipcc.ch/publications_and_data /ar4/syr/ en/contents.html]
27. MacDonald M, Simon J (2011) Climate, food security, & growth Ethiopia's complex relationship with livestock. Policy Brief 3. Brighter Green.
28. World Bank (2006) Ethiopia: Managing water resources to maximize sustainable growth (Report No 36000-ET) Washington, DC: The World Bank
29. Bezabih M, DiFalco S, Mekonnen A (2014) On the Impact of Weather Variability and Climate Change on Agriculture,Evidence from Ethiopia. Published by Sida, Environment for Development and Resource for the Future.
30. Bayrau A, Assefa B, Hagos A (2015) Productivity and welfare impact of climate change in sugarcane and cotton producing regions of Ethiopia. EDRI Research Report 23. Addis Ababa: Ethiopian Development Research Institute.
31. Afar National Regional State (2010) Programme of Plan on Adaptation to Climate Change.
32. Habtamu L (2012) Domestic Animal Biodiversity in Ethiopia and its Threats and Opportunities with Emphasis to Changing Climate: An Overview. Adv Life Sci Technol 6:33-39.
33. Iqubal A (2013) Livestock Husbandry: A Sustainable Livelihood in Ethiopia. International Journal of Economy, Management and Social Sciences 2: 603-607.
34. Gebreegziabher Z, Mekonnen A, Deribe R, Abera S, Kassahun MM (2014) Climate change can have significant negative impacts on Ethiopia's agriculture. EfD Discussion Paper 13-14.
35. Thornton P, Lipper L (2013) How does climate change alter agricultural strategies to support food security?. Background paper for the conference "Food Security Futures: Research Priorities for the 21st Century", 11-12 April 2013, Dublin.
36. Smith P, Martino D, Cai Z, Gwary D, Janzen H, et al. (2007a) Climate Change 2007: Mitigation. In: Metz B, Davidson OR, Bosch PR, Dave R, Meyer LA (eds.).Contribution of Working Group III to the Fourth Assessment Report of the Intergovernmental Panel on Climate Change. Cambridge University Press, Cambridge, United Kingdom and New York, USA.

37. FAO (2014a) Agriculture's greenhouse gas emissions on the rise. [http://www.fao.org/news/story/en/item/216137/icode/] Assessed on: Feb 25, 2016.
38. Marius K (2009) Climate Risks and Development Projects Assessment Report for a Community-Level Project in Guduru, Oromiya, Ethiopia.
39. IPCC. 2014. AR5 WGIII. Geneva. [http://www.ipcc.ch/report/ar5/]
40. David U, Lal R (2013) Soil emission of nitrous oxide and its mitigation. Dordrecht: Springer. BN 978-94-007-5364-8 (eBook). Springer Science+Business Media Dordrecht 2013.
41. Yibekal AT, Chanyalew SA and Getachew SE (2013) Understanding the process of adaptation to climate change by small-holder farmers: the case of east Hararghe Zone, Ethiopia A springer open journal of Agricultural and Food Economics
42. Busari AM, Kukal SS, Kaur A, Bhatt R, Dulazi AA (2015) Conservation tillage impacts on soil, crop and the environment. International Soil and Water Conservation Research. 3:119-129.
43. World Bank (2008b) Agriculture for development policy brief: adaptation and mitigation of climate change in agriculture: World Development Report 2008
44. IPCC (2007d) Mitigation of Climate Change. Working Group III Report of the Fourth Assessment Report of the Intergovernmental Panel on Climate Change. [www.ipcc.ch/publications_and_data/ar4/wg3/en/contents.html].
45. Denman KL, Brasseur G, Chidthaisong A, Ciais P, Cox PM, et al. Couplings between Changes in the Climate System and Biogeochemistry. In: Boonpragob K, Heimann M, Molina M (Eds.). Climate Change 2007: The Physical Science Basis. Contribution of Working Group I to the Fourth Assessment Report of the Intergovernmental Panel on Climate Change. Cambridge University Press, Cambridge, United Kingdom and New York, NY, USA, 2007.
46. Amdu B (2010) Analysis of Farmers' Perception and Adaptation to Climate Change and Variability: The Case of Choke Mountain, East Gojjam MSc thesis. Addis Ababa University.
47. Federal Democratic Republic of Ethiopia (2011) 'Ethiopia's Climate-Resilient Green Economy: Green Economy Strategy, Addis Ababa, Ethiopia.[http://www.uncsd2012.org/content/documents/287CRGE%20Ethiopia%20Green%20Economy_Brochure.pdf]
48. Ayana S, Beyene S, Nebil S, Sertse S, Tefera T (2011) Agriculture sector programme of plan on Adaptation to climate change. Federal Democratic Republic of Ethiopia.
49. FAO (2014b) Information brief agriculture/land at the Lima climate talks (1-12 December 2014). [http://www.fao.org/3/a-at620e.pdf] Assessed on: Feb 25, 2016.
50. UNFCCC (2009) About CDM. Clean Development Mechanism (CDM) section, United Nations Framework Convention on Climate Change (UNFCCC).
51. UNFCCC (2015) Draft Paris Agreement. Draft conclusions proposed by the Co-Chairs. Ad Hoc Working Group on the Durban Platform for Enhanced Action.
52. Initial National Communication of Ethiopia to the United Nations Framework Convention on Climate Change (UNFCCC) (2001) Report of the Federal Democratic Republic of Ethiopia, Ministry of Water Resources, National Meteorological Services Agency. [http://unfccc.int/resource/docs/natc/ethnc1.pdf]
53. FAO (2008) Organic agriculture and climate change. [http://www.fao.org/organicag/oa-specialfeatures/oa-climatechange/en/] Retrieved on: Dec 10, 2015.
54. FAO/IFAD (2008) Financing climate change adaptation and mitigation in the agriculture and forestry sectors. [http://unfccc.int/resource/docs/2008 /smsn/igo/015.pdf.] Assessed on: Feb 25, 2016.
55. FAO (2010a) Agriculture, food security and climate change in post Copenhagen processes: An FAO Information note. [http://foris.fao.org/static/data/nrc/InfoNote_PostCOP15_FAO.pdf] Assessed on: Feb 25, 2016.
56. FAO (2009) Food Security and Agricultural Mitigation in Developing Countries: Options for Capturing Synergies. [www.fao.org] Assessed on: Feb 25, 2016.
57. Robert J, Adrian M, Anne O (2009) High Sequestration, Low Emission, Food Secure Farming. Organic Agriculture - a Guide to Climate Change & Food Security, IFOAM.

58. Paul H, Ernsting A, Semino S, Gura S, Lorch A (2009) Agriculture and Climate Change: Real Problems, False Solutions. Report published for the 15thConference of the Parties of UNFCCC in Copenhagen, December.
59. FAO (2011) Strengthening Capacity for Climate Change Adaptation in the Agriculture Sector in Ethiopia. Proceedings from National Workshop held in Nazreth, Ethiopia.
60. Komba C, Muchapondwa E (2015) Adaptation to Climate Change by Smallholder Farmers in Tanzania. Environment for Development. Discussion Paper Series.
61. Action Aid (2008) The Time is NOW: Lessons From Farmers Adapting to Climate Change, Action Aid.
62. Lasco R, Pulhin F (2009) Agroforestry for Climate Change Adaptation and Mitigation. An academic presentation for the College of Forestry and Natural Resources (CFNR), University of the Philippines Los Baños, Philippines.
63. Rhodes ER, Jalloh A, Diouf A (2014) Review of research and policies for climate change adaptation in the agriculture sector in West Africa. Working Paper 090.
64. Muller A (2009) Benefits of Organic Agriculture as a Climate Change Adaptation and Mitigation Strategy for Developing Countries, Environment for Development Discussion Paper Series, EfD DP 09-09.
65. Ngeve MJ, Jalloh A, Ndjatsana M (2014) Review of Research and Policy for Climate Change Adaptation in the Agriculture Sector in Central Africa. Working Paper 098
66. FAO (Food and Agricultural Organization) (2006) Livestock Report 2006, Animal Production and Health Division, Food and Agriculture Organization of the United Nations, VialedelleTerme di Caracalla, 00100 Rome, Italy.
67. FAO (2010) Climate Smart Agriculture: Policies, Practices and Financing for Food Security, Adaptation and Mitigation, Rome, Italy: Food and Agriculture Organization [foris.fao.org/static/data/nrc/InfoNote_PostCOP15_FAO.pdf] Assessed on: Feb 25, 2016.
68. Takimoto A, Nair PKR, Nair VD (2008) Carbon Stock and Sequestration Potential of Traditional and Improved Agroforestry Systems in the West African Sahel. J Agriculture, Ecosystems and Environment 125:159-166.
69. Smith P, Martino D, Cai Z, Gwary D, Janzen H, et al. (2007b) Policy and Technological Constraints to Implementation of Greenhouse Gas Mitigation Options in Agriculture. Agriculture, Ecosystems and Environment 118: 6-28.
70. Nhemachena C, Hassan R (2007) Micro-level Analysis of Farmers Adaptation to Climate Change in Southern Africa. IFPRI Discussion Paper 00714.
71. Ishaya S, Abaje IB (2008) Indigenous People's Perception on climate Change and Adaptation Strategies in Jema'a Local Government area of Kaduna State in Nigeria. Journal of Geography and Regional Planning 1: 138-143.
72. Okolo W, Twyman J, Ampaire E, Acosta M (2015) Barriers to successful climate change policy implementation in Uganda: Findings from a qualitative policy study in Nwoya and Rakai Districts, Uganda. CCAFS Info Note. Copenhagen, Denmark: CGIAR Research Program on Climate Change, Agriculture and Food Security (CCAFS).
73. Bryan E, Deressa TT, Gbetibouo GA, Ringler C (2009) Adaptation to Climate Change in Ethiopia and South Africa: Options and Constraints. J Environmental Science and Policy 12: 413-426.

CHAPTER 3

Impacts of Climate Change on Agricultural Production in Enugu State, Nigeria

Enete IC*

Department of Geography & Meteorology, Nnamdi Azikiwe University, Awka, Nigeria

ABSTRACT

Climate Change is increasingly one of the most serious national security threats which will have significant impacts on natural resources, ecosystem and biodiversity. At the same time, it is likely to trigger food insecurity, human migration, economic, and social depression, environmental and political crisis, thereby affecting national development. To analyze the climate change impact on agricultural productivity in Enugu state, the study used primary data that have been collected from Enugu State Agricultural Development Programme (ENADP) on productivity for four decades. Also, rainfall data for thirty years (1981–2010) were collected from Nigeria Meteorological Agency to generate seasonality index and reliability index. The data have been analyzed using descriptive statistics and correlation analysis. Also, literatures were identified for review through a comprehensive search by using electronic and non-electronic databases. Related published literature and documents were searched in a systematic way using a range of key words relating to climate change impacts and agriculture. The study showed a general change in the seasonal rainfall regime. Rainfall regime in the state falls under a markedly seasonal regime with a long drier season; and a reliability of rainfall regime characterized by a more marked variability in the months of maximum rainfall. These changes are notable indicators of climate change. Again, the study showed that all the traditional crops with the exception of cassava and pepper had a significant field decrease as rainfall continued to be more erratic. The literature review indicates that climate change and agriculture both have double-barrel effect on each other. Reducing climate change - induced threats to agriculture that contributes to national development, there will be a need for farmers in Enugu to respond by planting different climate–adapted species, using pesticides and altering the dates of planting, harvesting and irrigation.

Keywords: Climate change; Rainfall; Seasonality index; Reliability index; Agriculture; Productivity

INTRODUCTION

The triple whammy problems of rapid increase in population, climate change and pressure on resources are major constraints to adequate agricultural production. Among the three, climate change is the most pressing challenges that the world faces today. Climate change is a defining phenomenon of the century. Given current atmospheric concentrations of Greenhouse Gases (GHGs), the world is already committed to significant warming. This is a serious challenge, given the wide range of expected climate impacts on natural systems, as well as on human societies, as assessed in the most recent report of the Intergovernmental Panel on Climate Change [1]. Climate change threatens to undermine the progress that has been achieved to date, especially in the agricultural sector [2].

It remains unclear; however, what the character of the new climates will be when they become fully established. While certain areas could become drier, others could be wetter. Also major shifts of climatic belts could occur. How, when and where these could happen remain largely unresolved [1]. The Stern Review, for example concluded that (i) the costs of climate change - induced extreme weather events could reach 0.5 - 1% of GDP by the middle of the century; (ii) a 2 to 3 degree Celsius rise in temperature could reduce global output by 3%; and (iii) if that rise should become 5 degrees, up to 10% of global output could be lost. The Review also concluded that it would need 1% of GDP to stabilize emissions at manageable levels.

In Nigeria, the climate change phenomenon manifests in many respects apart from the rising sea level and inundation of coastal lands by sea water. These include increased frequencies of extreme climatic events like strong storms, floods and above average daily minimum and maximum temperature [3]. There have also been marked intraand inter annual rainfall variability [4]. As shown by Ojo et al. [5], the durations and intensities of rainfall have increased in the last three decades, producing large runoffs, floods and water logging in many places. Also temperature has continued to be above normal with relatively higher figures in 1973, 1987 and 1998. Most stations have had temperature increases of 0.2 to 0.3°C per decade in Nigeria.

The actual and potential impacts of climate change in Nigeria are considerable and have far reaching effects. All sectors of our socioeconomic development, including agriculture are vulnerable to climate change. It presents significant threats to the achievement of the Millennium Development Goals especially those related to eliminating poverty and hunger and promoting environmental sustainability.

Productive agriculture is essential to feed a growing population and sustain modern civilization. Climate affects agriculture, a fact well known to every farmer. Year-to-year variations in harvest are largely due to variations in temperature and precipitation that can make the difference between bountiful "bumper" crops and economic ruin. The persistent sahelian drought of 1969 – 73 [6] and later the spell of 1979 – 83 [7] are outstanding examples.

The realization of these impacts, therefore, calls for a more detailed analysis of the threats posed by climate change to agriculture in Enugu State. The main thrust of this paper is to analyze and discuss the effects of climate change on agriculture in the state for application in crop production planning.

The Study Area

Enugu State is located on the Northwestern fringe of southeastern Nigeria between latitudes 6°30N and 7°10N and longitudes 6°35E and 7°30°E. It is located in the plains exceeding 200 metres above mean sea level and high lands, which is part of Nsukka – Udi – Okigwe Cuesta.

Enugu State has an estimated population of about 4,1671 million people [8]. The land area is approximately 16,727 square km. The soil type is mostly hydromorphic, which is heavy and composed of silt and clay.

Methodology

Rainfall data for thirty years (1981 – 2010) were collected and analyzed to obtain seasonality index and reliability index. Crop yield data for four years from the four decades were also obtained. Again literatures were identified for review through a comprehensive search by using electronic and non-electronic databases. Related published literature and documents were searched in a systematic way using a range of keywords relating to climate change impacts and agricultural production.

RESULTS AND DISCUSSION

Production pattern over the years

Rainfall departures from normal have been reported to be up to 30 percent in the Northern and Southeast parts of Nigeria [9]. Agu [10] reported a 10 percent departure in Enugu areas from 1991 to 1995; while Enete [11] and Enete and Okolie [12] observed a seasonality index of 0.814 (81.4%) for the period 1973-2003 for Nsukka and the Environs.

Table 1: Classification of Rainfall Regimes

S1 Class Limits	Type of Rainfall Seasonality
0.19 and above	Very equable
0.20 – 0.39	Equable with a definite wetter season
0.40 – 0.59	Rather seasonal with a short drier season
0.60 – 0.79	Seasonal
0.80 – 0.99	Markedly seasonal with a long drier season
1.00 – 1.19	Most rains in 3 months or less
1.20 and above	Extreme, almost all rain in 1-2 months

Source: Walsh and Lawler (1981)

Table 2: Crop yield Estimates.

S/No.	YEAR	1981			1991			2001			2010
	Annual Rainfall	3167.9mm			1824 mm			1393.9 mm			1247 mm
	Crop	Area HA	Prod MT	Yield HA	Area HA	Prod.MT	Yield HA	Area HA	Prod.MT	Yield HA	Area HA
		"000"	"000"	"000"	"000"	"000"	"000"	"000"	"000"	"000"	"000"
1	Cassava	157.51	1454.5	9.6	234. 9	2215.5	2215.5	9.4	293	3001.5	1o24
2	Yam	78.03	811.6	10.4	83.55	1044.4	14.2	71.1	1114.2	9.4	59.03
3	Rice	27.09	59.6	2.2	28.38	95.34	25	25.06	48.7	2	29.5
4	Coco Yam	106.66	725.3	6.8	106.04	901.4	8.5	95.02	797.7	6.6	52.82
5	Pepper	0.07	0.03	0.42	0.05	0.02	0.31	2.89	4.34	1.5	40.2

Source: ENADP (2010)

According to Walsh and lawler [13] classification (Table 1), rainfall regime in the State falls under a markedly seasonal regime with a long drier season; and a reliability of rainfall regime characterized by a more marked variability in the months of maximum rainfall. These changes according to summer [14] are notable indicators of climate change. It should be noted therefore, that the general change in the seasonal rainfall regime observed in the study appeared as a sufficient indicator of climate change in Enugu State. This by implication changed the farming rhythm in the State as well as productivity yield.

Table 2 shows the descriptive estimates of production pattern over the years in Enugu State. The highest yield for each decade was selected. These yields were correlated with annual rainfall pattern. It was observed from Table 2 that all the traditional crops with the exception of cassava and pepper had a significant field decrease as rainfall continued to be more erratic. The immediate response to this ugly trend has been increasing use of irrigation throughout the state and a shift from stable crops to economic crops such as pepper, vegetables as well as planting around flood plains of major rivers and streams.

Implications of climate change on agriculture

Changes in atmospheric CO2, temperature, precipitation, and soil moisture, individually or together, could alter crop production [15]. Higher Atmospheric CO2 levels could stimulate photosynthesis and crop production - a process called the CO2 fertilization effect.

In many crops, a significant increase in daytime temperature maxima during the growing season reduces photosynthesis and increase evapotranspiration, leading to a reduction in yield. However, if warming does occur primarily at night, rather than during the day, this could greatly reduce the negative impacts of climate change on crop productivity [16]. A decrease in water availability would results in decreased food production in regions where water becomes critical. Also, as crops are stressed by climate change they become more vulnerable to damaging pests and diseases [17].

Climate also affects animal husbandry. Indirect effects include climate-induced changes in the availability and price of feed grain and in pasture and forage crop yields. Extreme heat can affect the health of animals. For example, heat will kill poultry and decrease milk production in Cows. Also, climate controls the distribution of livestock pests and diseases [18].

Effects of agriculture on climate change

The global flux of several greenhouse gases is influenced by agriculture. Land clearing, much of it from agriculture is the second largest source of $CO2$ emissions after fossil fuel combustion, accounting for 10 to 30% of net global $CO2$ emissions [15]. Forests, grasslands, and soils store large quantities of carbon per unit area than most crops and when they are cleared for cultivation, much of this carbon is released to the atmosphere. Mean estimates of carbon loss from the conversion of terrestrial ecosystems to agriculture range from 21 to 46% [19].

Some agriculture produces Methane ($CH4$) – the second-most important greenhouse gas. Paddy rice cultivation is responsible for about 40% of global $CH4$ emissions. In flooded rice paddies, microbial decomposition of high organic aquatic sediments, under low oxygen conditions, release $CH4$ gas to the atmosphere. This source will continue to grow as rice cultivation expands in the future [15]. Livestock production is responsible for about 15% of global $CH4$ emissions. Ruminant animals (cattle, sheep, goats, etc) digest grasses and other cellulose forage in their stomachs and release $CH4$ to the air. Cattle represent about 75% of the total livestock $CH4$ emission.

Nitrous oxide ($N2O$) is another greenhouse gas closely linked to agricultural activities. Like carbon, Nitrogen in vegetation and soils is lost to the atmosphere during land clearing. Also, nitrogen fertilizers are applied to crops, and generally enhance growth. However, excess nitrogen from fertilizers is leached into the soil and, through microbial denitrification, converted to volatile $N2O$ and released into the atmosphere. Estimates of $N2O$ release from agricultural fertilizers range from 0.1 to 1.5% of applied nitrogen [15]. Again, different agricultural practices have different consequences for greenhouse gas emissions. Intensive agriculture uses large quantities of fossil fuel for tilling and harvesting and soil carbon can be depleted.

Adaptation Strategies

There are a number of influences that could mitigate the negative effects of climate change on crop production. First, and perhaps most important, is the potential for farming practices to adapt to climate change by planting different climate-adapted species, using pesticides, or altering the dates of planting, harvesting and mitigation. Such adoptions could minimize the impacts of climate change on crop yields.

CONCLUSION

Climate change will have direct impacts on agriculture in Enugu State. While some crops, like cassava or pepper may increase in their yield or range, climate change will increase existing risks of extinction of many threatened crops or lead to reduction in productivity. Climate change is expected to increase the frequency and intensity of severe weather events. Unfortunately, Enugu State like many other states in Nigeria largely lacks the infrastructure necessary to respond adequately to such events. Diseases such as malaria are likely to have wider ranges, impacting more poor farmers that are already most affected by such diseases. In the same manner, pest and crop diseases will expand affecting quantity and quality of crops.

Changing rainfall pattern in Enugu State could devastate the rain-fed agriculture on which Enugu farmers depend on for survival. Increased occurrence of drought may lead to reduced agricultural yields and diminished food security. Water supplies may also be altered, primarily through changes in temperature and rainfall. In general, climate changes have the potential to lead to large disruptions in agricultural sector in the state and have adverse impacts on food security.

REFERENCES

1. IPCC (2007) Climate change 2007: The Physical Science Basis. Contribution of working Group 1 to the Fourth Assessment Report of the Intergovernmental Panel on Climate Change, (edn). Solomon S, Qin D, Manning M, Chen Z, Marquis M, et al. Cambridge, U.K. Cambridge University Press.
2. World Bank (2010) World Development Report 2010: Development and Climate Change. Washington, D.C: World Bank.
3. Enete IC (2008) Potential Impacts of Climate Change on Coastal Settlements of Nigeria. A paper presented at the 15th National Conference of the Environment and Behaviour Association of Nigeria held at Babcock University, Ogun State.
4. Adefolalu DO (2000) "An Introduction to the Challenges of Climate Change and its Impact in Nigeria". Workshop on "Climate Change" and Natural Disasters in Nigeria.
5. Ojo O,Ojo K, Oni E (2001) Fundamentals of Physical and Dynamic Climatology. SEDEC Publishers, Lagos.
6. Glantz MH (1976) Le Sahel: Nine Fallacies of Natural Disasters.
7. Nicholson SE (1986) Sub-Saharan Rainfall in the Years 1976 – 80: Evidence of Continued Drought.
8. NPC (2007) National Population Commission, Nigeria.
9. Oguntoyimbo J, Richards P (1978) Drought and the Nigeria Farmer. J Arid Environ1:165 – 194.
10. Agu AN (1995) Sustainable Use of Min-Dams in Adapting to Climate Variation. A paper presented at the International Conference on the implications of Climate change. Global Warming and Environmental Degradation at Oshodi-Lagos.

11. Enete IC (2004) A study of Enugu Rainfall patterns from the Viewpoint of precipitation Dynamics. J Res Prod 5.
12. Enete IC, Okolie T (2007) Changing Farm Rhythm in Nsukka Plateau: Evidence of Climate Change. J Environ Stud 3: 16 – 19.
13. Walsh PD, Lawler DM (1981) Rainfall seasonality Description: Spatial patterns and change through time. Weather 36: 201-208.
14. Summer GN (1988) Precipitation: Process and Analysis, John Wiley & Sons, Limited, Australia.
15. Rosenzweig CR, Hillel D (1998) Climate Change and the Global Harvest: Potential Impacts of the Greenhouse Effect on Agriculture, Oxford University Press, Oxford.
16. Dhakhwa GB, Campbell CL (1998) Potential Effects of differential day-night warming in Global Climatic change on Crop Production. Climatic Change 40: 647-667.
17. Porter JH, Parry ML, Carter TR (1991) The Potential Effects of Climate Change on Agricultural Insect Pests. AgrForest Meteorol57:221-240
18. Rotter R, Van de Geijn SC (1999) Climate Change Effects on Plant Growth, Crop Yield and Livestock. Climatic Change 43:651-681.
19. Scholesinger WH (1986) Changes in Soil Carbon Storage and Associated Properties with Disturbance and Recovery. In Trabalka JR,Reichle DE (edn). The Changing Carbon Cycle: A Global Analysis. New York: Springer – Verlag 194-220.

CHAPTER 4

Transformational Adaptation: Agriculture and Climate Change

L. Rickards [A] [C] and S. M. Howden [B]

[A] *Melbourne School of Land and Environment, University of Melbourne, Level 1, 221 Bouverie Street, Carlton, Vic. 2010, Australia.*

[B] *Adaptive Primary Industries, Enterprises and Communities, Climate Adaptation National Research Flagship, CSIRO, Canberra 2911, ACT, Australia.*

ABSTRACT

Climate change presents the need and opportunity for what the Stern report called 'major, non-marginal change'. Such transformational adaptation is rapidly emerging as a serious topic in agriculture. This paper provides an overview of the topic as it applies to agriculture, focusing on the Australian situation. It does so by first defining transformational adaptation, distinguishing it from other more incremental but overlapping modes of climate change adaptation and positing its emergence in agriculture as a response to both drivers and opportunities. The multiple dimensions of transformational adaptation are highlighted before two types or cases are focussed upon in order to tease out issues and highlight two major examples of transformation in agriculture in the past. Four key issues about climate change adaptation in agriculture particularly pertinent for transformational adaptation are then reviewed: the identification, level, distribution and management of the costs of adaptation; the definition, potential for and need to avoid maladaptation; the capacity demands that this level of adaptation presents; and the role of government in adaptation. Overall, transformational adaptation poses potential great gains but also great risks. It reinforces the realisation that agricultural research can no longer remain insulated from off-farm, non-science or non-agricultural knowledge or processes. Support and guidance of transformational adaptation requires that we understand how Australian agriculture is currently, and could be, positioned within the landscape, rural communities, and broader social, political and cultural environment.

INTRODUCTION

Climate change adaptation is widely accepted as a necessity and opportunity but remains an ambiguous topic and task. Numerous definitions and types exist. An emerging focus is on 'transformational' change as an adaptive response (e.g. Howden *et al.* 2007; O'Brien 2011; Pelling 2011). This paper examines what transformational adaptation means in the context of agriculture. It defines transformational climate change adaptation in agriculture as major, purposeful action undertaken at the farm or supra-farm level in response to potential or actual climate change impacts and opportunities in the context of other drivers.

A recent comprehensive definition of climate change adaptation highlights the many questions that adaptation raises about what adapts, what is adapted to, how adaptation occurs and how effective it is:

- 'Adaptation involves changes in social-ecological systems in response to actual and expected impacts of climate change in the context of interacting nonclimatic changes. Adaptation strategies and actions can range from short-term coping to longer-term, deeper transformations, aims to meet more than climate change goals alone, and may or may not succeed in moderating harm or exploiting beneficial opportunities' (Moser and Ekstrom 2010).

In agriculture in Australia and elsewhere, research and action to date has been more in keeping with the Intergovernmental Panel on Climate Change definition, which frames adaptation as a question of 'adjustment' (Parry *et al.* 2007). Yet, as the definition above usefully highlights, adaptation also encompasses more radical change: 'longer-term, deeper transformations'. Such transformational change is increasingly on the agenda both by necessity and design (e.g. Cork *et al.* 2010; O'Brien 2011; Pelling

2011; WGBU 2011). For example, the Stern report in the UK highlights that climate change presents the need and opportunity for 'major, non-marginal' change (Stern 2007: 1). In agriculture, the scale of the climate change challenge and the growing influence of outside interests mean that transformation is both especially likely and desired. *Transformational* adaptation in and of agriculture was recently recognised as a priority by the Australian Primary Industries National Adaptation Research Plan (http://piarn.org.au/about-piarn/national-adaptation-research-plan, accessed 30 March 2012) and is an emerging field of research (e.g. Walker *et al.* 2009; Park *et al.* 2012).

This paper provides an introductory overview of transformational adaptation (TA) to climate change in agriculture. Given the nascent stage of thinking on the topic and space limitations, it is necessarily conceptual and exploratory. It consists of three main parts. First, the question of what TA is in relation to agriculture is discussed, highlighting the multiple dimensions involved, the relative nature of the concept and the need to think about both *in situ* and migratory change in land use. Second, two historical examples of transformation in agriculture are reviewed to highlight agriculture's experience with transformation and illustrate different types of transformation. Third, four inter-related issues of climate change adaptation of particular pertinence to TA of agriculture are discussed before conclusions and implications for further research are presented.

PRESSURES FOR TRANSFORMATION: TWO FRAMEWORKS OF ADAPTATION

Adaptation as 'fitting to' the environment: a focus on thresholds

A range of biophysical and social drivers are contributing to the rising interest in TA. These include how climate change adaptation is intellectually framed. Two major frameworks focus on need and opportunity, respectively. In terms of need, adaptation is typically conceived as a process of 'fitting to' an altered set of external circumstances, reflecting the bio-physical origins of much of the early literature on adaptation. Illustrated by the growing literature on 'limits', 'tipping points' and 'thresholds' (e.g. Russill and Nyssa 2009; Kwadijk *et al.* 2010; McNeall *et al.* 2011), this interpretation of adaptation has reminded society of humans' biological nature and the climatic, ecological and other natural limits we exist within.

Such limits and the potential for major changes in socio-natural systems have been investigated over many years in numerous fields including disturbance ecology and disaster risk management. But it is the growing evidence of the potential for substantial and/or rapid changes in future climate as a result of anthropogenic climate change that has most powerfully underlined that transformational change in human systems may be necessary. As the projected degree and pace of climate change accelerates (Rahmstorf *et al.* 2007; New *et al.* 2011) alongside the exacerbation of other biophysical-based limits such as less land, water, fossil fuels, phosphorus and biodiversity and rising demand for agricultural products, the likely need for more systemic, powerful adaptation of agriculture is increasingly being realised (e.g. Alexandra and Riddington 2007; Howden *et al.* 2007; Campbell 2008; Brussaard *et al.* 2010; Cork 2010*a*; Horlings and Marsden 2011; Sounness 2011). The aim of such transformational change is to maintain sets of activities, products (*sensu lato*), values and processes in the sector or region. In particular, transformative adaptators either seek to: (1) pro-actively avoid, uncertain and severe challenges in order to avoid transformational change of a more involuntary, uncontrolled and negative character; or (2) be 'first movers' so as to take full advantage of what they see as emerging opportunities (Howden *et al.* 2007; Park *et al.* 2012).

Adaptation as 'fitting with' the environment: a focus on proactive action

While the above is an important interpretation and driver of transformational change in agriculture, it does not capture the full situation. Reflecting the coupled nature of social-ecological systems such as agriculture, adaptation needs to be viewed not just as 'fitting to' biophysical limits of the sort discussed above, but 'fitting with': a process of co-evolving with one's system (Collins and Ison 2009*a*, 2009*b*; Ison 2010). This alternative framework – which in itself represents a mental adaptation – is particularly suitable for perceiving and conceiving of transformational change for three reasons. First, it better acknowledges ongoing 'background' change in the environment we are part of. Here, transformational change is a recognisable shift in the *type* of change occurring rather than the introduction of change to a stable setting. Second, in this frame, humans have a greater degree of agency over what can be influenced and changed, including the climate itself. A co-evolutionary perspective allows for a proactive as well as reactive approach to adaptation and broadens the scope of what can be changed, reframing accepted 'inevitabilities' such as infrastructure limitations, economic arrangements, agricultural policies or mental frameworks as a question of choice. This reframing helps move the debate from a negative perspective (cost-avoidance) into a potentially positive perspective (increased environmental or economic sustainability and opportunity) (Howden *et al.* 2007). As well as reducing harm, TA may be pursued as a way of creating new and better systems, consistent with the move in adaptation circles towards low-regrets strategies (where benefits exist regardless of whether projected impacts eventuate) and win-win options (in which all stakeholders benefit) (UNDP 2009). Adaptation in general is increasingly understood as a response to multiple drivers and a source of multiple benefits (e.g. Pittock 2009) and this is especially true of TA (Pelling 2011; Park *et al.* 2012).

A target of low-regrets strategies is to address the long-standing ecological, social and economic vulnerabilities that climate change exposes, such as soil degradation, biodiversity losses, dependence on declining oil reserves, low financial equity, and an aging farmer population (Green *et al.* 2003; Anderies *et al.* 2006; Walker and Salt 2006; Campbell 2008; Fresco 2009; PHAA 2009; Robinson 2009; Cribb 2010; Fedoroff *et al.* 2010). Research on agriculture's 'social licence' to operate suggests that there is a related growing pressure on agriculture to demonstrate it is a legitimate and preferred user of a land, water, biodiversity and atmospheric base progressively strained by climate change (Martin and Williams 2011) where there are alternative land uses such as for biofuel production, carbon sequestration, biodiversity conservation and other ecosystem services. This includes political and social pressure on agriculture to reduce greenhouse gas emissions. This brings us to the fourth advantage of a co-evolutionary interpretation of adaptation, which is that, like others, farmers need to adapt to the full suite of indirect and direct impacts of climate change, including the full range of its social, cultural, political, financial and physical effects, as well as other concurrent situations and threats. As a response to and source of significant flow-on effects, transformational change requires a perspective that accommodates this breadth and dynamism, as a co-evolutionary perspective does.

CONCEPTUALISING TRANSFORMATIONAL CLIMATE CHANGE ADAPTATION IN AGRICULTURE

Multiple dimensions of potential transformational change

Given that TA is being and increasingly will be demanded of agriculture in the face of climate change, work is needed to clarify the concept. This requires accepting its heuristic, subjective and relative character. Different perspectives on what counts as transformational change exist and are not reducible to one another. This is for two reasons. First of all, there are many dimensions over which TA to climate change can be conceived (Table 1). While a large spatial scale is one possible dimension, TA can also be thought of differently in relation to smaller scales, from an individual to a family to a community, or from a single farm business or property, to a sector or region.

Table 1. Possible characteristics of transformational adaptation (TA) across different dimensions of change at the farm and regional scales

Dimension of change	Characteristic of TA	Example for farm system	Example for regional system
Depth of change	Relatively profound change in the system (may be interpreted as the creation of a 'new system')	Move from cereals to farm forestry for carbon sequestration	Change from being a dryland region to an irrigated region, or vice versa
Generality of change	General or intangible change in direction	Shift in goals from optimising productivity to more sustainable, long-term systems	Rezoning of land from agricultural to residential or conservation
Spatial scale of change	Change across whole system of interest	Conversion of a farm from a conventional to an organic system	Establishment of conservation corridors across all farmland in a region to facilitate species migration
Effect on the system	Even if action taken seems minor, has a relatively profound effect on the system	Adoption of substantial off-farm work by farmer to increase income, affecting many aspects of farm management and life	Closure of local abattoir due to reduced level and reliability of local production
Permanence of change	Change is not necessarily permanent, but difficult to reverse	Revegetate pasture with native vegetation Large financial investment in new land	Extension of rail system into new cropping region

Second, across any one dimension, the level of change that counts as 'transformational' is subjective and relative. It is relative in two ways. First it is a question of the difference between the deliberate versus background change. Climate change adaptation is essentially about 'persistence through change', which raises the questions of what persists (subject to pre-existing trajectories or variability) and what changes (in a desired, deliberate or obvious way). In TA, this ratio between persistence and change is low; more of the system is changed than is continued as is. Second, this ratio is relative to less transformational forms of change. Adaptation as adjustment (incremental adaptation) infers that the ideal 'ratio' of what remains relatively constant to what is deliberately changed is high. Most of the existing system remains on its pre-existing trajectory. Park *et al.* (2012) note that the central aim of incremental adaptations is to maintain 'the essence and integrity of an incumbent system or process at a given scale' (p. 5).

This second element of relativity is emphasised in the diagram by Howden *et al.* (2010) (Figure 1) that brought the concept of TA to prominence in the agricultural sector. It presents 'transformational adaptation' as the more radical end of a spectrum of change that begins with incremental adaptation (changes in practices and technologies within an existing system such as planting times or row spacing) and extends through systems adaptation (changes to an existing system, such as major change in the focus between livestock and cropping, new crop types or the adoption of precision agriculture), mapped against an increasing degree of climate change.

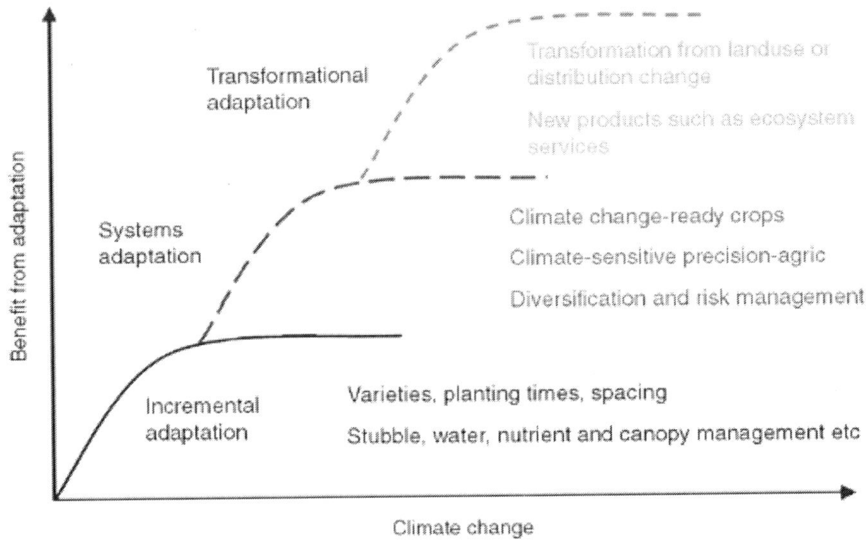

Figure 1. Levels of adaptation in relation to benefits from adaptation actions and degree of climate change, with illustrative examples (from Howden *et al.* 2010).

The third point to note about TA is that it does not exist in a vacuum: the three types of adaptation in Figure 1 overlap conceptually and in practice. The adaptation being pursued in any situation depends in part on the type of decision being faced. For some decisions (e.g. the amount of fertiliser to apply to a crop), reducing the decision horizon will likely always be the more rational way of proceeding. For other decisions (e.g. where to purchase a property), TA will be needed. Here, incremental adaptation alone may act as a blockage for necessary change by increasing investment in the existing system or locale and narrowing down alternatives for change: what the resilience, transition and policy literatures refer to as 'lock in trap', 'incrementalism' and 'negative resilience' (e.g. Handmer and Dovers 1996; Allison and Hobbs 2004; Anderies *et al.* 2006).

While the need for some degree of TA will likely become more indispensable and involuntary as climate change increases, it is not always necessary or desirable to wait for such pressures to accumulate. TA is already being pursued earlier to capture various benefits, address pre-existing structural issues, pre-empt possible problems or to accommodate the long lead times on associated decisions and actions (Howden *et al.* 2010; Park *et al.* 2012). TA will also always need to be accompanied by more incremental forms of change to allow for adjustments to fine-tune systems to ongoing change. The earlier decision makers act the more uncertainty and variability needs to be accommodated. This requires making decisions 'robust' across a range of possible future conditions future (Wilby and Dessai 2010; Stafford Smith *et al.* 2011) rather than focusing too narrowly on one climatic signal such as dryness. It also requires building in capacity for monitoring, learning and adjustment (Howden *et al.* 2007). If robustness and flexibility are taken seriously, deliberately acting early on some dimensions to alter the system (e.g. addressing pre-existing vulnerabilities, or adaptation to climate change) is likely to allow for a greater range of options and more influence over circumstances than non-action (Smit *et al.* 1996). How the different types of adaptation relate to each other in practice is a topic of ongoing research (Park *et al.* 2012). Depending on how TA is defined, incremental adaptations may accumulate on one farm or across a region to create or trigger a transformational change. Park *et al.* (2012) provide examples from the wine industry about how incremental and TA by actors operating at different scales interact, with transformational change at either the farm scale or sectoral scale enabling incremental adaptation at the opposite scale.

Two illustrative types of TA in agriculture

Further insight into TA in agriculture can be gained by considering two illustrative and potentially influential cases:

1. Changes in goal (for example resulting in a major change in land use and/or employment through trying to do more or something different); and/or
2. Changes in location (of an agricultural activity and/or agriculturalists) (Figure 2).

Both strategies are directed at reducing adaptors' vulnerability to climate change impacts (broadly defined), where vulnerability is understood as a combination of sensitivity and exposure to impacts and capacity to adapt (Smit *et al.* 1996). The main difference between them is their relative focus on sensitivity and exposure. Where the anticipated change is negative, an *in situ* change in land use or occupation primarily aims to reduce the adaptors' sensitivity to impacts by shifting to a less climate-sensitive mode of operating, while a spatial relocation primarily aims to reduce the adaptors' exposure to impacts by seeking out a new area anticipated to be more amenable to the continuance of their original activity or occupation. These transformational processes of replacement and displacement may result in the maintenance of agricultural activity, even if there is a shift from food- to non-food-based agriculture in some places. They may also involve a shift out of agriculture, particularly when combined. In contrast, if the perception of climate change is net positive, then the change in land use or location could aim to increase the sensitivity of the operation to capitalise on the anticipated benefits.

The choices involved in such changes highlight the importance of willingness to adapt as well as capacity to adapt (Moser and Ekstrom 2010). The strategies contrast those farmers who are strongly attached to their land and will go to great lengths (including working off-farm, planting it with Blue Gums etc.) to maintain ownership of it, and those who are strongly attached to their farming occupation (and perhaps particular enterprise types such as livestock or cropping) but do not care deeply about where they practice it. In this and other ways, TA intersects with identity issues within farming (e.g. see Marshall 2010), including the different ideals of agricultural masculinity dominant in different settings (e.g. Ferrell 2011).

Changes in goal and location presuppose and intersect with other adaptations and broader change processes, both in the landscape and socio-economically, technically and politically. This includes adaptations at the sub-farm level, which are defined here as system changes and not considered in detail. That said, a farmer purchasing an additional property in a new area in order to avoid projected climate impacts in the original area (i.e. partially relocating) could count as transformation, as would substituting a significant amount (not necessarily all) of their agricultural work with off-farm employment. Evidence is emerging of both of these adaptation strategies at work, some in explicit response to climate change (e.g. Park *et al.* 2012; Rickards 2012). Like more definitive changes in goal or location, these adaptation processes at the farm level may aggregate over time and space to create large-scale change. Such a high level view is now taken to describe major transformations within agriculture in the past.

PAST AND PRESENT TRANSFORMATIONAL PROCESSES IN AGRICULTURE

Not only are agriculturalists familiar with the idea of and need for climate adaptation (Kandlikar and Risbey 2000; Howden and Stokes 2009), but so too are they familiar with major transformations (Lobao and Meyer

2001; Barr 2009). These past and ongoing transformations boost the slim evidence base for transformational climate change adaptation by providing temporal and spatial analogues. As for climate change adaptation in general (Ford *et al.* 2010; McLeman and Hunter 2010), these can be used to gain insights into change processes and identify their possible characteristics and constraints.

Eras of agricultural production: an example of changes in goal

At a broad scale, an analogue of the first form of transformational change in agriculture identified above – a change in goal – is provided by the emergence of different eras of agriculture. At a very coarse scale, there was first a shift from traditional agriculture to 'modern' science and market-based agriculture, substituting at least at the sectoral level subsistence-based goals and values with ones associated with productivity and professionalism (Rickards 2006). More recently, 'productivist' agriculture has been replaced in part by 'multifunctional' or 'post-productivist' agriculture in Britain and the European Union through alterations to the Common Agricultural Policy arising from changing societal values (Potter and Tilzey 2005; Wilson 2007, 2008). Seen to a far lesser extent in Australia, this shift represents the formal addition of goals associated with land stewardship, biodiversity conservation and ecosystem services to those to do with productivity (Cocklin *et al.* 2006). This insertion of environmental sustainability goals and policies into the agricultural arena is likely to increase under climate change, especially in relation to the mitigation and sequestration imperative that agriculture increasingly seems expected to respond to (Bebbington *et al.* 2011). More generally, the farm business has been decoupled to a degree from farm production *per se* as the goal of increasing productivity has been replaced by that of profitability. This has led to a range of diversifications of farm income, in the most extreme cases involving a complete replacement of farm production income and an exit from farming (in most cases involving a concomitant shift in location for the farm members). As growing numbers of farmers have taken the latter option, farmer numbers have declined significantly in Australia and elsewhere. The resultant 'Great Transformation' of agriculture, as Lobao and Meyer (2001) call it in relation to the United States, provides a further analogue of potential TA under climate change, not just as a generic transformational change but as one that is likely to specifically feature among climate change adaptations.

The importation of British agriculture to Australia: an example of a change in land use/location

'British' agriculture in its increasingly scientific form was exported to Australia with colonialism in the eighteenth and nineteenth centuries. Indeed, expansion of agricultural land was a key tool of British colonialisation of the 'empty landscape' of *Terra Nullius* (Anderson 2003). This relocation or expansion of sedentary agriculture provides an analogue of the second form of transformational change in agriculture identified above and can illustrate significant lessons about such changes in location. First, in contrast to the frontier mentality that accompanied the clearing of land for agriculture, we now know that rather than inserting agriculture into a vacuum, the process was one of replacing existing land uses, people and values, including indigenous 'fire stick' farming (Jones 1969). The same holds for any relocation of agriculture today: it will always be in replacement of another set of attributes, values and goals, and thus necessitates careful consideration of costs, trade-offs and power issues. In this way, the processes of displacement and replacement are intricately linked. They are also linked to the extent that no agricultural system can be 'imported' to a new location wholesale without significant modifications. Pointing to the interaction of the TA process of relocation with incremental and systems adaptations, farmers and extension agents are acutely aware of the need to adjust generic advice to the local biophysical context (Pannell *et al.* 2006). While the point of relocating agriculture under climate change is to better match an existing agricultural system to a local climate elsewhere, this match will not be perfect, the climate will change in the new location also, and numerous other biophysical factors will be different from the original location (Park *et al.* 2012), requiring *in situ* changes to the agricultural system in its new location, as illustrated in Figure 2 above. Understanding how the 'new' agricultural system interacts with the local environment is essential for identifying, avoiding and managing negative externalities such as habitat loss and water pollution, which is another critical lesson that has emerged from the importation of British agriculture into the Australian context.

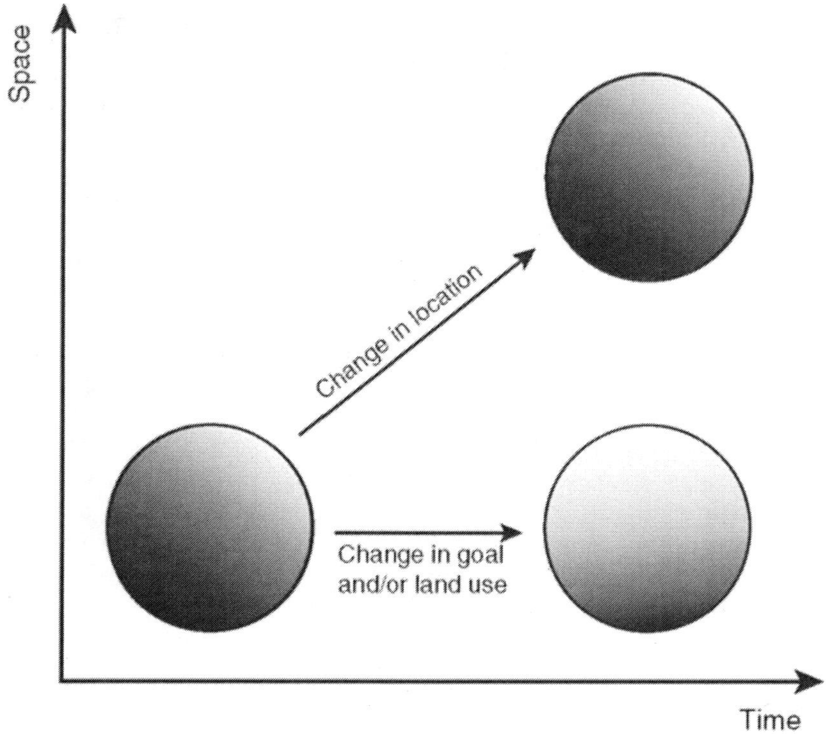

Figure 2. Schematic diagram of the two main forms of transformational adaptation in agriculture: (*a*) a major *in situ* change; (*b*) a change in spatial location, with minor accompanying *in situ* changes.

ISSUES FOR TRANSFORMATIONAL ADAPTATION

The above discussion of some major past transformational changes in agriculture illustrate that such change can involve serious challenges, risks and benefits. It also reinforces that climate change and adaptation are not happening in isolation; there is already a complex collection of change processes in play. Climate change impacts, adaptation efforts, and the impacts of adaptation itself will interact with these processes, and will likely exacerbate the complexity and uncertainty involved.

In this section, we turn to the climate change adaptation literature to present four general climate change adaptation issues of particular pertinence to TA in agriculture. Bridging literatures in this way is important for facilitating interdisciplinary, social learning of the sort needed to respond to climate change (Collins and Ison 2009*a*; Ison 2010). These issues – costs, maladaptation, adaptive capacity, and role of government – are discussed in turn below.

The costs of transformational adaptation

Common to all change, adaptation to climate change can involve some significant costs. Five are considered here. First are transaction costs, which is the toll on resources (mental, emotional, physical, financial, social etc) that the *process* of change exacts. Transaction costs are largely independent of the positive value or rationality of the change; a point that agricultural extension has come to realise following decades of frustration at farmers' seemingly illogical recalcitrance (Pannell *et al.* 2006). For a farmer, even learning about and considering the applicability of a potential change takes precious time and energy (Pannell *et al.* 2006). Climate change adaptations can involve particularly significant transaction costs, in part because of the uncertainty surrounding what adaptation is needed in response to what potential changes, what the options are and what is most appropriate: the contingency cost can be high. Transaction costs and risks can also be high because the expected benefits of action may not be realised for a long time, or even ever, and the actions can contest with existing norms and values (Howden *et al.* 2008; Orlove 2009). In the case of TA, the retarding influence of uncertainty is greatly exacerbated as the longer time scale, larger spatial and social scale, and greater magnitude of action increases the degree of uncertainty and the deliberation needed. As Stafford Smith *et al.* (2011) discuss, adaptation decisions with long life times (long lasting effects) usually also have long lead times.

The second type of cost to note is opportunity cost. The path dependency of adaptation decisions and especially TA decisions means that this form of cost is also particularly significant in the context of climate change. Path dependency refers to the way in which decisions taken now – particularly transformative ones – may constrain the options available at a later point (Inderberg and Eikeland 2009). As the uncertainty surrounding conditions under climate change demands a flexible or robust approach in which options are kept open (Hallegatte 2009; Wilby and Dessai 2010), this path dependency is especially problematic and requires that strenuous efforts are made to reduce 'lock in' (Allison and Hobbs 2004) and other opportunity costs. TA also highlights the potential for positive change and wider benefits under the guise of adaptation. A further opportunity cost is that which emerges when action is not taken to seize such potential. This appears to be a key motivation for current transformative climate change adaptations in the wine, peanut and rice industries (Park *et al.* 2012).

Opportunity costs can be accepted consciously or unconsciously. The latter unintended consequences are the third form of cost of relevance to climate change adaptation: those costs arising out of the unexpected, often long-term and long distance risks created by our actions, epitomised by climate change itself (Beck 2006) that can be systematically hidden from view if a narrow conceptualisation of value is used in economic framing (Costanza 2001). As Eriksen *et al.* (2011) note, 'not every response to climate change is a good one' (p. 7). For example, a climate adaptation that increases greenhouse gas emissions would be difficult to justify as being internally consistent (Howden *et al.* 2007). TA is particularly vulnerable to creating unintended consequences because the complexity of change involved is beyond the limits of our scientific capacity to predict or trace, as illustrated in the cases of transformational change in agriculture above. That said, avoidance of the unintended consequences of inadequate action is also a major motivator for TA (see Handmer and Dovers 1996).

The fourth type of cost is that which arises out of the incompleteness of adaptation, as adaptation will never fully compensate for climate change losses where the change is negative (Easterling *et al.* 2007). There will therefore always be residual losses and in some cases enhanced gains. Some of these losses may be intended. Economists focussed on cost-efficiency, for example, advocate for *a priori* acceptance of a certain level of damage (Fankhauser 2010). The rationale is that there is an economically optimal level of adaptation, represented by where the cost of adaptation action intersects with the projected cost of climate change impacts at a particular point in time (the latter being the fifth type of cost to note). Although seriously marred by the ignorance within and static nature of the predictions on which the calculations are based (Hulme *et al.* 2009), this approach has popularised the notion of 'over-adaptation': the idea that it is possible to act too soon or too vigorously. TA is especially vulnerable to criticism of over-adaptation, given that the long lead time and opportunity-focus of much such adaptation can require acting in advance of a necessitating degree of climate change (Stafford Smith *et al.* 2011). To the extent that TA involves preventative action, it is further vulnerable to this criticism, as the very successfulness of prevention can diminish public awareness of the reality of the problem at which such action is directed, leading to assertions that such actions are 'over the top' (Zaalberg *et al.* 2009). Nevertheless, there are already examples of farmers making transformative changes in response to the climate trends that are already occurring (Park *et al.* 2012). Ongoing evaluation of the success of these will hopefully inform the discussions in relation to over-adaptation. Hindsight will also inform us whether the costs incurred by reactive, incremental adaptation justified an earlier, staged approach to more systemic and transformational change.

The upshot of the above discussion is that where climate change is negative, adaptation is likely to reduce the size of the losses rather than remove them. Decision makers therefore need to address the types, level and distribution of losses they want to aim for. While TA carries the risk of significant losses, it also has a unique potential to avert potentially greater losses. Rising calls for radical change and action (Bailey and Wilson 2009) indicate that, as discussed by (Jones 2010) and implied by Figure 1 above, across sectors people are increasingly willing to take a risk in order to try to reduce the extent to which we are at risk under climate change. Agriculture is no exception.

Maladaptation

Whether society will succeed in reducing the extent to which we are at risk from climate change is an open question. Adaptation efforts that fail in this way, or involve excessive costs in the process, are a form of maladaptation. In addition to how effective and efficient adaptation actions are, 'good adaptation' is increasingly assessed, at least in theory, according to other more searching criteria, notably: equity (distributional justice, including future generations); social legitimacy (procedural justice); and sustainability (including inter-species justice) (Adger *et al.* 2005; Barnett and O'Neill 2010; O'Brien *et al.* 2009; Orlove 2009; Eriksen and Brown 2011) as well as risk of increased greenhouse gas emissions which would operate as a positive feedback to climate change (Howden *et al.* 2007).

The systems perspective associated with TA is crucial for attempting to trace the consequences of potential and actual adaptation actions across time, space and social settings in order to try to avoid maladaptation for one's self or others. The high order change that TA involves is also posited as a way of avoiding maladaptation, given the existing problems of inequity and unsustainability that currently exist and potentially may be exacerbated by climate change (Leichenko and O'Brien 2008). It is also in response to concerns about the perceived inadequacy of current proposed adaptation and mitigation actions that calls for TA have been initiated. Calls for consideration of TA are escalating, whether based on the ongoing failure to initiate effective mitigation (e.g. Ayers and Huq 2009), a concern about the long-term ineffectiveness of specific adaptation approaches (e.g. Stokes and Howden 2010), or a desire to address pre-climate change vulnerability (e.g. O'Brien and Wolf 2010).

Of particular pertinence to agriculture is debate about how adaptive the existing responses to climate variability are in the long term. Responding to the impacts of climatic extremes and natural disasters is unavoidable and therefore a stepping stone to longer term adaptation (e.g. McKeon et al. 1993). Failure to acknowledge this can lead to unrealistic and callous expectations of those recovering from disasters (McKeon *et al.* 2004). Recovery processes can also involve many of the same approaches needed for climate change adaptation, particularly when the climatic extreme involved is aligned with changes in average conditions, as is the case for drought in Australia (Howden *et al.* 2010). Yet, research also indicates that responses to variability may in the longer term prove maladaptive. In a study of graziers in Queensland, for example, Marshall (2010) concludes that: 'Enhanced strategies for coping with climate variability will provide a way for encouraging gradual, incremental adjustments for climate adaptation' but may also lead to over-confidence among producers and 'in fact, make them vulnerable to more extreme and frequent climate events predicted for the future (p. 40). Importantly, the consequences of these extreme events may be effectively irreversible and thus reactive, incremental adaptation based on historical experience of extremes and post-hoc learning will potential degrade future production options (McKeon *et al.* 2004). Similarly Handmer and Dovers (1996) stress that 'change at the margins' can create a false sense of security and delay needed transformational change and (Davies 2009) argues that perpetuating a state of mere 'coping' is no substitute for genuine adaptation. Overall, the relationship between adaptation to climate variability and climate change is complex, highly pertinent to agriculture, and in need of in-depth research.

Adaptive capacity

TA requires a higher level of adaptive capacity than incremental or systems-scale adaptation due to the greater risks and complexity arising from the change. To the extent that it can involve collective, voluntary action (a common mode of change in agriculture), it also requires an especially high level of willingness to adapt, which is determined in part by the potential benefits to be gained and the perceived legitimacy of the adaptation process. In many cases TA also requires substantial resources.

The attributes required for transformation may be somewhat different from those required for incremental adaptation. Cork (2010*b*) lists the characteristics of transformability as: experimentation; support for change; trust (social capital); human, built and natural capital; and cross-scale awareness. These characteristics are important in incremental or systems adaptations as well as in change processes more generally, but it could be argued that they are more crucial in transformative climate adaptation. More generally, the five types of capital identified by Nelson *et al.* (2010*a*, 2010*b*) – human capital (e.g. health, education), social capital (within-group and between-group connectedness), financial capital (assets, cash), physical capital (infrastructure, technology) and natural capital (soil, water, biodiversity) – are needed for the perpetuation and improvement of rural livelihoods in general and adaptation to climate change in particular. Empirical research on farming families' decision making experiencing drought and other pressures (Greenhill *et al.* 2009; King *et al.* 2009; Rickards 2012) suggests that the accumulative effects of climate change and other issues erode their net level of these 'capitals' or assets and may encourage maladaptive substitution between them (e.g. exploiting one's human health and energy to substitute for a lack of financial capital by taking on more work). At the level of human capital, transformative learning research with producers suggests that critical reflection on the relationship between practices, the environment, roles, responsibilities and social norms can lead to the kind of transformative change needed for major adaptation to climate change (Tarnoczi 2011). A wealth of research also highlights the importance of social and human capital in enabling the sort of positive, enduring, collective action TA will involve (e.g. Meert *et al.* 2005; Gooch and Rigano 2010; Burton and Paragahawewa 2011).

Role of government

The final issue to highlight in relation to TA in agriculture is the role of government. This is shaped by broader debates about the role of government both in relation to climate change mitigation and adaptation, agriculture and society more generally. In terms of the former, it is acknowledged that climate change can be framed as an unprecedented market failure (Stern 2007), and therefore legitimates government intervention. Yet, the form this intervention could most effectively take is less certain, especially in relation to adaptation which is often misconceived as a mere local-level, private-good concern that will happen autonomously (Nelson *et al.* 2011). Sensitivity to the private benefits of adaptation is enhanced in Australian agriculture, where since the 1970s there has been a move to more precisely distinguish private good from public good activities in agriculture in order to help reduce and justify government expenditure in the sector (Simpson and Dargusch 2010), including in relation to climate-related disasters (Botterill 2003). While this stance remains strongly influential, climate change and associated food security concerns are prompting growing calls to redress some of the consequent reduction in government investment in agricultural research, development and extension (RDE), both in Australia and beyond (e.g. Cribb 2010; PMSEIC 2010). The case for such investment seems particularly strong in support of transformational adaptation, with Pannell (2010) for example arguing that government RDE investment should be focussed on supporting 'decisions that are larger and indivisible, such as land purchase or the decision to exit from agriculture' (unpaginated). Nelson *et al.* (2011) argue that there is a key role of government in identifying and implementing policies that foster pro-active and anticipatory adaptation especially addressing market failures such as moral hazard and adverse selection of risk management approaches.

The appropriate type as well as quantum of agricultural RDE is needed to support TA in agriculture. Information is needed that provides flexible options for farmers and others to use, monitor and evaluate. To be effective, particularly at a transformational level, it is crucial that the underpinning research is cross scale, cross sectoral and integrated with mitigation (Nelson 2009), the latter being a major policy and societal shift that agriculture needs to adapt to. Research to support transformational change therefore needs to be interdisciplinary and engage with relevant insights from diverse areas, including futures thinking (Inayatullah 2008), innovation studies and socio-technical transitions studies (e.g. Geels and Schot 2007; Dolata 2009; Leggewie and Welzer 2010), rural and political geography (e.g. Gibson *et al.* 2010), resilience thinking (e.g. Cork 2010*b*; Gelcich *et al.* 2010), and the mainstream climate risk management literature, including that on migration and climate change adaptation (e.g. Black *et al.* 2011). As pointed out elsewhere, conventional insular agricultural research is increasingly inadequate in the face of growing complexity and uncertainty (see, for example, PMSEIC 2010). More than interdisciplinary, such research also needs to be transdisciplinary, which refers to the integration of non-academic knowledge through participatory processes and in developing systems of adaptive governance (Nelson *et al.* 2008). CSIRO, for example, is starting to work with farmers undertaking transformational changes to better understand and help assess the consequences of different transformation options, in part to avoid problems of the past and the issues of over- or maladaptation discussed above (Park *et al.* 2012). Policy makers, agricultural organisations, agribusiness, agricultural advisors and rural community groups are other important 'end users' and shapers of research into TA of and in agriculture. Effort is needed to bring these different perspectives together to discuss the cross-scale and cross-sectoral shifts that may eventuate under climate change and increase the social learning that is needed for appropriate adaptation.

CONCLUSION

TA is a crucial and increasingly relevant aspect of agriculture's adaptation to climate change. It is also little studied and poorly understood. This paper has sought to provide an introductory overview of the topic in relation to agriculture. It has highlighted that it is a multidimensional, relative and critical concept, which can be understood more easily by focusing on two major types of change: changes in the goals and/or location, including the option for farmers of exiting the industry. Historical and ongoing examples of other transformational change processes in agriculture provide important insights into what TA under climate change may involve, and further research is required in this area. Further research is also needed to engage the agricultural research community with the broader learning emerging out of the mainstream climate change adaptation literature and other disciplines. TA reinforces the realisation that agricultural research can no longer remain insulated from off-farm, non-science or non-agricultural knowledge or processes. Just as support and guidance of TA requires that we understand how Australian agriculture is currently, and could be, positioned climatically, we need to understand its position within the landscape, rural communities, and broader social, political and cultural environment. It is only by building our capacity in adopting such a broad systems perspective that TA will avoid the maladaptations of the past and enable opportunities for positive change to be realised.

ACKNOWLEDGEMENT

The authors would like to thank the Primary Industries Adaptation Research Network for facilitating the production of this paper.

REFERENCES

1. Adger WN, Arnell NW, Tompkins EL (2005) Successful adaptation to climate change across scales. Global Environmental Change 15, 77–86.
2. Alexandra J, Riddington C (2007) Redreaming the rural landscape. Futures 39, 324–339.
3. Allison HE, Hobbs RJ (2004) Resilience, adaptive capacity, and the 'Lock-in trap' of the Western Australian agricultural region. Ecology and Society 9. Available at: www.ecologyandsociety.org/vol9/iss1/art3 (accessed 30 March 2012)
4. Anderies J, Ryan P, Walker P (2006) Loss of resilience, crisis and institutional change: lessons from an intensive agricultural system in southeastern Australia. Ecosystems 9, 865–878.
5. Anderson K (2003) White natures: Sydney's Royal Agricultural Show in post-humanist perspective. Transactions of the Institute of British Geographers 28, 422–441.
6. Ayers J, Huq S (2009) The value of linking mitigation and adaptation: a case study of Bangladesh. Environmental Management 43, 753–764.
7. Bailey I, Wilson GA (2009) Theorising transitional pathways in response to climate change: technocentrism, ecocentrism, and the carbon economy. Environment & Planning A 41, 2324–2341.
8. Barnett J, O'Neill S (2010) Maladaptation. Global Environmental Change 20, 211–213.
9. Barr N (2009) 'The house on the hill: the transformation of Australia's farming communities.' (Land and Water Australia and Halstead Press: Canberra)
10. Bebbington J, Asaduzzaman M, Fernandez A, Clark M, Guillou M, Jahn M, Erda L, Mamo T, Van Bo N, Nobre CA, Scholes RJ, Sharma R, Wakhungu J (2011) Achieving Food Security in the Face of Climate Change: summary for policy makers from the

Commission on Sustainable Agriculture and Climate Change. CGIAR Research Program on Climate Change, Agriculture and Food Security (CCAFS), Copenhagen, Denmark. Available at: www.ccafs.cgiar.org/commission (accessed 30 March 2012)

11. Beck U (2006) Living in the world risk society. Economy and Society 35, 329–345.
12. Black R, Kniveton D, Schmidt-Verkerk K (2011) Migration and climate change: towards an integrated assessment of sensitivity. Environment & Planning A 43, 431–450.
13. Botterill LC (2003) Uncertain climate: the recent history of drought policy in Australia. The Australian Journal of Politics and History 49, 61–74.
14. Brussaard L, Caron P, Campbell B, Lipper L, Mainka S, Rabbinge R, Babin D, Pulleman M (2010) Reconciling biodiversity conservation and food security: scientific challenges for a new agriculture. Current Opinion in Environmental Sustainability 2, 34–42.
15. Burton RJF, Paragahawewa UH (2011) Creating culturally sustainable agri-environmental schemes. Journal of Rural Studies 27, 95–104.
16. Campbell A (2008) 'Paddock to plate: food, farming and Victoria's progress to sustainability. ACF's Future Food & Farm Project.' (Australian Conservation Foundation: Melbourne)
17. Cocklin C, Dibden J, Mautner N (2006) From market to multifunctionality? Land stewardship in Australia. The Geographical Journal 172, 197–205.
18. Collins K, Ison R (2009a) Jumping off Arnstein's ladder: social learning as a new policy paradigm for climate change adaptation. Environmental Policy & Governance 19, 358–373.
19. Collins K, Ison R (2009b) Living with environmental change: adaptation as social learning. Environmental Policy and Governance 19, 351–357.
20. Cork S (2010a) 'Resilience and transformation: preparing Australia for uncertain futures.' (CSIRO Publishing: Melbourne)
21. Cork S (Ed.) (2010b) Resilience of social-ecological systems. In 'Resilience and transformation: preparing Australia for uncertain futures'. pp. 131–142. (Australia 21 and CSIRO: Melbourne)
22. Cork S, Eckersley R, Walker B (2010) 'Rapid and surprising change in Australia's future: anticipating and preparing for future challenges and opportunities on the way to a sustainable Australia.' (Australia 21: Canberra). www.australia21.org.au/oldsite/pdf/Tipping2007.pdf
23. Costanza R (2001) Visions, values, valuation, and the need for an ecological economics. Bioscience 51, 459–468.
24. Cribb J (2010) 'The coming famine: the global food crisis and what we can do to avoid it.' (CSIRO Publishing: Melbourne)
25. Davies S (2009) Are coping strategies a cop out? In 'The Earthscan Reader on adaptation to climate change'. (Eds ELF Schipper, I Burton) pp. 99–116. (Earthscan: London)
26. Dolata U (2009) Transformative capacity, adaptability, patterns of change: an analytical framework. Research Policy 38, 1066–1076.
27. Easterling WE, Aggarwal PK, Batima P, Brander K, Erda L, Howden SM, Kirilenko A, Morton J, Soussana JF, Schmidhuber J, Tubiello FN (2007) Food, fibre and forest products. Climate Change 2007: impacts, adaptation and vulnerability. In 'Contribution of Working Group II to the Fourth Assessment Report of the Intergovernmental Panel on Climate Change'. (Eds ML Parry, OF Canziani, JP Palutikof, PJ van der Linden, CE Hanson) pp. 273–313. (Cambridge University Press: Cambridge, UK)
28. Eriksen S, Brown K (2011) Sustainable adaptation to climate change. Climate and Development 3, 3–6.
29. Eriksen S, Aldunce P, Bahinipati CS, Martins RD, Molefe JI, Nhemachena C, O'Brien K, Olorunfemi F, Park J, Sygna L, Ulsrud K (2011) When not every response to climate change is a good one: identifying principles for sustainable adaptation. Climate and Development 3, 7–20.

30. Fankhauser S (2010) The costs of adaptation. WIREs Climate Change 1, 23–30.
31. Fedoroff NV, Battisti DS, Beachy RN, Cooper PJM, Fischhoff DA, Hodges CN, Knauf VC, Lobell D, Mazur BJ, Molden D, Reynolds MP, Ronald PC, Rosegrant MW, Sanchez PA, Vonshak A, Zhu JK (2010) Radically rethinking agriculture for the 21st Century. Science 327, 833–834.
32. Ferrell AK (2011) Doing masculinity: gendered challenges to replacing burley tobacco in central Kentucky. Agriculture and Human Values
33. Ford JD, Keskitalo ECH, Smith T, Pearce T, Berrang-Ford L, Duerden F, Smit B (2010) Case study and analogue methodologies in climate change vulnerability research. WIREs Climate Change 1, 19
34. Fresco LO (2009) Challenges for food system adaptation today and tomorrow. Environmental Science & Policy 12, 378–385.
35. Geels FW, Schot J (2007) Typology of sociotechnical transition pathways. Research Policy 36, 399–417.
36. Gelcich S, Hughes TP, Olsson P, Folke C, Defeo O, Fernandez M, Foal S, Gunderson LH, Gundersong LH, Rodríguez-Sickert C, Scheffer M, Steneck RS, Castilla JC (2010) Navigating transformations in governance of Chilean marine coastal resources. Proceedings of the National Academy of Sciences of the United States of America 107, 16794–16799.
37. Gibson K, Cahill A, McKay D (2010) Rethinking the dynamics of rural transformation: performing different development pathways in a Philippine municipality. Transactions of the Institute of British Geographers 35, 237–255.
38. Gooch M, Rigano D (2010) Enhancing community-scale social resilience: what is the connection between healthy communities and healthy waterways? The Australian Geographer 41, 507–520.
39. Green K, Harvey M, McMeekin A (2003) Transformations in food consumption and production systems. Journal of Environmental Policy and Planning 5, 145–163.
40. Greenhill J, King D, Lane A, MacDougall C (2009) Understanding resilience in South Australian farm families. Rural Society 19, 318–325.
41. Hallegatte S (2009) Strategies to adapt to an uncertain climate change. Global Environmental Change 19, 240–247.
42. Handmer J, Dovers S (1996) A typology of resilience: rethinking institutions for sustainable development. Organization & Environment 9, 482–511.
43. Horlings LG, Marsden TK (2011) Towards the real green revolution? Exploring the conceptual dimensions of a new ecological modernisation of agriculture that could 'feed the world'. Global Environmental Change21, 441–452.
44. Howden SM, Stokes CJ (2009) Introduction. In 'Adapting agriculture to climate change: preparing Australian agriculture, forestry and fisheries for the future'. (Eds CJ Stokes, SM Howden) pp. 1–12. (CSIRO: Melbourne)
45. Howden SM, Soussana JF, Tubiello FN, Chhetri N, Dunlop M, Meinke H (2007) Adapting agriculture to climate change. Proceedings of the National Academy of Sciences of the United States of America 104, 19691–19696.
46. Howden SM, Crimp SJ, Stokes CJ (2008) Climate change and Australian livestock systems: impacts, research and policy issues. Australian Journal of Experimental Agriculture 48, 780–788.
47. Howden SM, Crimp S, Nelson RN (2010) Australian agriculture in a climate of change. In 'Managing Climate Change: Papers from GREENHOUSE 2009 Conference'. (Eds I Jubb, P Holper, W Cai) pp. 101–112. (CSIRO Publishing: Melbourne)
48. Hulme M, Pielke R, Dessai S (2009) Keeping prediction in perspective. Nature 3, 2
49. Inayatullah S (2008) Six pillars: futures thinking for transforming. Foresight 10, 4–21.
50. Inderberg TH, Eikeland PO (2009) Limits to adaptation: analysing institutional constraints. In 'Adapting to climate: thresholds, values and governance'. (Eds W Adger, I Lorenzonin, K O'Brien) pp. 433–447. (Cambridge University Press: Cambridge, UK)

51. Ison R (2010) 'Systems practice: how to act in a climate change world.' (The Open University and Springer: London)
52. Jones R (1969) Fire-stick farming. Australian Natural History 16, 224
53. Jones R (2010) The use of scenarios in adaption planning: managing risks in simple to complex settings. Report for the Victorian Centre for Climate Change Adaptation Research. Scenarios for Climate Adaptation Program. Available at: www.vcccar.org.au/files/vcccar/Jones%20scenarios%20presentation%2011-11-10.pdf
54. Kandlikar M, Risbey J (2000) Agricultural impacts of climate change: if adaptation is the answer, what is the question? Climatic Change 45, 529–539.
55. King D, Lane A, MacDougall C, Greenhill J (2009) The resilience and mental health and wellbeing of farm families experiencing climate variation in South Australia. Report for the National Institute of Labour Studies Incorporated, Adelaide, Australia.
56. Kwadijk JCJ, Haasnoot M, Mulder JPM, Hoogvliet MMC, Jeuken ABM, van der Krogt RAA, van Oostrom NGC, Schelfhout HA, van Velzen EH, van Waveren H, de Wit MJM (2010) Using adaptation tipping points to prepare for climate change and sea level rise: a case study in the Netherlands. WIRES Climate Change1, 729–740.
57. Leggewie C, Welzer H (2010) Another 'great transformation'? Social and cultural consequences of climate change. Journal of Renewable and Sustainable Energy 2, 031009
58. Leichenko RM, O'Brien KL (2008) 'Environmental change and globalization: double exposures.' (Oxford University Press: Oxford, UK)
59. Lobao L, Meyer K (2001) The great agricultural transition: crisis, change, and social consequences of Twentieth Century US farming. Annual Review of Sociology 27, 103–124.
60. Marshall NA (2010) Understanding social resilience to climate variability in primary enterprises and industries. Global Environmental Change 20, 36–43.
61. Martin P, Williams J (2011) Renegotiating farmers' social licence. In 'Defending the social licence of farming: issues, challenges and new directions for agriculture'. (Eds J Williams, P Martin) pp. 195–202. (CSIRO Publishing: Melbourne)
62. McKeon GM Howden SM Abel NOJ King JM 1993 Climate change: adapting tropical and subtropical grasslands. In 'Grasslands for our world'. (Ed. MJ Baker) pp. 426–435. (SIR Publishing: Wellington, New Zealand)
63. McKeon G, Hall W, Henry B, Stone G, Watson I (Eds) (2004) 'Pasture degradation and recovery in Australia's Rangelands: learning from history.' (Queensland Department of Natural Resources, Mines and Energy: Brisbane)
64. McLeman RA, Hunter LM (2010) Migration in the context of vulnerability and adaptation to climate change: insights from analogues. WIRES Climate Change 1, 450–462.
65. McNeall D, Halloran PR, Good P, Betts RA (2011) Analyzing abrupt and nonlinear climate changes and their impacts. WIRES Climate Change 2, 663–686.
66. Meert H, van Huylenbroeck G, Vernimmen T, Bourgeois M, van Hecke E (2005) Farm household survival strategies and diversification on marginal farms. Journal of Rural Studies 21, 81–97.
67. Moser SC, Ekstrom JA (2010) A framework to diagnose barriers to climate change adaptation. Proceedings of the National Academy of Sciences of the United States of America 107, 22026–22031.
68. Nelson GC (2009) Agriculture and Climate Change: An Agenda for Negotiation in Copenhagen, 2020 Vision for Food. Agriculture and the Environment, International Food Policy Research Institute, Washington, DC.
69. Nelson R, Howden M, Smith MS (2008) Using adaptive governance to rethink the way science supports Australian drought policy. Environmental Science & Policy 11, 588–601.
70. Nelson R, Kokic P, Crimp S, Martin P, Meinke H, Howden M (2010a) The vulnerability of Australian agriculture to climate variability & change: Part I –

reconciling the supply and demand for integrated assessments. Environmental Science & Policy 13, 8–17.

71. Nelson R, Kokic P, Crimp S, Martin P, Meinke H, Howden M, Devoil P, McKeon G, Nidumolu U (2010b) The vulnerability of Australian agriculture to climate variability & change: Part II – vulnerability assessments that support adaptation. Environmental Science & Policy 13, 18–27.
72. Nelson R, Byron N, Stafford Smith M (2011) Towards economic and policy analysis of adaptation to climate change. In 'Paper presented at the Australian Economics Society Annual Conference'. 11–14 July, Canberra, ACT. Available at: http://publicpolicy.anu.edu.au/coombs/workshops/climate_change_adaptation_in_australia/Adaptation%20-%20towards%20eco%20&%20policy%20analysis.pdf
73. New M, Liverman D, Schroder H, Anderson K (2011) Four degrees and beyond: the potential for a global temperature increase of four degrees and its implications. Philosophical Transactions of the Royal Society A 369, 6–19.
74. O'Brien K (2011) Global environmental change II: from adaptation to deliberate transformation. Progress in Human Geography
75. O'Brien KL, Wolf J (2010) A values-based approach to vulnerability and adaptation to climate change. WIRES Climate Change 1, 232–243.
76. O'Brien K, Hayward B, Berkes F (2009) Rethinking social contracts: building resilience in a changing climate. Ecology and Society 14. Available at: www.ecologyandsociety.org/vol14/iss2/art12/
77. Orlove B (2009) The past, the present and some possible futures of adaptation. In 'Adapting to climate change: thresholds, values, governance'. (Eds WN Adger, I Lorenzoni, K O'Brien) pp. 131–163. (Cambridge University Press: Cambridge, UK)
78. Pannell DJ (2010) Policy for climate change adaptation in agriculture. In 'Paper presented at the 54th Annual Conference of the Australian Agricultural and Resource Economics Society'. Adelaide, 10–12 February 2010.
79. Pannell DJ, Marshall G, Barr N, Curtis A, Vanclay F, Wilkinson R (2006) Understanding and promoting adoption of conservation practices by rural landholders. Australian Journal of Experimental Agriculture 46, 1407–1424.
80. Park SE, Marshall NA, Jakku E, Dowd AM, Howden SM, Mendham E, Fleming A (2012) Informing adaptation responses to climate change through theories of transformation. Global Environmental Change 22, 115–126.
81. Parry ML Canziani OF Palutikof JP van der Linden PJ Hanson CE 2007 Climatic change: impacts, adaptation and vulnerability. Contribution of Working Group II to the Fourth Assessment Report of the Intergovernmental Panel on Climate Change, Cambridge, UK.
82. Pelling M (2011) 'Adaptation to climate change: from resilience to transformation.' (Routledge: London)
83. PHAA (2009) 'A future for food: addressing public health, sustainability and equity from paddock to plate.' (Public Health Association of Australia: Canberra) www.phaa.net.au/documents/PHAA%20Report.pdf
84. Pittock J (2009) Lessons for climate change adaptation from better management of rivers. Climate and Development 1, 194–211.
85. PMSEIC (2010) 'Australia and food security in a changing world.' (The Prime Minister's Science, Engineering and Innovation Council: Canberra)
86. Potter C, Tilzey M (2005) Agricultural policy discourses in the European post-Fordist transition: neoliberalism, neomercantilism and multifunctionality. Progress in Human Geography 29, 581–600.
87. Rahmstorf S, Cazenave A, Church JA, Hansen JE, Keeling RF, Parker DE, Somerville RCJ (2007) Recent climate observations compared to projections. Science 316, 709
88. Rickards L (2006) Capable, enlightened and masculine: constructing English agriculturalist ideals in formal agricultural education, 1845–2003. PhD Thesis, School of Geography and Environment, University of Oxford, UK.
89. Rickards L (2012) Critical breaking point? The effects of climate variability and other pressures on farm households. Final Report for the Birchip Cropping Group,

Birchip, Australia. Available at: www.bcg.org.au/cb_pages/ Social_and_Comm_Dev.php

90. Robinson GM (2009) Towards sustainable agriculture: current debates. Geography Compass 3, 1757–1773.
91. Russill C, Nyssa Z (2009) The tipping point trend in climate change communication. Global Environmental Change 19, 336–344.
92. Simpson S, Dargusch P (2010) Classifying public benefit in Australian agricultural research. Australasian Agribusiness Review 18, 13
93. Smit B, McNabb D, Smithers J (1996) Agricultural adaptation to climatic variation. Climatic Change 33, 7–29.
94. Sounness C (2011) Five big questions: assisting broadacre agriculture to face the need for transformational change. In 'Climate Change Research Strategy for Primary Industries Conference'. Feburary 2011, Melbourne. Available at: www.theccrspiconference.com.au/viewStory/Presentations
95. Stafford Smith M, Horrocks L, Harvey A, Hamilton C (2011) Rethinking adaptation for a 4°C world. Philosophical Transactions of the Royal Society A 369, 196–216.
96. Stern N (2007) 'Stern Review: the economics of climate change.' (Cambridge University Press: Cambridge, UK)
97. Stokes CJ, Howden SM (Eds) (2010) 'Adapting agriculture to climate change: preparing Australian agriculture, forestry and fisheries for the future.' (CSIRO Publishing: Melbourne)
98. Tarnoczi T (2011) Transformative learning and adaptation to climate change in the Canadian Prairie agro-ecosystem. Mitigation and Adaptation Strategies for Global Change 16, 387–406.
99. UNDP (2009) A climate risk management approach to disaster reduction and adaptation to climate change. In 'The Earthscan Reader on adaptation to climate change'. (Eds ELF Schipper, I Burton) pp. 229–248. (Earthscan: London)
100. Walker B, Salt D (2006) 'Resilience thinking: sustaining ecosystems and people in a changing world.' (Island Press: New York)
101. Walker B, Abel N, Anderies J, Ryan P (2009) Resilience, adaptability and transformability in the Goulburn-Broken catchment, Australia. Ecology & Society 14, No. 12
102. WGBU (2011) World in transition: a social contract for sustainability. Report by the German Advisory Council on Global Change (WGBU), Berlin.
103. Wilby RL, Dessai S (2010) Robust adaptation to climate change. Weather 65, 180–185.
104. Wilson G (2007) 'Multifunctional agriculture: a transition theory perspective.' (CABI: Wallingford, UK)
105. Wilson GA (2008) From 'weak' to 'strong' multifunctionality: conceptualising farm-level multifunctional transitional pathways. Journal of Rural Studies 24, 367–383.
106. Zaalberg R, Midden C, Meijnders A, McCalley T (2009) Prevention, adaptation, and threat denial: flooding experiences in the Netherlands. Risk Analysis 29, 1759–1778.|

CHAPTER 5

Climate Change Impacts, Agroforestry Adaptation and Policy Environment in Sri Lanka

Mangala De Zoysa[1], Makoto Inoue[2]

[1]*Department of Agricultural Economics, Faculty of Agriculture, University of Ruhuna, Kamburupitiya, Sri Lanka*

[2]*Global Forest Environmental Studies, Department of Global Agricultural Sciences, Graduate School of Agricultural and Life Sciences, The University of Tokyo, Tokyo, Japan*

ABSTRACT

Despite its low emissions of greenhouse gasses (GHGs), Sri Lanka is considered as a vulnerable small island nation under climate change. Agroforestry, which uniquely integrates trees into land use systems, has historically contributed to climate change adaptation in Sri Lanka. Hence, the promotion of agroforestry practices is vitally important to enhance the resiliency of the country to future climate change. This paper reviews the literature and discusses the adverse impacts of climate change on agriculture and forestry, the effects of adapting agroforestry on climate changes, and important policies for promoting agroforestry adaptation in Sri Lanka. The adverse impacts of climate change on agriculture and forestry are identified as follows: endangerment of natural assets; prevalence of pests, diseases and invasive species; agriculture and forest damage; and high levels of food insecurity. The adaptation of agroforestry impacts climate change by increasing the tree cover outside forests, enhancing forest carbon stocks, conserving biodiversity, reducing risks and damage intensity, maintaining health and vitality, and scaling up multiple benefits. Changing regulations and legislation, developing awareness and capacity, planning for climate-smart agroforest landscapes, offering no-regret options and implementing climate change adjustment programs are discussed as important policy measures. It is concluded that agroforestry, along with favourable policy, has an important role in climate change adaptation by enhancing resilience to climate impacts on farming systems.

Keywords: Climate Change Impacts, Agroforestry Adaptation, Policy Environment

1. INTRODUCTION

The quantity of greenhouse gasses (GHGs) emitted in Sri Lanka does not contribute to global warming (Yamane, 2009). However, climate change impacts in Sri Lanka involve the increasing variability of rainfall intensity and regimes, increasing frequency of storms, increasing mean ambient temperature and sea level rise (Climate Change Secretariat, 2010). The World Bank Group (2011) predicts further increases of climate variability and extreme events across Sri Lanka in the future. The country's mean air temperature increases by 0.016°C per year during the period of 1961-1990 (Chandrapala 1996), and it is predicted that the mean temperature may increase by approximately 0.9°C to 4°C by the year 2100 (Basnayake et al., 2007). The annual average rainfall in Sri Lanka decreases by 144 mm from 1961 to 1990; this is a decrease of approximately 7% compared with the period of 1931 to 1960 (Baba, 2010). According to the World Bank Group (2011), historical records from 1974- 2004 indicate that floods and droughts are increasing. The rainfall intensity, amount of rainfall per rainy day and the average rainfall per spell have increased in most parts of the country (Ratnayake & Herath 2005). El Nino events caused severe drought in Sri Lanka from January to March 1983. Approximately 55% of the land area, mostly in the dry and intermediate zones, received less than 10% of the normal rainfall (Yamane, 2009). A flood triggered by 730 mm of rain was an extreme event and severe climate anomaly in the southern province of Sri Lanka on 17 May 2003 (Cruz et al. 2007).

Sri Lanka is considered as a vulnerable small island nation that is under serious threat from various climate change impacts according to the United

Nations Framework Convention on Climate Change (UNFCCC) and the Intergovernmental Panel on Climate Change (IPCC). The impacts of climate change are manifested in all socio- economic sectors, including agriculture, forest resources, biodiversity, health, energy, and human settlements. The Sri Lankan government ratified the UNFCCC in 1993 and the Kyoto Protocol in 2002. The Ministry of Environment has developed the National Climate Change Adaptation Strategy (NCCAS) for a climate-change-re- silient future. The concept of adaptation is closely associated with vulnerability. The UNFCCC and the IPCC have considered the vulnerability and adaptation to climate change as integral components (Yamane, 2009). Farming and forestry have been threatened by the complexity and magnitude of weather-related phenomena (www.Copa-Cogeca.eu). Strengthening adaptation mechanisms by combining agriculture and forestry through sustainable agroforestry management reduces the impacts of climate change. Agroforestry has an important role in climate change adaptation through diversified land-use practices, protected livelihoods and sources of income, enhanced agricultural productivity, and mitigated weather-related production losses, which enhance a region's resilience to climate impacts on farming systems (www.fao.org/forestry2010). Sri Lanka has a long history of integrating trees into land use systems, and many diverse agroforestry systems are in use due to the cultural, climatic and topographic variations of the country.

The authors intend to determine the use of agroforestry in Sri Lanka as a promising land use option for adapting to climate change. Further attempts are made to ascertain the problem of climate change, and agroforestry practices that are restructured from traditional practices to ecologically sound and farmer-based solutions can meet location-specific requirements of policy and receive institutional support. Therefore, this paper reviews the literature and discusses the impacts of climate change on forest and agriculture, the agroforestry adaptation for climate changes and policy environment required to promote agroforestry for climate change adaptation in Sri Lanka.

2. ADVERSE IMPACTS OF CLIMATE CHANGE ON AGRICULTURE AND FORESTRY

2.1. Endangering Natural Assets

The impacts of climate changes on natural assets both directly and indirectly affect the agricultural and forestry sectors. The mean annual and seasonal daytime maximum and mean annual and seasonal night time

Table 1. The damage of climatic disasters in Sri Lanka.

Year	Type of Disaster	Number of deaths	Number of affected families	Expenditure for Relief (Rs. In Millions)
1986	Floods & Landslides	40	118,494	13.8
	Droughts	-	5303	2.4
1987	Floods & Landslides	-	5053	0.6
	Droughts	-	484,925	72.9
1988	Floods & Landslides	03	26,373	2.9
	Droughts	-	652,363	28.5
1989	Floods & Landslides	325	86,176	49.1
	Droughts	-	238,426	24.6
1990	Floods & Landslides	37	157,427	38.7
	Droughts	-	203,794	6.7
1991	Floods & Landslides	34	55,491	78.9
	Droughts	-	203,794	230.2
1992	Floods & Landslides	25	71,080	154.6
	Droughts	-	-	-
1993	Floods	06	210,874	30.0
	Landslides	29	870	4.42
	Droughts	-	16,383	7.6
1994	Floods	-	353,409	37.4
	Landslides	08	284	0.63
	Droughts	-	-	0.62
1995	Floods	-	353,409	37.4
	Landslides	-	484	2.9
	Droughts	-	-	-
1996	Floods	03	8238	12.2
	Landslides	-	75	0.05
	Droughts	-	199,535	424.9

Source: http://www.iedm.ges.kyoto-u.ac.jp/workshop/IEDM%20Forum2006/rekha%20nianthi.pdf

minimum air temperatures in Sri Lanka have increased with trends of 0.026°C and 0.017°C per year, respectively during the period of 1960-2001 (Zubair et al. 2005). The damages of climatic disasters mainly floods, landslides and droughts occurred during 1986-1996 in Sri Lanka are shown in Table 1. According to the National Disaster Management Centre, drought is the most expensive and prolonged hazard in Sri Lanka. The mean rainfall is projected to decrease by 4% in terms of quantity and spatial distribution compared to historical records (Responsible Tourism Partnership Sri Lanka, 2008). Far more people are severely affected by periodic droughts, especially those in the dry zone of Sri Lanka. The culture-based agriculture that is influenced by the unique hydraulic civilisation in Sri Lanka and that has been practiced for centuries has also been affected. Vulnerability mapping performed by the Ministry of Environment in 2011 shows that the irrigation sector's vulnerability to drought exposure is widespread, but it is more concentrated in the dry zone. Approximately 55% of the land area, mostly in the dry and intermediate zones, received less than 10% of the normal rainfall. Climate change impacts are critical for the continuation of traditional cultural practices at the local level due to the loss of species. Biodiversity in most areas of Sri Lanka is vulnerable to the impacts of climate change. From the 58 assessed dipterocarps plant species which are remarkably endemic in Sri Lanka, 42 species mostly confined to the Wet Zone are threatened (Climate Change Secretariat, 2010). Forests play an important role in regulating water in Sri Lanka, which critically depends on the staple foods, power generation and drinking water. Forest resources in Sri Lanka decreased from 36% of the total land area in 1990 to 29% in 2006 (Responsible Tourism Partnership Sri Lanka, 2008). The total forest cover of 1.94 million hectares in 1999 is estimated to have decreased by 2% - 3% in 2012 (Climate Change Secretariat, 2010). In contrast, heavy rains erode top soil and wash away fertilisers and other chemicals. It has been estimated that approximately 30 cm of soil has already been eroded from upland tea plantations (Wijeratne, 1996). The projected sea level rise in Sri Lanka, i.e., 0.3 m by 2010 and 1.0 m by 2070, would adversely influence the coastal environment, agriculture, forest, biodiversity, health, energy, and human settlements (Natcom, 2000). It has been estimated that for a rise of sea level in 0.3 m and 1 m, the inundation will be in the range 41 km^2 and 91.25 km^2respectively for lowlands along the southwest coast (Weerakkody, 1996). The Indian Ocean tsunami of 26 December 2004, an extreme event, affected over two-thirds of the coastline and destroyed assets, infrastructure, agriculture and forests; the disaster claimed over 35,000 human lives in Sri Lanka (Mattsson et al., 2009).

2.2. Prevalence of Pests, Diseases and Invasive Species

Sri Lanka was self-sufficient regarding food and practiced agriculture with low external inputs prior to introducing modern agriculture. Under traditional diverse system of cultivation, mixed cropping was more popular in rain-fed conditions and was unlikely to promote pests or spread disease. Traditionally grown crop varieties were well adapted to the environment, and pest attacks did not occur at significant levels. Currently, crop vulnerability to disease, pest infestations, and choking weeds have been increased by the altered weather patterns. Less diverse farms and forests face a major environmental challenge regarding the climate-induced risk of pests and diseases. Warmer temperatures due to climate change generally develop conditions for the proliferation of plant pests and diseases. Climate change affects pest and disease incidence and virulence because diseases and insect populations are strongly dependent upon temperature and humidity. Increases in the pest and disease problems of coconuts in Sri Lanka due to increasing air temperatures increase the investment in pest control (Peiris et al. 2004). Climate change impacts are expected as abrupt invasions and changes in population dynamics. Invasive alien species have spread at an unprecedented rate and dominate over native and natural species; this trend has severely threatened natural ecosystems and biodiversity in Sri Lanka (Mattsson et al., 2009). The accelerated spread of invasive alien species as a result of climate change impacts is affecting agricultural lands and diminishing agricultural production in many areas of Sri Lanka. Climate change has increased the risk of diseases also in some tree species that have grown for many years. Forest species, mainly Eucalypt species and acacias, and many other perennial crops are affected by the C. quinqueseptatum pathogen (Booth et al., 2000).

2.3. Agriculture and Forest Damage

The potential impacts of climate changes on the agricultural and forestry sectors include increased damage to crops, trees and soils due to drought, flooding and land degradation. The broad impacts of climate change on agriculture in Sri Lanka are the impacts of temperature and precipitation on crop growth. With the increase in temperature, crop injuries are possible because crops are cultivated in Sri Lanka at optimum temperature ranges. Exceeding ambient temperature over 35°C even for just 60 - 90 minutes at the flowering stage, can cause crop damages for paddy. Increased temperatures have negatively affected high-value crops, mainly vegetable and potato cultivations of the country (Eriyagama, et al., 2010). Harvesting crops according to the ancient seasons of "Yala" and "Maha" based on the timing of monsoons in Sri Lanka has shifted in recent decades. The variability of rain from monsoons and convection has increased significantly in recent decades due to climate change and has become a recurrent problem in crop production (Punyawardena, 2007). Paddy farming output is estimated to decrease by 20% - 30% in the next 20 to 30 years due to the negative impact of monsoonal changes (Baba, 2010). Major irrigation in Sri Lanka offers protection from drought; however, during severe droughts, there is no water in the irrigation channels and crops fail (Yamane, 2009). Prolonged droughts also affect plantation crops such as tea, rubber and coconut and minor export crops such as cocoa and coffee, which account for slightly more than 700,000 hectares of agricultural land in the wet and intermediate zones. Empirical findings show that the growth and yield of tea are adversely affected by the increases in temperature, soil moisture deficits, and saturation vapour pressure deficits at low elevations (Sivakumar et al., 2005). Except in the upper wet zone, rising temperatures and diminishing rainfall reduce tea yields in many tea-growing regions. The optimum rainfall in different tea growing regions varies from 223 to 417 mm per month. The productivity of made tea could reduce by 30 - 80 kg with the reduction of monthly rainfall by 100 mm (Wijeratne et al. 2007). Climatic factors, mainly rainfall and temperature, influence the yields of coconuts, which are grown as a rain-fed crop in Sri Lanka. Extended dry spells and excessive cloudiness during the wet season can reduce coconut yields and cause annual losses between US$32 and US$73 million (Fernando et al. 2007). Somaratne & Dhanapala (1996) have estimated that increased temperature and rainfall would result in a decrease of the tropical rain forest of 2% - 11% and an increase of the tropical dry forest of 7% - 8%. Shifting the Dry Zone spread into the Intermediate Zone and the Intermediate Zone spread into the Wet Zone from 10 m band where they could become more water scarce and drier is predicted due to climate change. These forests and the species that would contain by the forest area of 1248 km^2 within the intermediate zone/dry and 544 km^2 within the intermediate zone/wet zone could be severely affected (Climate Change Secretariat, 2010). The existing ambient temperature is at or very near the upper margins of the optimum temperature range for many wild relatives of crops in the dry zone, which is also threatened by climate change. It is predicted that under extreme drought conditions, fire hazards increase in forests. Wild relatives, including five species of wild pepper, eight species of wild cinnamon, ginger, cardamom and nutmeg, that occur in the natural forests of the wet zone are vulnerable to the threat of climate change. Medicinal plants, ornamental plants, timber species and wild relatives of non-food crop species are also among the threatened wild varieties in the natural forests of Sri Lanka (Dela, 1998). An increase in the volume of sea water entering estuaries or lagoons due to climate change adversely affects the peripheral agricultural lands and crops in Sri Lanka. A tsunami in 2004 seriously damaged the agricultural and forestry sectors in Sri Lanka via seawater intrusion, salinisation of lands and water body contamination (Mattsson et al., 2009).

2.4. High Levels of Food Insecurity

Sri Lanka's vulnerability to climate-change-induced natural disasters, such as floods, droughts, cyclones, landslides, and epidemics, substantially threatens food security. Sri Lanka is predicted to be one of the hotspots of food insecurity in the Asia-Pacific region. Food production in South Asia is alarming with the shifts in the monsoonal rainfall patterns due to global warming. Food security is often directly or indirectly dependent on agricultural and forest ecosystem services. The vulnerability of water resources and agricultural output in Sri Lanka is the key challenge to national food security in the wake of climate change. Climate change impacts in the form of prolonged and more frequent droughts in Sri Lanka reduce the availability of water for rain-fed agriculture and often reduce crop production. Nellemann et al. (2009) predict a further decrease of 0% - 15% in agricultural productivity of the country by 2080. Climate change impacts can be large in the agriculturally intensive areas located in the dry zone where some production is already experiencing water stress. Field crops, including coarse grains, grain legumes, oil seeds and condiments that are grown on rain-fed upland areas in the dry and intermediate zones have also been adversely affected. The irrigation sector's vulnerability to drought exposure is widespread but more concentrated in the dry zone where there is high dependency on irrigation for agriculture. The annual cropping intensities of major irrigation schemes of the country are low and generate lower annual incomes for farmers (Dharmasena, 2010). Agriculture under small-scale irrigation is confined to one season only in some years due to the drought. Farmers face problems regarding the availability of groundwater for agriculture when droughts are frequent (Jinapala, 2010). Impacts of droughts and saline water intrusion in the coastal areas, combined with the increasing demands on freshwater resources due to expanding human settlements in Sri Lanka, are likely to reduce water availability and affect both irrigated and rain-fed agriculture (The World Bank Group, 2011). In Sri Lanka, a large proportion of the poor population who are living in vulnerable areas are dependent on weather-reliant industries, such as agriculture, forestry and fisheries. Climate change impacts on water resources and the agriculture sector may seriously effect food production, the livelihoods of farmers and the economy in Sri Lanka. The communities that already have high levels of food insecurity, environmental degradation and limited options for coping with adverse weather conditions are severely affected. Severe droughts impacted over 3.5 million people of the country who required emergency assistance in the form of food, water, and medical assistance between 2000 and 2005 (UNESCAP, 2007). Preliminary agricultural vulnerability mapping suggests that the Nuwara Eliya, Badulla, Moneragala, Ratnapura and Anuradhapura are as of the country are more sensitive to climate change due to their heavy reliance on primary agriculture. Further, they have a lower adaptive capacity and face higher vulnerability due to their lower levels of infrastructural and socioeconomic assets.

3. ADAPTATION OF AGROFORESTRY TO CLIMATE CHANGE

3.1. Increase Tree Cover Outside Forests

All of the technically suitable land areas for forestation cannot be devoted to plantation forestry because they are agricultural lands that support local populations. A more appropriate land use system could be agroforestry, specifically, growing trees in conjunction with agricultural crops. Agroforestry systems in Sri Lanka provide corridors that connect distant reserves through the matrix effect on species diversity in landscape mosaics with native tree cover and through the persistence and movement of species across landscapes. Agroforestry in the rural landscape contributes to environmental sustainability and benefits climate change adaptation by storing carbon, halting land degradation, and fixing nitrogen. Home gardens, which are widespread and vary in species composition and tree density, are the best developed agroforestry system in Sri Lanka. Home garden systems cover 22% of the land area and are considered forest analogues that supply more than 70% of the timber and 80% of the fuel wood outside natural and planted forests in Sri Lanka (UN-REDD Program, 2012). Home gardens in Sri Lanka offer great potential for restoring and increasing forest cover and connectivity. The district-wise changes of the forest cover in home gardens during 1983-1999 are shown an increase except Mulathive and Ratnapura Districts where forest cover shows a decrease (Figure 1). In many parts of the wet zone, home gardens represent an important type of land use at the margins of larger rainforest fragments and still occur in landscapes where natural forest is scarce. The Kandyan home gardens (KHGs) are traditional systems of agroforestry found around the Kandy District in the central highland of Sri Lanka that have evolved into unique ecosystems and provide numerous goods and services. KHGs cover approximately 40% of the Kandy District and primarily contain perennial and semi-perennial trees, shrubs and up to 30 different crops (Kallesoe & De Alvis, 2004). Presently, approximately 9,851,250 home gardens cover 9701 sq km comprising 2220 sq km of tea estates, 1158 sq km of rubber plantations, and 3948 sq km of coconut plantations within the coconut triangle; privately held woodlots also provide additional tree cover outside forests (UN-REDD Program 2012).

The government of Sri Lanka implemented the Participatory Forestry Project (PFP) during 1993-2000 by promoting co-management and agroforestry with the intention of reducing deforestation and improving house hold livelihoods. The PFP included the farmers' woodlots scheme, protective woodlots, home garden development, and the village reforestation system for benefit sharing from agroforestry (UN-REDD Program, 2012). The project aimed to establish a total of 14,750 hectares of reforested land, 9000 hectares of homestead gardens, 4000 hectares of farm wood lots (FWL), 1500 hectares of protective wood lots (PWL) and 250 hectares of mis- cellaneous plantings over the 7-year project period (Kallesoe & De Alvis, 2005). The Upper Watershed Management Project (UWMP) was initiated in 1998 and implemented until the year 2005 to address management issues in four major watersheds in Sri Lanka, i.e., Uma, Kirindi, Walawe and Kalu, and to suggest sustainable management approaches. Maintaining and increasing forest cover, applying soil conservation measures, and integrating homestead garden developments into the broader basin management are the main activities related to

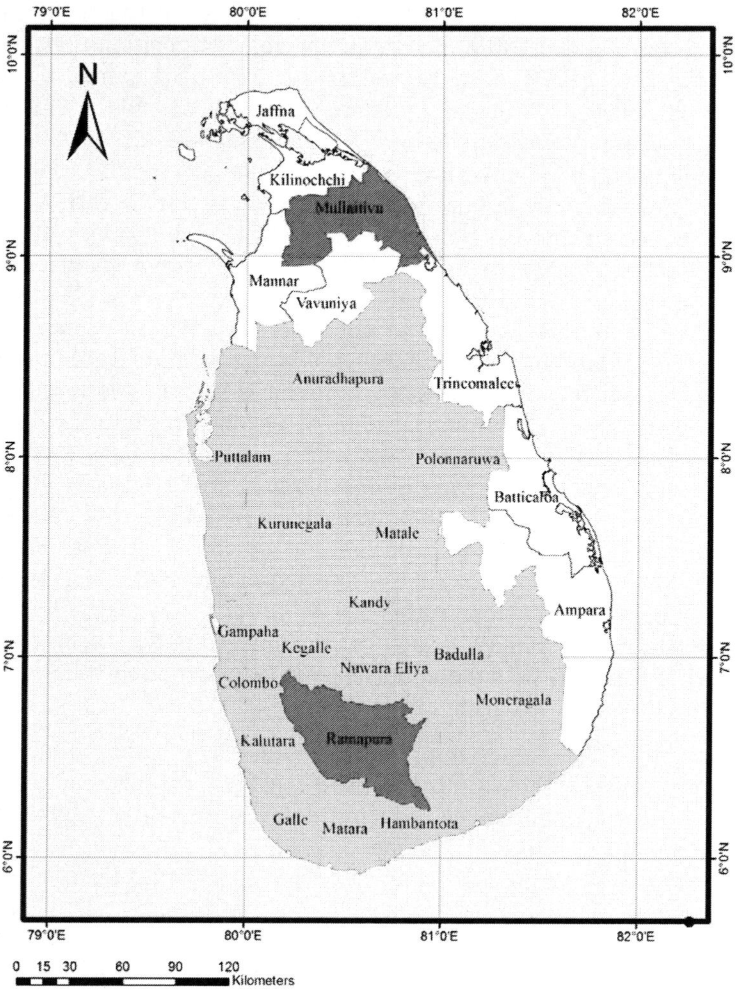

Figure 1. District-wise changes of forest cover in home gardens 1983-1999. Green colour symbolizes an increase, red colour symbolizes a decrease and white colour symbolizes no data of forest cover of Home gardens Source: Näsström, R. & Mattsson, E., 2009.

watershed conservation and rehabilitation (Kallesoe & De Alvis, 2005). The government of Sri Lanka declared Gliricidia sepium as the fourth main plantation crop to fall under the purview of the Ministry of Plantation Industries. Popularising Gliricidia plantations would increase the present forest cover of 1.7 million ha to a more acceptable level. Gliricidia is grown in extensive areas of scrub land and has minimal value for food crops due to inadequate rainfall and low fertility, but it is suitable for growing short rotation fuel wood (www.scribd.com/doc/95318104). Help Green (Pvt.) Ltd. initiated in 1988 has developed 600 ha of teak woodlots in non-forest lands and expected to increase to 5000 ha in which private investments are attracted by offering substantial economic returns from timber sales and by contributing towards environmental health. They have expanded their business model to include agroforestry by intercropping cash crops, such as soybeans, chilies, sweet potatoes, ginger and sunflowers, with teak trees to increase revenue and stimulate tree and eco-tourism services (Kallesoe & De Alvis, 2005). Some other private companies are planting the trees of timber species such as teak, mahogany, agar-wood and sandalwood on non-forest lands, for the general public to invest. Presently, touchwood has invested for about 1400 ha of plantations, while Sadaharitha Plantations has invested over 405 ha of teak and 30 ha of sandalwood (UN-REDD Program 2012). In line with the Green villages and "Dayata Sevena" national tree planting programs, local authorities, schools, home garden owners and the private sector are involved in agroforestry systems to enhance timber resources, canopy cover, and carbon stocks, which can be considered under REDD+ in Sri Lanka (UN-REDD Program 2012). Integrating agroforestry characteristics into Agro-well-based agriculture in the Yan Oya in dry zone of Sri Lanka has led to the development of canopy cover and helped to conserve soil moisture, reduce soil erosion, and maintain bio-diversity. Areas with perennial canopies have been increasing over all well-based lands from a mere 0.3% in 1994 to 6% in 2004 with the decrease of shifting cultivation lands from 30% to 22% during the same period (Perera, 2010).

3.2. Enhance Forest Carbon Stocks

Sri Lanka is not producing GHGs that induce global warming and climate change. The CO2 emission in Sri Lanka is less than 0.1 percent of the total global CO_2 emissions and is marginal relative to other countries (Baba, 2010). Green House Gas (GHG) emission levels in Sri Lanka are low in both absolute (26.1 Mt CO_2) and per capita terms (1.3 t CO_2), excluding land use change in 2005 (Mattsson et al., 2012). It has been estimated that the baseline deforestation emissions in Sri Lanka amounted to 17 Mt CO_2/year in the 1992-1996 period. According to the Responsible Tourism Partnership Sri Lanka (2008), 2 million hectares of forest in Sri Lanka store approximately 21 Mtons of CO_2 per hectare for a total of approximately 40 million tons of carbon. Establishing forestry and vegetation on non-forested lands and promoting mixed cropping to ensure fertile soil in the long-run reduce the emissions in Sri Lanka. Moving from low-biomass land-use systems, such as grasslands, agricultural fallows and permanent shrub lands, to agroforestry systems will produce a net gain of C stocks by sequestering CO_2 via biomass C storage (Roshetko et al., 2007). Tree species and management of agroforestry systems also influence carbon storage. A substantial increase of agroforestry area will have a large impact on the flux and long-term storage of C in the terrestrial biosphere. The energy in Sri Lanka also mainly originates from clean sources, such as biomass and hydro power. The use of biomass, which is produced sustainably for energy generation, does not lead to a net build-up of CO_2 levels in the atmosphere. Biomass that is produced in non- forest land is cheap and easy to attain; therefore, it is a popular option and viable for the mostly poor rural population of the country. As a clean source, primary solid biomass produces 54.2% of the energy requirement of rural population in Sri Lanka (Baba, 2010).

Non-forest lands, such as home gardens, tea estates, rubber plantations, coconut plantations within the coconut triangle and privately held woodlots, provide additional tree cover and are a major source of fuel wood and timber. Current data on the scale of agroforestry planting or of trees outside forests and their contributions to carbon stocks are not available in Sri Lanka. However, different forms of agroforestry, home gardens and boundary plantings have been well recognised as potential long-rotation systems that mitigate CO2 and sequester sizeable quantities of C in plant biomass (Albrecht & Kandji, 2003). Home gardens in Sri Lanka contain a significant fraction of the total above-ground biomass carbon stock in the terrestrial system, and this proportion has increased from almost one-sixth in 1992 to nearly one-fifth in 2010. Home gardens store significant amount of carbon, with above ground biomass carbon stocks with a mean value of 35 $Mg{\cdot}C{\cdot}ha^{-1}$ in dry zone while 87 $Mg{\cdot}C{\cdot}ha^{-1}$ in wet zone in the terrestrial system in Sri Lanka (Mattsson et al., 2013). The tree density per hectare ranged from 338 in dry zone home gardens to 2108 in wet zone home gardens; these values are reflected in the higher carbon stock estimates within the wet zone home gardens (Mattsson et al., 2012). Agroforestry, the tree- bearing agricultural land-use systems, sequester carbon at higher rates than systems containing only annual crops, pastures, or grasslands; the agroforestry sequestration rate is generally 0.2 - 3.1 $t{\cdot}C{\cdot}ha^{-1}$ per year (Pandey, 2002). Home garden agroforestry systems in Sri Lanka are a valuable part of the country's tree cover and act as important carbon sinks. Large coconut trees that are mixed with many small trees are common in rich, diverse home gardens in southern Sri Lanka. Coconut (194.4 t·C/ha) and home garden (107.9 t·C/ha) store a high volume of carbon stocks (Table 2). In addition to providing C inputs, such as root biomass, litter and prunings to the soil, agroforestry trees in agricultural fields often reduced soil erosion, which is a crucial process in the soil C dynamics. The majority of agroforestry systems produce non-wood products, such as fruits, vegetables, spices, oils, and resins, and are harvested with a negligible impact on the C stock of a system compared to the harvest of wood products (Roshetko et al., 2007). Hence, changes in carbon stocks within home gardens, forest analogue systems, are considered to be of paramount importance for a national REDD+ Programme in Sri Lanka (UN- REDD Program, 2012). Agroforestry in Sri Lanka, which introduces trees into existing agricultural systems, falls into the Clean Development Mechanism-Afforestation/Reforestation (CDM-AR) eligible categories. The clean development mechanism (CDM) assists developing countries with the multiple goals of poverty reduction, environmental benefits and cost-effective emissions reductions.

3.3. Conserve Biodiversity

The need for biodiversity conservation in human-dominated landscapes is greater today as species are forced to persist in degraded and secondary habitats outside forest reserves. Agroforestry impinges on biodiversity by reducing exploitation of protected areas, increasing biodiversity in a human-dominated or working landscape, and increasing the species and within-species diversity of trees in farming systems (Garrity, 2004). Traditional farming systems in Sri Lanka have been developed by farmers who manage production systems to best suit their local conditions. The main crops habitats are rice fields, crop plantations, vegetable plots, chena (similar to shifting cultivation) plots and home gardens. The traditional species of these land types are clearly resistant to diseases and insect pests, suitable for variable soil and climate conditions and well adapted to their new environments, but they are also still genetically similar to their wild relatives (MoENR, 2009a). An average of 40% of the tree species in traditional forest gardens in Sri Lanka is endemic. Ariyadasa (2002) found 400 different woody species in home gardens in Sri Lanka covering 87% of the country where the majority of tree species were indigenous or endemic with multi-purpose uses. An enormous potential for adaptation to progressive climate change is provided by both inter- and intra-specific agricultural biodiversity. Agroforestry conserves biodiversity, even if the species composition and biodiversity of the lost natural forests are not the same, depending on how the agroforests are managed and how close they are to the natural forests in terms of time and space. Presently, the forest gardens in Sri Lanka are structurally and functionally analogous to the natural forest and are established in alleys with different combinations of seasonal crops, cash intercrops, fodder-dairy, and woodlots (Table 3). Based on variations in soil, annual rainfall and altitude, 46 agro-ecological regions have been identified in Sri Lanka. Maintaining agroforests is therefore a good means to maintain a certain level of biodiversity. The change in species composition may sometimes be an advantage if the new species are better adapted to changing conditions. Home gardens in southern Sri Lanka have a rich diversity with many smaller fruit trees mixed with larger trees, such as coconut. A study of home garden vegetation conducted in semi-urban home gardens in the Western Province revealed as many as 101 species, including coconut, rubber, fruit and timber, comprising over 2817 individuals; the stem density was over 400 trees/ha (Dela, 1998). The sustainable management of Kandyan Home Gardens (KHGs), the traditional systems of agroforestry found in the central high-

Table 2. Mean values of basal area, standing biomass and carbon stock of the different forest types.

	Natural Forest	Coconut	*Casuarina*	Home garden	Mangrove
BA (m^2/ha)	75.6	62.4	36.3	31.2	26.4
Standing biomass (m^3/ha)	702.4	457.5	285.2	253.8	222.5
Carbon stock (t·C/ha)	294.8	194.4	121.5	107.9	94.6

Source: Mattsson et al., (2012).

Table 3. Species list of an 0.5 ha analog forest.

Forest layer	Crop/Tree	Botanical Name	Number
Emergent	Areca nut	*Areca catechu*	54 (6.9)
	Eucalyptus	*Eucalyptus* sp.	6 (0.8)
	Fishtail palm (kitul)	*Caryota urens*	15 (1.9)
Canopy	Avocado	*Persea americana*	16 (2.1)
	Breadfruit	*Artocarpus incisa*	5 (0.6)
	Mango	*Mangifera indica*	15 (1.9)
	Ceylon olive (weralu)	*Elaeocarpus glandifuler*	2 (0.3)
	Papaw (papol)	*Carica papaya*	18 (2.3)
	Jak fruit (kos)	*Artocarpus integrifolia*	2 (0.3)
	Sapu	*Michelia champac*	7 (0.9)
Sub-canopy	Banana	*Musa sapientum*	30 (3.9)
	Cherimoya (anoda)	*Anona cherimolia*	9 (1.2)
	Curry leaf tree (karapincha)	*Murraya koengii*	9 (1.2)
	Orange	*Citrus sinensis*	20 (2.6)
	Lime	*Citrus aurantifolia*	20 (2.6)
Shrub	Coffee	*Coffea arabica & robusta*	270 (34.7)
	Cardamom	*Elletaria cardamomum*	280 (36.0)
Total			778 (100)
Ground cover	Ginger Turmeric Yam "Arrowroot" (hulankeriya)	*Zingiber officinale* *Curcuma domestica* *Dioscorea* spp.	18 plots of 4 m^2 each

(Percentages are in parenthesis); Source: After Nuberg, I. K., & Evans, D. G. (1993).

land of Sri Lanka that cover approximately 40% of the Kandy District, conserves a wide range of biodiversity areas containing more than 30 different crops, primarily perennial and semi-perennial trees and shrubs (Kallesoe & De Alvis, 2005). Beverage crops, including tea, coffee and cacao, are successfully cultivated under a large number of leguminous and other shade trees. Pepper under shade trees, mixed spices and beverage crops under shade trees and cardamom grown beneath forest cover are the three main agroforestry subsystems that grow spice crops under shade trees. Except cinnamon, all perennial spices are cultivated in agroforestry types of farming systems. Spices have been cultivated over 50,000 ha of wet zone accounting for about 6% of the land under all perennials crops (Lindara et al., 2004). Horticultural plantings of native species in landscaping, gardens, roadside areas, and parks are classic examples of species responding in different locations of the country as the climate changes. Climate Change Secretariat (2010) has recognised that the conservation of watersheds and wetland biodiversity are of crucial importance when adapting to climate change in Sri Lanka.

According to the Climate Change Secretariat (2010), functionally diverse vegetation in Sri Lanka tends to adapt to climate change and climatic variability better than impoverished vegetation, while genetically diverse species appear to have more potential for long-term persistence. The resources in biodiversity areas include food, fuel, fibre, wood products, timber, medicines, biomedical materials, ornamental species of commercial value, raw materials for industry, and areas for recreation and aesthetic enjoyment, which are vital to meet the consumptive and economic needs of the country. Hence, conserving and sustainably managing biodiversity of agroforestry is of critical importance to reduce the negative impacts of climate change on biodiversity, ecosystem services, and economic advancement in Sri Lanka.

3.4. Reduce Risk and Intensity of Damage

Reducing the risk and intensity of damage has become a major concern for many farmers and foresters in Sri Lanka under changing climatic conditions. From the elderly farmer's descriptions of farming patterns, there is evidence that the climate is getting hotter and drier. One possibility to achieve analogue forestry is to use cropping systems that mimic the structure and successional processes of a tropical forest to adapt to the climate change impacts. Resilience can be increased through the reduced exposure of cultivated plants to environmental stresses by planting combination of trees and crops as agroforests. Agroforestry systems in Sri Lanka with multi-layer canopy, which cover the ground, improve soil structure by incorporating tree biomass into the soil and reducing soil erosion to create a multi-layer defence mechanism against rain and wind erosion and to increase the infiltration capacity. Traditional agroforestry systems in Sri Lanka have been widely practiced throughout history as a means of slowing soil and other environmental degradation. Shadow crops also contribute to additional protection against extreme rainfall and are less exposed to soil erosion. The complex and diverse landscape structure created by agroforestry systems is important for biological pest control by supporting populations of natural enemies of agricultural pests (Pandey, 2002). Planting multipurpose trees on farms offers an alternative to adopting organic practices and integrated pest management and can reduce dependence on pesticides. Diversified productions of agroforestry systems are a form of risk avoidance and ensure food production even under the current agricultural crises. The government program "Api Wawamu Rata Nagamu" (Let us grow, and uplift the nation) launched in 2007 to develop 375,000 home gardens. The national tree planting program called the "Divi Neguma" (Livelihood Development) aimed to establish 1.5 million home gardens to achieve self-sufficiency of vegetables and to reduce vegetable prices; the number has further increased to 2.5 million home gardens (Kumari et al. 2009). Due to the complexity of measures that may have contradictory effects, there is a tendency for integrated management practices to include different treatments that help maintain the vitality of the agroforests, including timely agricultural and silvicultural practices and species mixing. The increasing salinity of low-lying coastal areas due to sea level rise has devastating effects on productive agricultural land. Introducing strips of vegetation has reduced the salinity level and created favourable conditions for crop diversification. Use of several adaptation strategies by home gardeners in Sri Lanka not only functions as a way to reduce pressure on natural forest land but also resists climate change impacts. Home garden structures in Sri Lanka exhibited resilience against and protection from the 2004 tsunami. In the Hambantota District, houses with home gardens or in the vicinity of neighbours' home gardens received less damage than those without home gardens during the tsunami (Mattsson et al., 2009).

The economies of tropical small island countries are based on diverse agricultural systems and are subject to fewer impacts from climate change. Due to the high sensitivity of agriculture to the prevailing climate variability, the dependency of households on home gardens to secure their livelihoods is increasing. The objectives of the Climate Change Enabling Activity Project (CCEAP) in Sri Lanka initiated in 2001 were to develop an adaptation strategy and to recommend mitigation measures needed to offset and limit potential impacts of climate change on the different sectors of the national economy (Kallesoe & De Alvis, 2005). Agroforestry systems provide higher cash incomes and greater purchasing power for farmers, especially when agriculture is not practiced or when crops fail. A study of intercropping of coconuts in the wet and intermediate wet zones has revealed that farmers prefer bananas, black pepper, coffee, and ginger (in order) as intercrops. The main factors that influenced the farmers' preference for crops are profitability, marketing facilities and convenience (MoENR, 2009b). Considering the urgent national challenge, the goal of the "Api Wawamu Rata Nagamu" programs are to reduce living costs and imports of food products, enhance food security, maintain environmentally friendly traditional agriculture methods and identify avenues to test new technologies (Kumari et al. 2009). Dendro plantations in semi-utilised lands, lands used for tea and coconut, and largely unutilised lands are encouraged to supply fuel wood for alternative power generation (www.scribd.com/doc/95318104).

3.5. Maintain Health and Vitality

The main threats to health and vitality of agriculture and forests are extreme weather events, pests and diseases. The diversity of agroforestry usually enhances biodiversity and is useful for maintaining health and vitality. Promoting diverse age classes, species mixes, structural diversities, and genetic diversity in agroforestry systems in Sri Lanka provide the most practical opportunities for resetting ecological trajectories. Other intrinsic attributes of the land use activity of agroforestry include the low inputs and ecological security. Most of the agroforestry systems of the country are characterised by high levels of on-site nutrient conservation, similarly to self- nourishing natural forest ecosystems. Kandyan Home Gardens (KHGs), the traditional systems of agroforestry in Kandy, have evolved into unique ecosystems that provide numerous goods and services and maintain promising health and vitality. In these complex agroforestry systems, the decline in soil fertility is considered minimal despite the continuous harvests of crops, wood and other products. Soil physical and chemical properties are improved by returning the abundant litter and pruned biomass to the soil, along with the decay of roots. Smallholder farmers in ecologically marginal areas of Sri Lanka are practicing alley cropping as an appropriate agroforestry technology restores the fertility of degraded soil and provide other services, such as erosion control (Nanayakkara, 1991). Boundary plantings and live fences in home gardens in Sri Lanka, which are more densely spaced, comprise a number of different species that modify microclimates, improve water conditions, and improve soil characteristics positively effect crop production. Extending agroforestry, especially to uplands and very steep slopes, increases climate resiliency, provides different sources of income, and reduces land erosion. The UWMP addressed management issues, such as applying soil conservation measures, integrating homestead garden development activities related to watershed conservation, and rehabilitating the four major watersheds (Kallesoe & De Alvis, 2005). A structure of alleys created by hedgerows of nitrogen-fixing woody species planted along the contour is characterised by a sequence of two alleys of seasonal crops and one alley of perennial crops in watersheds. A block of alleys is dedicated to fodder production for a livestock system or a woodlot system in which farmers are persuaded to maintain perennial vegetative cover on as much of the land as possible (Nuberg & Evans, 1993). The KHGs are often found on steep slopes, and the ground cover of agroforests protects lands from severe erosion and floods (Kallesoe & De Alvis 2005). The organic production of fruits, tea and spices in agroforestry systems has expanded rabidly in Sri Lanka over the last few decades for producing environmentally friendly products, and limiting the use of chemicals and artificial fertilizer. Over 3300 farmers have cultivated 14,250 ha of land with organic products and another 4550 ha were under conversion in 2000 now generating more than 31 million USD a year while providing environmental services (Kallesoe & De Alvis (2005). A study conducted by the Coconut Research Institute (CRI) has proved that Gliricidia improves soil fertility and the microclimate in sustainable agriculture projects. Leaf litter covers the soil and reduces the soil temperature underneath the Gliricidia canopy, thereby increasing root activity, reducing soil moisture losses and reducing soil carbon oxidation. Soil from the Gliricidia-Coconut inter-cultivation has a greater proportion soil nutrients than in Coconut mono-cropping system (Table 4). The trials conducted by the CRI have identified Gliricidia as the best tree species for fixing biological nitrogen in the soil, enhancing soil fertility and encouraging the soil rehabilitation capacity (Anon, 2004). Soil from the mixed cropping area of coconut palms and Gliricidia inter-cultivation had a greater proportion of soil carbon, and high N and Ca levels were identified over leaves. Compared to coconut monoculture, the incorporation of Gliricidia into the farming system produces the highest wood yield in the range of 15 to 30 m.t. per ha for 10,000 trees/ha densities (www.scribd.com/doc/95318104). According to research conducted at the Tea, Rubber and Coconut Research institutions and the Department of Export Agriculture, a considerable diversification of cash crops has occurred from the originally introduced germplasms selected by growers. These crops have served to produce high- yielding varieties that are also resistant to pests, disease and adverse climatic conditions (MoENR, 2009b).

Table 4. Effect of Gliricidia inter-cultivation on soil N, P, K, Ca and Mg.

Situation	N (ppm)		P (ppm)		K%	Mg%
	0 - 15	16 - 30	0 - 15	16 - 30	0 - 15 16 - 30	0 - 15 16 - 30
Coconut only	490	327	3.4	Trace	0.15 - 0.17	0.8 - 0.7
Coconut + *Gliricidia*	644	471	8.1	5.6	0.23 - 0.26	1.1 - 1.1

Sources: http://projects.nri.org/biomass/conference_papers/sustainable_biomass_production_in_srilanka.pdf

3.6. Scale up Multiple Benefits

Multiple benefit approaches of the sustainable management of agroforestry increase yields, incomes, food secu- rity and climate resilience, protect biodiversity, and reduce greenhouse gas emissions, often simultaneously. The direct forms of production and the indirect mechanisms that promote enhanced and sustained production are fundamental to the traditional agroforestry systems in Sri Lanka. The ancient and locally adapted agroforestry system of home gardens, which presently occupy nearly 22% of the land area in Sri Lanka, offer wide range of products and several ecosystem services to users. The wood production from home gardens could reach 681,400 m^3 in 2020 as expected in the Forestry Sector Master Plan of the country (MALF, 1995). Home gardens provide not only diverse fruit, vegetable, and spice products but also sources of mineral nutrients, particularly for women and children. Traditional systems of agroforestry of Kandyan Home Gardens (KHGs) that primarily contain perennial and semi-perennial trees, shrubs and more than 30 different crops contribute to local livelihoods and yield cash crops and products for subsistence use (Kallesoe & De Alvis, 2005). The estimated mean technical efficiency for the spice based agroforestry production in Matale district in Sri Lanka is 84.32%, which indicates that output could be further increased substantially (Lindara et al., 2004).

With the decline of village forests and access to local forest resources, farmers in Sri Lanka have learned to utilise agroforestry systems to meet household needs, reduce risks, diversify income streams and efficiently use the limited labour and financial resources. Indigenous fruit tree cultivation in agroforestry can have a significant effect on the health and nutrition of the rural communities, particularly the quality of child nutrition. Depending on the species, trees that shade crops to reduce evapotranspiration, erosion control and nutrient cycling are important in agroforestry systems (Albrecht & Kandji, 2003). Smallholders in the Upper Mahaweli catchment of Sri Lanka use different production systems to occupy the alleys of contour hedgerows to control soil erosion. The Participatory Forestry Project (PFP) was implemented during 1993-2000 in Sri Lanka with the intent of reducing deforestation and improving household livelihoods by promoting co-management and agroforestry. The homestead garden subcomponent of the PFP was to improve livelihoods and the health status of poor families by providing alternative livelihoods in the forms of growing and selling timber and fruit (Kallesoe & De Alvis 2005). According to the Forestry Sector Master Plan (FSMP) (1995), the projected wood production of peeler logs, saw logs and poles from home gardens and other non-forest tree resources would rapidly increase during the period from 2000 to 2020 in Sri Lanka (Table 5). The agroforestry programs implemented under the PFP created employment opportunities, raised income, reduced poverty and rehabilitated degraded areas. Gliricidia in agroforestry systems improves the soil fertility, improves the microclimate, and is ideal for fuel wood production under sustainable agriculture. Gliricidia presently produces fuel wood in the agro-based energy sector, such as for tea factories and desiccated coconut mills, to meet the overall energy security for the country. The production of leguminous soil enhancement, provision of animal fodder and production of biogas from Gliricidia coppice crops provide families with a fully integrated farming system. Growing leguminous trees and grass in mixed forage systems enhances the yield and quality of the understory grass and total output of the system. Gliricidia, on average, produces 23 m.t./ha of foliage that can be used as a soil mulch for improving soil fertility (Anon, 2004). Gliricidia, as a wood-based energy producer, is able to alleviate poverty, reduce foreign exchange and receive sufficient support from local technology (www.scribd.com/doc/95318104). Growing trees in home gardens and other agroforestry systems in Sri Lanka is promoted as a main strategy to supply wood and other forest products for meeting household and market needs. Sample surveys conducted on intercropping with coconuts listed the factors such as profitability, marketing facilities and convenience as the major reasons for farmers' preference for their selected crops for intercropping. Bananas, black and pineapple (in order) compose pepper, coffee, ginger (in order) are the most preferred intercrops of farmers, while turmeric, betel, vegetables the second group (Liyanage et al., 1984).

Table 5. Projected wood production from home gardens and other non-forest tree resources ('000 m^3).

Source/year	2000	2005	2010	2015	2020
Peeler logs (rubber)	7.9	8.3	8.4	8.2	8.2
Sawlogs					
Home gardens	582.7	631.3	681.4	731.7	769.2
Rubber	256.3	269.9	270.0	263.3	265.8
Coconut and palmyra	202.4	220.0	210.9	166.4	154.3
Trees on tea lands	75.9	75.9	75.9	75.9	75.9
Other perennials	68.6	71.7	74.9	78.3	81.9
Roadside planting	4.8	5.0	5.1	5.2	5.2
Poles					
Home gardens	831.4	853.7	873.4	889.5	910.8
Other perennials	47.7	49.9	52.1	54.5	57.0

Source: FSMP (1995).

4. POLICIES FOR AGROFORESTRY ADAPTATION

4.1. Changing Regulations and Legislation

Sri Lanka has no strong obligations to the international climate change policy arena because the country neither emits a large quantity of greenhouse gasses (GHGs) that contribute to global warming nor is regarded as vulnerable to climate change (Yamane, 2009). However, there is a need to revise existing agriculture policy to take into account climate change impacts. New approaches to national policy for climate change adaptation under National Adaptation Programs of Action (NAPAs) are now being developed. It is suggested that the net emissions could be reduced by changing the current land uses to adopt better management practices of crop production and livestock farming. The introduction of trees into existing agriculture systems as agroforestry and the rehabilitation of degraded areas through tree planting are classified as eligible CDM projects under the Bonn Agreement. Food insecurity in Sri Lanka has also become an important policy issue that should be addressed due to its vulnerability to climate change impacts. The National Climate Change Adaptation Strategy (NCCAS) of Sri Lanka includes minimising climate change impacts on food security and safeguarding natural resources and biodiversity from climate change impacts. The Biodiversity Conservation Action Plan of 1999 in Sri Lanka recognised maintaining a high biodiversity of agricultural lands, especially in traditional farming systems, to increase future national agricultural productivity and to meet the challenges posed by climate change. The Commission on Sustainable Agriculture and Climate Change in 2011 has identified the policy changes and actions needed to achieve sustainable agriculture that contribute to food security and poverty reduction and to respond to climate change adaptation and mitigation goals. Farmers' dependency on evolving modern agriculture and government policies should be gradually replaced by self-confidence, self-motivation and the empowerment of integrating the sustainability of neglected centuries-old wisdom of traditional agriculture (Dharmasena, 2010). The framework of REDD+ reduces the unequal treatment of forests compared to agricultural crops and renders forest management more competitive with other forms of land use. The national REDD+ Program may help to demonstrate the value of agroforestry systems, gain the interest of potential investors, and popularise the enhancement of carbon stocks in home gardens and crop plantations (UN-REDD Programme, 2012). Inclusion of agriculture in climate change policies and inclusion of climate issues in agricultural policies are required to ensure cost-effective investments and offer benefits to the rural poor (Vermeulen et al., 2012). Adaptation measures and policies should be developed with the full participation of local communities and other stakeholders. The appropriate policy packages should cover aspects such as harvesting, processing, and utilisation of farm- grown wood and that ensure credit and extension services to popularise agroforestry among smallholder producers.

4.2. Awareness Creation and Capacity Building

Many farmers are aware of climate change but only few have a good understanding of the degree of uncertainty of climate change and impact projections. Unless they are educated in agroforest and climate change, they are susceptible to wrongly interpreting those projections. Scientifically backed technical strengths and community-based knowledge are the keys to developing effective adaptation strategies and reducing vulnerability of the natural resources and communities. Raising awareness and the development capacity of communities are vital tools for improving their understanding of the concept of climate change, their ability to effectively manage forest ecosystems and their livelihoods on a sustainable basis. The potential for carbon sequestration and an evaluation of the socio-economic benefits of carbon-offset projects of the forestry sector must be examined. Funding for research and advisory and vocational training schemes on agroforestry are of the utmost importance to develop ideas and new technologies that help farmers and forest producers prepare for the coming challenges. Outcomes of climate change research must be packaged and communicated via training programs, updates to grower field days, farm-forestry conferences, and farm-forestry project steering groups. Identification of the specific user needs through direct participation of farming communities is essential to determine the agro-mete- orological services and actual value of forecasts for the systems. By developing a range of deliberate manage- ment techniques for agroforestry systems and enhancing management skills through participatory demonstration trials, farmers can produce quality products for specific market opportunities (Roshetko et al., 2007).

4.3. Planning for Climate-Smart Agroforest Landscapes

Climate-smart agriculture is one of the sustainable agriculture and rural development objectives of the Millennium Development Goals (MDGs) of reducing hunger and improving environmental management. The Commission on Sustainable Agriculture and Climate Change under the United Nations Framework Convention on Climate Change (FCCC) was set up to help develop climate-smart agricultural practices and achieve a food-se- cure world in the face of climate change. Developing countries require support of smallholder farming systems and the associated food systems to achieve food security due to the complications of climate change. Agroforestry is a set of innovative practices in which smallholder farmers may combat climate change by reverting to more natural productive systems that provide improved ecological and social functions. To accelerate the process of climate-smart production systems, science and traditional knowledge need to be integrated with necessary institutional and policy adjustments. According to the Secretariat to the Convention on Biological Diversity, the people in Sri Lanka continue to retain a reasonable stock of traditional knowledge and accumulated wisdom, which can be utilised to promote the sustainable use of biodiversity and to combat the problems of climate change. Interest in agroforestry is increasing due to its potential to address various on-farm adaptation needs with food shortages and the increased threat of climate change. Agroforestry products may generate financial capital beyond subsistence levels, thereby aiding capital accumulation and reinvestment at the farm level to strengthen the system's ability to cope with adverse impacts of a changing climate. Agroforestry under integrated landscape management approaches support food production, ecosystem conservation, and rural livelihoods across entire landscapes. Climate-smart agroforestry requires actions beyond the farm scale and working at the landscape scale to ensure inter-sectoral coordination and cooperation. The climate-smart agroforestry approach provides the opportunity for all relevant stakeholders to become involved in landscape planning, program implementation and progress monitoring (Scherr et al., 2012)

4.4. No-Regret Options

To address uncertainties, taking actions that offer significant development benefits under a range of climate scenarios are called "no-regret" options. These approaches, including promoting agroforestry systems, help maintain agricultural production and have obvious benefits with or without climate change (climate change building smallholder resilience.mht). New agroforestry models can help reduce uncertainty in local assessments of vulnerability to climate change. "No-regret" options provide economic benefits, simultaneously help to reduce GHG emissions, and provide additional environmental benefits, such as improvements in biodiversity protection, water management and soil quality (www.Copa-Cogeca.eu). In the Yan Oya watershed, which is in the dry zone of Sri Lanka, farmers have provided a unique opportunity to explore new land use patterns. Adaptation strategies consisting of no-regret options would be sustainable regardless of the magnitude of climate change. Both rainwater harvesting and restoration of existing tanks to promote agroforestry systems in Sri Lanka are "no regrets" adaptation methods that simultaneously deliver climate resilience and address current development needs. Integrating agroforestry characteristics into agro-well-based agriculture by using shallow groundwater provides the opportunity to cultivate perennial crops and seasonal crops throughout the year and is a new trend with farmers. Agroforestry as an adaptation option with lower life-cycle costs is known as a "no-regret" option because it will lower the cost of providing economic benefits and environmental services in Sri Lanka (Bhattacharya et al., 2003). Adaptation alternatives of agroforestry systems for climate change are "no regret" options for Sri Lanka because they are cost-effective and simultaneously address sustainable development goals. Some farmers maintain animal husbandry, such as rearing cows, under perennial crop systems. Shade and grass under trees can be converted to "agroforestry systems" that are more productive and effective land use systems compared to the conventional systems. These systems contribute unquantified and unvalued benefits to a farming household via vegetables, fruits, spices, firewood, timber, medicines, milk, shade, organic matter, fodder, soil moisture, wind breaks, and aesthetic value. Adaptation of these approaches by agroforestry systems support com- munities both in building resilience to a range of potential surprises and in adjusting to long-term climatic trends.

4.5. Climate Change Adjustment Programs

The Millennium Development Goals (MDG), which aim to reduce the number of hungry and desperately poor by at least half by 2015, could be supported by promising agroforestry pathways that increase on-farm food production and income (Garrity, 2004). Agroforestry in Sri Lanka contribute to the rehabilitation of degraded agricultural lands, increased yields of small holder farmers, and CO_2 sequestration. Planning programs across the entire agroforestry system enable farmers to plan and prepare for challenges, such as prolonged drought, climate variability, biosecurity threats, emergency management and variable market conditions, and to sustainably manage their natural resource base. Forecasting the seasonal climate provides the opportunity for farmers to choose whether to opt for lower risk, lower return strategies or to adopt new technologies and intensify production (Vermeulen et al., 2012). Under the Climate Change Adjustment Program, farmers are provided with advice and a grant to develop a climate change action plan that outlines the aims and steps required to manage the impacts of climate change and improve the farm's long-term prospects (www.daff.gov.au/climatechange/australias-farming-future/climate-change-adjustment-assistance). Developing countries are more vulnerable to climate change impacts due to the lack of economic and institutional capacities (Yamane, 2009). High investment costs of dedicated plantations and low biomass productivity are included in the main technical barriers to biomass production for energy in Sri Lanka (Bhattacharya et al., 2003). Management, in the case of climate change, of the agroforest environment must seek new models that enable fast and precise decisions. Sri Lanka has long been interested in the Clean Development Mechanism (CDM) financing of emission reduction projects; however, these projects pay more attention to vulnerable aspects of the country and adaptation to climate change (Yamane, 2009). Agroforestry farmers need mechanisms to provide incentives and reward their efforts to adapt as they face the uncertainties of climate change. The use of marginal cropland and degraded non-crop land for biomass production and energy, which lead to the co-benefits of soil and water conservation and land reclamation in Sri Lanka, are justified for financial incentives (Bhattacharya et al., 2003). Improved agroforestry systems that produce social and environmental benefits meet the conditions for eligible afforestation/reforestation (A/R) activities according to the Clean Development Mechanism (CDM) (Verchot et al., 2007). Management-intensive agroforestry systems that incorporate traditional chena farming in Sri Lanka would qualify for CDM projects.

5. CONCLUSIONS AND RECOMMENDATIONS

Mangala De Zoysa, Makoto Inoue Climate change impacts by complex weather-related phenomena have threatened agricultural and forest ecosystems and the livelihood of agricultural and local communities in Sri Lanka. Agroforestry in Sri Lanka has an important role in climate change adaptation through diversified land-use practices, sustainable livelihoods, sources of income, enhanced forest and agricultural productivity, and reduced weather-related production losses, which enhance resilience against climate impacts. Sustainable agroforestry management, along with raising awareness, strengthening technical capacities, creating policies for actions that encourage cross-sectoral and landscape approaches, is a key to helping agriculture- and forest-dependent people adapt to new conditions caused by climate change in Sri Lanka. Agroforestry activities on the ground that are aimed at climate change adaptation need to be developed and implemented through the involvement of a range of related sector agencies at local, national and international levels. The cooperation of those agencies is very important for the success of the strategies that are adapted to local circumstances.

REFERENCES

1. Albrecht, A., & Serigne, T. K. (2003). Carbon Sequestration in Tropical Agroforestry Systems. Agriculture, Ecosystems and Environment, 99, 15-27.
2. Anon (2004). Annual Report of the Coconut Research Institute (CRI) of Sri Lanka. Lunuwila: CRI.
3. Ariyadasa, K. P. (2002). Assessment of Tree Resources in the Homegardens of Sri Lanka. Bangkok, ECFAO Partnership Program on Information and Analysis for Sustainable Forest Management (2000-2002).
4. Basnayake, B. R. S. B., Abjhyasinghe, K. R., Sumathipala, W. L., Punyawardene, B. V. R. P., Perera, N., & Joseph, P. G. (2007). Climate Change in Sri Lanka: Impacts, Adaptation & Mitigation. Proceedings of the National Conference on Climate Change, Centre for Climate Change Studies, and Kobe-Asia-Pacific Network for Global Change Research, Department of Meteorology, Colombo, Sri Lanka.
5. Baba, N. (2010) Sinking the Pearl of the Indian Ocean: Climate Change in Sri Lanka. Global Majority E-Journal, 1, 4-16.
6. Bhattacharya, S. C., Salam, P. A., Pham, H. L., & Ravindranath, N. H. (2003). Sustainable Biomass Production for Energy in Selected Asian Countries. Biomass and Bioenergy, 25, 471-482.

7. Booth, T. H., Jovanovic, T., Old, K. M., & Dudzinski, M. J. (2000). Climatic Mapping to Identify High-Risk Areas for Cylindrocladium quinqueseptatum Leaf Blight on Eucalypts in Mainland South East Asia and around the World. Environmental Pollution, 108, 365-372.
8. Chandrapala, L. (1996). Long Term Trends of Rainfall and Temperature in Sri Lanka. In Y. P. Abrol, S. Gadgil, & G. B. Pant (Eds.), Climate Variability and Agriculture (pp. 150-152). New Delhi: Narosa Publishing House.
9. Climate Change Secretariat (2010) Strengthening Capacity for Climate Change Adaptation, ADB TA 7326 (SRI), Climate Change Secretariat Ministry of Environment, Colombo, Sri Lanka.
10. Cruz, R. V., Harasawa, H., Lal, M., Wu, S., Anokhin, Y., Punsalmaa, B., Honda, Y., Jafari, M., Li, C., & Huu, N. (2007). Asia Climate Change, 2007, Impacts, Adaptation and Vulnerability. In M. L. Parry, O. F. Canziani, & J. P. Palutikof (Eds.), Contribution of Working Group II to the 4th Assessment Report of the Intergovernmental Panel on Climate Change (pp. 468). Cambridge: Cambridge University Press.
11. Dela, J. D. S. (1998). The Ecology and Social Biology of a Selected Population of the Western Purple-Faced Leaf Monkey. Ph.D. Thesis, Peradeniya: University of Peradeniya.
12. Dharmasena, P. B. (2010). Traditional Rice Farming in Sri Lanka. Economic Review, 36, 48-53.
13. Eriyagama, N., Smakhtin, V., Chandrapala, L., & Fernando, K. (2010). Impacts of Climate Change on Water Resources and Agriculture in Sri Lanka: A Review and Preliminary Vulnerability Mapping. IWMI Research Report 135, Colombo: International Water Management Institute, 51-55. http://www.iwmi.cgiar.org/.../iwmi-research-report-135
14. Forestry Sector Master Plan of Sri Lanka (FSMP) (1995). Ministry of Lands, Agriculture and Forestry, Sri Lanka.
15. Garrity, D. P. (2004). Agroforestry and the Achievement of the Millennium Development Goals. Agroforestry Systems, 61, 5-17.
16. Jinapala, K. (2010) Overview: The Volume on Irrigation for Food Security. In P. Weligamage, G. G. A. Godaliyadda, & K. Ji-napala (Eds.), Proceedings from the National Conference on Water, Food Security and Climate Change in Sri Lanka. www.iwmi.cgiar.org/...ations/Other/PDF/SLWC_vol-1.pdf
17. Kallesoe, M., & De Alvis, D. (2005) Financial Incentives for Ecosystem Conservation: A Review of the Development of Markets for Environmental Services in Sri Lanka. IUCN Water, Nature and Economics Technical Paper No. 4. IUCN— The World Conservation Union, Ecosystems and Livelihoods Group Asia, Colombo.
18. Kumari, M. A. S., Kansuntisukmongkol, K., & Brockelman, W. Y. (2009). Plant Diversity in Home Gardens and Its Contribution to Household Economy in Suburban Areas in Sri Lanka. Environment and Natural Resources Journal, 7, 12-30.
19. Lindara, M. J. K. L., Johnsen, F. H., & Gunatilake, H. M. (2004) Technical Efficiency in the Spice Based Agroforestry Sector in Matale, Sri Lanka. Noragric Working Paper No. 34, Noragric Agricultural University of Norway. http://www.eldis.org/vfile/upload/1/document/0708/DOC12243.pdf
20. Liyanage, M., Tejwani, K. G., & Nair, P. K. R. (1984). Intercropping Under Coconuts in Sri Lanka. Agroforestry Systems, 2, 215-228.
21. Mattsson, E., Ostwald, M., Nissanka, S. P., & Marambe, B. (2013). Homegardens as a Multi-Functional Land-Use Strategy in Sri Lanka with Focus on Carbon Sequestration. AMBIO, 42, 892-902.
22. Mattsson, E., Ostwald, M., Nissanka, S. P., Holmer, B., & Palma, M. (2009). Recovery and Protection of Coastal Ecosystems after Tsunami Event and Potential for Participatory Forestry CDM—Examples from Sri Lanka. Ocean & Coastal Management, 52, 1-9.
23. Mattsson, E., Persson, U. M., Ostwald, M., & Nissanka, S. P. (2012). REDD+ Readiness Implications for Sri Lanka in Terms of Reducing Deforestation. Journal of Environmental Management, 100, 29-40. http://www.sifi.se/wp-content/uploads

/2012 /02/ REDD±-readiness-implications-for-Sri-Lanka-in-terms-of-reducing-deforestation.pdf http://dx.doi.org/10.1016/j.jenvman.2012.01.018
24. MALF (1995). Sri Lanka Forestry Sector Master Plan. Ministry of Agriculture, Lands and Forestry, Sri Lanka.
25. MoENR (2009a). National Workshop on Mainstreaming Climate Change for Sustainable Development in Sri Lanka.
26. MoENR (2009b). Fourth Country Report to the Convention on Biological Diversity.
27. Nanayakkara, V. (1991). Agroforestry Systems in Sri Lanka. In W. Mellink, Y. S. Rao, & K. G. Mac Dicken (Eds.), Agroforestry in Asia and the Pacific (pp. 304). Bangkok: Winrock International.
28. National Communication (Natcom) (2000). Democratic Socialist Republic of Sri Lanka. Initial National Communication under the United Nations Framework Convention on Climate Change. http://unfccc.int/national_reports/non-annex_i _natcom /items/2979.php
29. Näsström, R., & Mattsson, E. (2009). Country Report Sri Lanka. Land-Use Change and Forestry at the National and Sub- National Level. Focali Report 2011:04, Gothenburg.
30. Nellemann, C., MacDevette, M., Manders, T., Eickhout, B., Svihus, B., Prins, A. G., & Kaltenborn, B. P. (2009). The Environmental Food Crisis—The Environment's Role in Averting Future Food Crises. A UNEP Rapid Response Assessment. United Nations Environment Program, GRID-Arendal. http://www.grida.no
31. Nuberg, I. K., & Evans, D. G. (1993). Alley Cropping and Analog Forests for Soil Conservation in the Dry Uplands of Sri Lanka. Agroforestry Systems, 24, 247-269.
32. Pandey, D. N. (2002). Carbon Sequestration in Agroforestry Systems. Climate Policy, 2, 367-377. http://www.climatepolicy.com
33. Peiris, T. S. G., Wijeratne, M., Ranasinghe, C. S., Anandacumaraswamy, A., Fernando, M. T. N., Jayakody, A., & Ratnasiri, J. (2004). Proceedings from 2nd AIACC Regional Workshop for Asia and the Pacific, 2-5 November 2004, Traders Hotel, Pasay City.
34. Perera, M. (2010). Integrating Agro-Forestry Characteristics into Agro-Well-Based Agricuture. In P. Weligemege, G. G. A. Godaliyadda, & K. Jinapala (Eds.), Irrigation for Food Security. Proceedings from the National Conference on Water, Food Security and Climate Change in Sri Lanka. Vol. 1. http://www.iwmi. cgiar.org /Publications/Other/PDF/SLWC_vol-1.pdf
35. Punyawardena, B. V. R. (2007). Impacts of Climate Change on Agriculture in Sri Lanka and Possible Response Strategies: Impacts, Adaptation and Mitigation. Proceedings from National Conference on Climate Change 2007. www.ips.lk/.../308-climate-change-issues-in-sri-lanka
36. Ratnayake, U., & Herath, G. (2005). Changes in Water Cycle: Effect on Natural Disasters and Ecosystems. Sri Lanka National Water Development Report. In N. T. S. Wijesekera, K. A. U. S. Imbulana, & N. B. Paris (Eds.), World Water Assessment Program. Paris. www.ircwash.org/...ources/sri-lanka-national-water...
37. Responsible Tourism Partnership Sri Lanka (2008) Carbon-neutral Sri Lanka—Longhaul Destination of the Developing World, Colombo: Responsible Tourism Partnership Sri Lanka. www.allconferences.com/...erences/2011/20110124022808
38. Roshetko, J. M., Lasco, R. D., & Angeles, M. S. D. (2007). Smallholder Agroforestry Systems for Carbon Storage. Mitigation and Adaptation Strategies for Global Change. World Agroforestry Center, 12, 219-242.
39. Scherr, S. J., Shames, S., & Friedman, R. (2012). From Climate-Smart Agriculture to Climate-Smart Landscapes. Agriculture & Food Security, 1, 12. http://www. agricultureandfoodsecurity.com/content/1/1/12
40. Sivakumar, M. V. K., Brunini, O., & Das, H. P. (2005). Impacts of Present and Future Climate Variability on Agriculture and Forestry in the Arid and Semi-Arid Tropics. Climate Change, 70, 31-72.
41. Somaratne, S., & Dhanapala, A. H. (1996). Potential Impact of Global Climate Change on Forest Distribution in Sri Lanka. Water, Air, and Soil Pollution, 92, 129-135.

42. The World Bank Group (2011). Vulnerability, Risk Reduction, and Adaptation to Climate Change, Sri Lanka. Climate Risk and Adaptation Country Profile April 2011. The World Bank Group, Washington. http://sdwebx.worldbank.org /climateportalb /doc/GFDRRCountryProfiles/wb_gfdrr_climate_change_country_profile_for_LKA.pdf
43. UNESCAP (2007) Statistical Yearbook for Asia and the Pacific 2007. Thailand: United Nations Publication. http://www.unescap.org/stat/data/syb2007/ESCAP-SYB2007.pdf
44. UN-REDD Programme (2012) Sri Lanka REDD+ Readiness. Preparation Proposal, UN-REDD Programme Eighth Policy Board Meeting. http://www.unredd.net/index.php?option=com_docman&task=doc
45. Verchot, L. V., Van Noordwijk, M., Kandji, S., Tomich, T., Ong, C., Albrecht, A., Mackensen, J., Bantilan, C., Anupama, K. V., & Palm, C. (2007). Climate Change: Linking Adaptation and Mitigation through Agroforestry. Mitigation and Adaptation Strategies for Global Change, 12, 901-918.
46. Vermeulen, S. J., Aggarwal, P. K., Ainslie, A., Angelone, C., Campbell, B. M., Challinor, A. J., Hansen, J.W., Ingram, J. S. I., Jarvis, A., Kristjanson, P., Lau, C., Nelson, G. C., Thornton, P. K., & Wollenberg, E. (2012). Options for Support to Agriculture and Food Security under Climate Change. Environmental Science & Policy, 15, 136-144.
47. Weerakkody, U. (1996). Impact of Sea Level Rise on Beach Erosion and Inundation of Wetlands in the South-West Coast of Sri Lanka. Regional Workshop on Climate Change Vulnerability and Adaptation in Asia and the Pacific.
48. Wijeratne, M. A. (1996). Vulnerability of Sri Lanka Tea Production to Global Climate Change. Water, Air, and Soil Pollution, 92, 87-94.
49. Wijeratne, M. A., Anandacoomaraswamy, A., Amaratunge, M. K. S. L. D., Ratnasiri, J., Basnayake, B. R. S. B., & Kalra, N. (2007). Assessment of Impact of Climate Change on Productivity of Tea (Camellia sinensis L.) Plantations in Sri Lanka. Journal of the National Science Foundation of Sri Lanka, 35, 119-126.
50. Yamane, A. (2009). Climate Change and Hazardscape of Sri Lanka. Environment and Planning, 41, 2396-2416.
51. Zubair, L., Hansen, J., Chandimala, J., Siraj, M. R. A., Siriwardhana, M., Ariyaratne, K., Bandara, I., Bulathsinghala, H., Abeyratne, T., & Samuel, T. D. M. A. (2005). Current Climate and Climate Change Assessments for Coconut and Tea Plantations in Sri Lanka. IRI, FECT, NRMS and UoP Contribution to AS12 Project Report to Be Submitted to Start, Washington DC.

CHAPTER 6

Greenhouse Gas Emissions from Agricultural Food Production to Supply Indian Diets: Implications for Climate Change Mitigation

Sylvia H. Vettera, *, Tek B. Sapkotab , Jon Hilliera , Clare M. Stirlingc , Jennie I. Macdiarmidd , Lukasz Aleksandrowicze,f , Rosemary Greene,f , Edward J.M. Joye,f , Alan D. Dangoure,f , Pete Smitha

[a] *Institute of Biological and Environmental Sciences, University of Aberdeen, Aberdeen AB24 3UU, UK*

[b] *International Maize and Wheat Improvement Centre (CIMMYT), Sustainable Intensification Program, NASC Complex, New Delhi 110012, India*

[c] *International Maize and Wheat Improvement Centre (CIMMYT), Sustainable Intensification Program, Apdo, Postal 6-641, 06600 Mexico, Distrito Federal, Mexico*

[d] *Rowett Institute of Nutrition and Health, University of Aberdeen, Aberdeen AB25 2ZD, UK*

[e] *Faculty of Epidemiology and Population Health, London School of Hygiene & Tropical Medicine, Keppel Street, London WC1E 7HT, UK*

[f] *Leverhulme Centre for Integrative Research on Agriculture and Health, London WC1H 0PD, UK*

ABSTRACT

Agriculture is a major source of greenhouse gas (GHG) emissions globally. The growing global population is putting pressure on agricultural production systems that aim to secure food production while minimising GHG emissions. In this study, the GHG emissions associated with the production of major food commodities in India are calculated using the Cool Farm Tool. GHG emissions, based on farm management for major crops (including cereals like wheat and rice, pulses, potatoes, fruits and vegetables) and livestock-based products (milk, eggs, chicken and mutton meat), are quantified and compared. Livestock and rice production were found to be the main sources of GHG emissions in Indian agriculture with a country average of 5.65 kg CO_2eq kg^{-1} rice, 45.54 kg CO_2eq kg^{-1} mutton meat and 2.4 kg CO_2eq kg^{-1} milk. Production of cereals (except rice), fruits and vegetables in India emits comparatively less GHGs with <1 kg CO_2eq kg^{-1} product. These findings suggest that a shift towards dietary patterns with greater consumption of animal source foods could greatly increase GHG emissions from Indian agriculture. A range of mitigation options are available that could reduce emissions from current levels and may be compatible with increased future food production and consumption demands in India.

Keywords: Agriculture; Cool Farm Tool; Greenhouse gas emissions; Indian diets; Sustainability

1. INTRODUCTION

Agriculture is an important sector of the economy in India, contributing about 20% of national gross domestic product, and providing a livelihood for nearly two-thirds of the population (ICAR, 2015). Equally important is the contribution of agriculture to national food security. India achieved self-sufficiency in food production after the Green Revolution (GR), but retaining this success has been challenging due to the increasing scarcity of resources, including labour, water, energy, and rising costs of production (Saharawat et al., 2010). Increased use of production inputs, such as mineral fertiliser, has made Indian agriculture more greenhouse gas (GHG)-intensive. Agricultural production is a major emitter of GHGs, currently accounting for 18% of total GHG emissions in India (INCCA, 2010). Recent estimates report that global food production must increase by 70% to meet the projected food demand of the estimated 9 billion global population by 2050 (CTA-CCAFS, 2011). With a population of ~1.3 billion, it is evident that the food system in India will be central to the global challenge of providing sufficient nutritious food while minimising GHG emissions. However, given the increasing population and shifting dietary patterns, GHG emissions from agricultural production in India are expected to change.

Quantifying GHG emissions associated with the production of food items in India is an important stage in quantifying GHG emissions associated with diets. It allows us to (i) identify variation in GHG emissions between typical dietary patterns within India; (ii) forecast the effect of changes in diets on GHG emissions; and (iii) identify options to minimise GHG emissions from food production, either through production-side changes or through dietary changes. For example, a number of countries have experienced a 'nutrition transition' associated with greater disposable incomes, urbanisation and globalisation. The transition is typified by increasing consumption of animal products, edible oils and sweetened beverages and decreasing consumption of cereals and pulses (Drewnowski and Popkin, 1997 ; Popkin *et al.*, 2012). There is evidence that a similar trend is emerging among some population groups in India, although cultural preferences for lacto-ovo-vegetarian diets suggest that India's experience will differ from other countries including China (Baker and Friel, 2014 ; Misra *et al.*, 2011). The implications of dietary changes in India for GHG emissions have not been quantified.

In India, the majority of agricultural GHG emissions occur at the primary production stage (Pathak et al., 2010), and are generated through the production and use of agricultural inputs, farm machinery, soil disturbance, residue management and irrigation. These practices are used to increase yields and improve harvests. Due to its direct contribution to global GHG emissions, agriculture can also serve as an important climate change mitigation strategy (Smith *et al.*, 2013 ; Smith *et al.*, 2008), both by reducing GHG emissions to the atmosphere, and by sequestering atmospheric carbon into plant biomass and soil, though the role of some soil carbon sequestration practices for climate mitigation has been questioned (Powlson et al., 2014). India's Intended Nationally Determined Contributions (INDCs) to the United Nations Framework Convention on Climate Change (UNFCCC, http://unfccc.int/2860.php [accessed 19.05.2016]) place emphasis on mitigation from agriculture, and various mitigation strategies (particularly concerning methane, CH_4, and nitrous oxide, N_2O) have been proposed (Smith *et al.*, 2014 ; Smith *et al.*, 2008). Quantification of GHG emissions from the production of different food commodities helps farmers, researchers and policymakers to understand and manage these emissions, and identify mitigation responses that are consistent with the food security and economic development priorities of countries (Hillier *et al.*, 2011 ; Whittaker *et al.*, 2013).

Various methods exist to estimate GHGs from agriculture, ranging from simple Tier 1 methods (IPCC 2006) to complex process-based models, which simulate the soil carbon and nitrogen cycles in some detail (Ogle et al., 2013). Several tools and calculators have been developed for estimating GHG fluxes from farm activities and to support decision making in terms of identifying informed interventions. Here, we used a modified version of The Cool Farm Tool (Hillier et al., 2011), which integrates several empirical models into one tool for GHG estimation from farm activities. The tool recognises context-specific factors that influence GHG emissions such as pedo-climatic characteristics, production inputs, and other management practices at farm level. GHG emissions from livestock products are calculated using the comprehensive data from the 19th Livestock Census of the Government of India (GOI, 2012) following the approach of Herrero et al. (2013).

The objective of the study is to analyse and compare farm-level GHG emissions of major food commodities at a national scale in India. The study gives an overview of emission-related hotspots, and discusses the implications for low carbon development in relation to changing diets in India.

2. MATERIALS AND METHODS

2.1. Food items

We calculated GHG emissions from agricultural production of several crops and livestock products in India (Table 1). Items for analysis were chosen from the total list of consumed foods recorded in the Indian Migration Study (IMS), a regional survey that measured dietary intake in 2005–2007 (Bowen et al., 2011). Based on project logistics, we set an objective of analysing a set of 20 food items. We used two criteria to choose items for analysis: having at least one item from each broad food group (i.e., cereals, pulses, tubers, vegetables, etc.); and within each group, selecting items reported to be consumed in the greatest quantity within the study (in kg $capita^{-1}$ d^{-1}). In total, 17 single food items were analysed, including 13 single crops, and four animal-sourced products (Table 1), plus three crop groups (other cereals, other pulses and crops used for vegetable oils). These items represent about 72% of reported consumption of food (kg) in the IMS, and about 75% of consumption when assessed in a 2012 nationally-representative household expenditure survey (National Sample Survey Organisation, 2007). The livestock products included in the analysis are milk, eggs, poultry and mutton meat. Fish and seafood were excluded from the study. This is because the consumption of these groups was low within the IMS sample which meant that the additional methodological effort and data acquisition required to conduct the analysis was not justified. Our study therefore focusses on agricultural produce.

Table 1. Major crops and livestock products by % of total intake in India, number of data points available with management information, averaged data and standard deviation for each product, nitrogen input and GHG emissions for different scales.

crop/ livestock prod.	group	subgroup	% of consumption from total food in Indian diets	nr. of data points	yield [tonnes/ ha]	std dev	N [kg/ ha]	std dev	GHG [kg ha^{-1}]	std dev	GHG [kg kg^{-1} product]	std dev	GHG [kg 1000 kcal^{-1}]	std dev
Milk	Livestock product	Dairy-lo-fat	18.17	105							2.42	0.90	3.97	1.48
Wheat	Cereals	Cereals	9.42	6017	3.26	1.14	139.41	51.47	977.15	439.70	0.34	0.21	0.12	0.07
Paddy Rice	Rice	Cereals	8.97	11993	3.61	1.51	114.37	54.21	8447.59	4754.41	5.65	4.59	1.21	0.96
Mango[a]	Fruit	Fruit	4.60	/	10.4	/	11.7	/	750.00		0.07		0.16	
Onion	Other	Spices	3.72	48	19.55	8.59	192.57	98.24	1599.65	969.44	0.10	0.07	0.39	0.29
Tomato[a]	Other	Vegetable	3.67	/	130	/	360	/	3000.00		0.15		0.88	
Potato	Potato	Tuber	2.69	394	23.83	9.27	236.01	181.24	3406.46	2727.19	0.22	0.23	0.33	0.35
Orange[a]	Fruit	Fruit	2.57	/	10.3	/	113	/	1300.00		0.13		0.37	
Sugarcane	Other	Other	1.90	1312	79.35	33.49	258.84	122.67	3954.34	3975.21	0.09	0.22	0.73	2.07
Lentil	Pulses	Pulses	1.89	425	0.90	0.39	16.03	14.96	292.17	303.45	0.38	0.38	0.13	0.13
Spinach	Other	Vegetable	1.29	/	21	/	33.5	/	1100.00		0.05		0.30	
Peas	Pulses	Pulses	1.17	128	1.39	0.75	41.41	38.11	540.09	250.37	0.42	0.21	0.81	0.84
Poultry	Livestock product	Chicken	0.74	69							2.59	0.08	1.40	0.04
Egg	Livestock product	Egg	0.45	69							2.59	0.08	1.87	0.06
Groundnut	Pulses	Nuts and oils	0.39	629	1.36	0.73	50.66	44.71	383.58	295.60	0.38	0.47	0.10	0.13
Mutton	Ruminant meat	Meat	0.38	280							45.54	11.89	17.32	4.52
Spices (Cumin Seed)	Other	Nuts and oils	0.08	/	2	/	100	/	1500.00		0.75		0.25	
Other Cereals[b]	Cereals	Cereals	2.76	3520	1.94	1.41	64.43	53.65	707.32	377.03	0.43	0.50	0.06	0.13
Other Pulses[c]	Pulses	Pulses	3.8	3720	0.82	0.57	25.14	36.50	490.32	359.31	0.75	1.59	0.14	0.38
Crops for vegetable oils[d]	Other	Nuts and oils	2.84	2569	1.30	0.62	40.66	36.20	532.12	632.39	0.54	0.93	0.12	0.18

[a] No plot/farm data were available; typical management and statistical information were used to generate management information.
[b] Includes bajra, barley, maize, ragi and jowar.
[c] Includes black, red and green gram.
[d] Includes coconut, rapeseed, soybean, safflower, sesamum, sunflower.

2.2. Management data

Agricultural input and management information, including yield of major crops grown in India, were obtained from the Directorate of Economics and Statistics of the Government of India (http://eands.dacnet.nic.in [accessed 01.10.2015]). The Government of India conducts cost of cultivation surveys at the Indian district level using multi-stage sampling. Districts within states, and villages within districts, formed the first and second stage unit of sampling with the ultimate unit of data collection being the household (CSO, 2002). The district and villages were selected in order to cover the major crops grown in the country. Figure 1 shows the locations of households selected for the survey, which forms the foundation of the activity data used in this study. In total, there were 34,577 data points across India used in the study. Of these, 53% of data points were for paddy rice and wheat, representing the proportionate area under rice and wheat cultivation in

India. Data on temperature and rainfall were obtained from the WorldClim global climate database (http://worldclim.org/ [accessed 01.10.2015]), and soil data (soil texture, soil organic carbon, soil pH, bulk density) were obtained from Shangguan et al. (2014). The water management system before and during rice cultivation was determined from databases at national and state levels (Gupta et al., 2009), and expert opinion (experts from CIMMYT). The analysis includes a representative distribution of irrigation management strategies for rice, from flooded to alternative wetting and drying systems. In India, agricultural residues left in the field after harvest are sometimes burnt *in-situ* to facilitate cultivation of subsequent crops, or used for other purposes off-site. The information on residue management of different crops, including burning, was obtained from Gadde et al. (2009) and Jain et al. (2014) at state level. The area under different crop cultivation in each state and union territory were obtained from state agriculture departments, the Directorate of Economics and Statistics of the Government of India, and FAOSTAT (FAOSTAT, 2015).

State-wise details of livestock by breed, age, sex and management type were obtained from the 19th Livestock Census of the Government of India (GOI, 2012). The information on livestock body weight, feed consumption and *per-capita* production of meat and milk (Table 2) were based on Singhal and Mohini (2002) and on expert judgement from the National Dairy Research Institute (NDRI) following relationships outlined in Herrero et al. (2013).

Management information for crops not included in the data set from the Directorate of Economics and Statistics of the Government of India, was generated from another source of general management information (http://www.haifa-group.com/knowledge_center/recommendations/fruit_trees/ [accessed 01.06.2015]), and statistics from FAOSTAT (2015).

2.3. Model and greenhouse gas emissions

GHG emissions from crops were calculated using the Cool Farm Tool (CFT) (Hillier et al., 2011; CFT: https://www.coolfarmtool.org/ [accessed 01.10.2015]). The CFT is a GHG emission calculator which allows users to estimate annual GHG emissions associated with the production of crops or livestock products from production to the farm gate (Hillier et al., 2011). It comprises a generic set of empirical models that are used to estimate full farm-gate product emissions constituting a mix of Tier 1, Tier 2, and simple Tier 3 approaches (see IPCC, 1997 for definitions of tiers for GHG estimation in national GHG inventories). GHG emissions were estimated from inputs including general information about soil and climate, and the set of management options on the farm: fertilisation, pesticide and herbicide use, residue management, machinery and energy use.

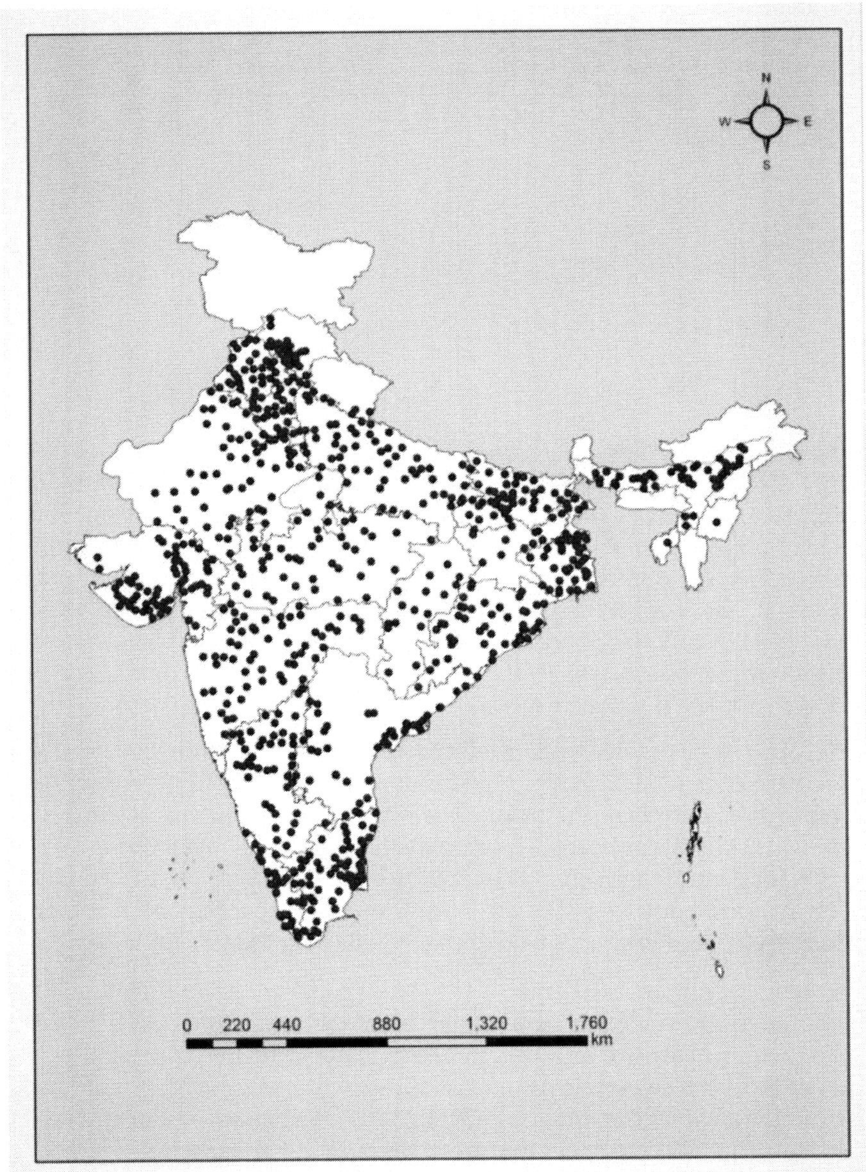

Figure 1. Location of sampled villages for the cost of production survey in India, from which activity data were derived for this study.

Table 2. Livestock body weight, feed consumption and per-capita production of meat and milk used in the analysis.

						Feed type		
livestock	breed	age/category	body wt (kg)	product (meat/ milk)	meat (kg/animal/ year)	dry fodder (kg/animal/ year)	green fodder (kg/animal/ year)	concentrates (kg/animal/ year)
sheep/ mutton	exotic	under 1 year	24	meat	14	365	1460	109.5
sheep/ mutton	local	under 1 year	18	meat	10	365	1277.5	116.8
sheep/ mutton	exotic	more than 1 year	42	meat	23	547.5	2190	146
sheep/ mutton	local	more than 1 year	38	meat	20	365	1642.5	109.5
cattle	exotic	milk	350	milk	2555	1460	8030	730
cattle	local	milk	300	milk	1095	1460	6570	730
poultry			1.8	meat	10 kg/cycle			30 kg/cycle
poultry			1.5	egg	315 egg/cycle			82 kg/cycle

For the current analysis, a version of the CFT implemented in Matlab (R2012a [7.14.0739], MathWorks, USA) was used to calculate the emissions for on-farm plots across India. The exception was for rice production where the method of Yan et al. (2005) was preferred to the Cool Farm Tool (which uses the Tier 1 method of IPCC (2006)), due to the greater granularity of the Yan et al. method, which bases estimates of CH_4emissions on several variables (i.e. soil pH, climate, organic amendment, pre-water regime, water regime) which were available at plot level in this study but were not factored in to the IPCC tier 1 method (IPCC, 2006).

GHG emissions from livestock products were calculated using the approach of Herrero et al. (2013) which provides data on GHG emissions from enteric fermentation and manure management for several animal groups (i.e. ruminants, small ruminants, pigs and poultry) using data for India on livestock systems and feed. National GHG emissions were calculated based on the average body weight of the livestock for different regions. Additional emissions for feed production were calculated using the CFT for feed crops.

We account only for GHG emissions related to farm management, and do not account for processing or transport after the farm-gate. GHG emissions up to the farm gate are reported in CO_2 equivalent (CO_2eq) *per* ha of crops and *per* head for livestock using the 100 year global warming potentials used in national GHG accounting (IPCC, 2006). For comparison, all results are also presented on a *per* kg production basis. GHG emissions were also converted to kg CO_2eq kcal^{-1} using FAO (2001) data for the energy content of the food commodities.

3. RESULTS

3.1. GHG emissions of food items up to the farm-gate

The analysis of GHG emissions from farm management in India presents the variability of emissions across India based on different management practices (Figure 2, Table 1). There are more data for the most widely consumed crops (i.e. wheat and rice; Table 1) than other products. The variability in GHG emissions for wheat is less than that for rice or potato. For paddy rice, >10,000 plots were available for analysis with a wide range of management practices, which is reflected in the GHG emission results. The main reason for the wide range in GHG emissions seen in rice is water management, which is the main determinant of CH_4 emissions. In particular, continuous flooding generates the highest CH_4 emissions, while longer and more frequent periods of water drainage reduces emissions. For example, changing the water regime from continuously flooded to multiple drainage periods, reduces CH_4 emissions by 9-fold (data not shown). High emissions on a *per*-ha-basis correspond with high GHG emissions *per* kg rice.

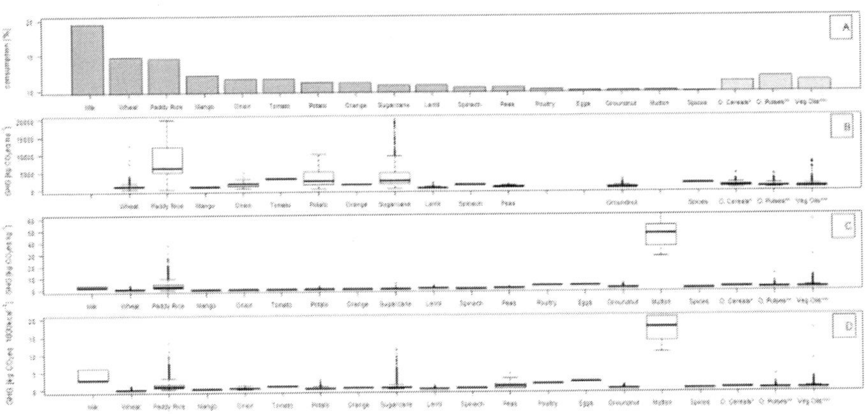

Figure 2. (A) mass of food group consumption as a%-age of total consumption reported in the Indian Migration Survey (reference); (B) GHG emissions associated with production of food groups in India per hectare, (C) per kg yield, (D) per 1000 kcal. (*Other Cereals: includes bajra, barley, maize, ragi and jowar; ** Other Pulses: includes black, red and green gram; *** Crops for vegetable oils: includes coconut, rapeseed, soybean, safflower, sesamum, sunflower).

With the exception of flooded rice, the major source of variation in GHG emissions for crops is due to variation in fertiliser application. The results show a wide range of GHG emissions for rice, potato and sugarcane on a *per*-ha-basis, and a narrow range for other crops (Figure 2B). The groups "other cereals" (i.e. bajra, barley, maize, ragi and jowar), and "other pulses" (i.e. black, red and green gram), had broadly similar GHG emissions. Emissions from vegetable oil crops showed more variation across the different crops (i.e. coconut, rapeseed, soybean, safflower, sesamum, sunflower).

GHG emissions *per* kg of livestock product (Figure 2C) varies markedly between livestock types. GHG emissions are highest for mutton meat (as the example for ruminant meat), followed by other livestock production such as poultry and dairy (milk). GHG emissions *per* kg of product were greater for livestock products than for crops, with the exception of rice. Mean GHG emissions were <1 kg CO_2eq kg^{-1} product for all crops except rice, with decreasing emissions across the categories of spices, pulses and nuts, wheat, fruits, vegetables and roots, and sugarcane, respectively (Table 1).

GHG emissions *per* kcal show a different ranking, although products from ruminant animals have the highest emissions using all metrics. GHG emissions *per* kcal show a small increase across cereals, pulses, vegetables, fruits and animal-source foods. Rice in particular had higher emissions than other crops, and mutton meat had markedly high emissions, several times more than other animal-source products.

3.2. GHG emissions from food consumption in India

Figure 3A shows relative reported consumption by weight of commodities in the IMS, while 3B shows their relative contribution to emissions. Rice and livestock products contribute the most to total dietary GHG emissions, with the third contributor being ruminant meat. Although ruminant meat had the greatest GHG emissions *per* unit product, it contributed less to overall GHG emissions (12.5%) as consumption is low, accounting for only 0.4% of total intake. Cereals other than rice and fruit products account for 12.9% and 22.5% of reported consumption by weight, yet as their emissions *per* unit of product are low, they make a relatively small contribution to total dietary GHG emissions, representing only 3.2% and 1.1% of total emissions, respectively. The group "other" (including various crops from the subgroups nuts and oils, spices, and vegetables) also contributed little to total dietary GHG emissions.

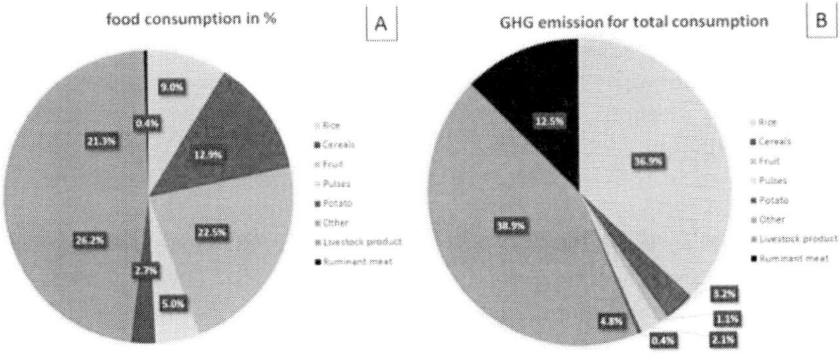

Figure 3. Proportion of consumption of food groups in Indian diets (A), and distribution of GHG emissions from agricultural production of this diet (B).

4. DISCUSSION

4.1. GHG emissions from crop production

GHG emission calculations for agricultural production of the commonly consumed food items in India are based on a substantial dataset with representative plots across the country. This analysis gives an overview of GHG emissions produced through on-farm management at different locations, representing diverse soil types and climatic conditions, and encompasses major drivers of variation within the country. Using the same model to calculate GHG emissions for all of the major food groups allows emissions from different crops and products to be compared.

On a *per* ha basis, GHG emissions for major food crops in India are generally lower than those in Europe and North America, with GHG emissions for cereals 2–3-fold greater in Europe (2000–3000 kg CO_2eq ha^{-1} yr^{-1}) (Carlton et al., 2012). A study in Canada estimated GHG emissions for spring wheat of 600 and 1400 kg CO_2eq ha^{-1} yr^{-1} (Gan et al., 2012). GHG emissions for potatoes and peas show a similar range in Europe of ~3000 kg CO_2eq ha^{-1} yr^{-1} and ~660 kg CO_2eq ha^{-1} yr^{-1}, respectively (Carlton et al., 2012). A Swedish study reported GHG emissions of wheat of 0.2–0.6 kg CO_2eq kg^{-1}production (Röös et al., 2011); the calculated GHG emissions for wheat in India are towards the lower end of that range. A comparison of rice production in a Chinese study showed a range from 2000 kg CO_2eq ha^{-1} yr^{-1} for upland rice up to 20000 kg CO_2eq ha^{-1} yr^{-1} for paddy rice production, and similar values have been reported for Indian rice (Li et al., 2006).

The reported estimates of GHG emissions from farm management differ across the above studies partly because each uses different boundary conditions. In our study, the calculated GHG emissions on a *per*-ha-scale follow the same methods for all crops and differ mainly because of changes in management and fertiliser use for crops. For cereals in general, less fertiliser is used in India than in Europe. These differences are also partly reflected in yield. The yields for cereals, pulses and potatoes have increased over recent years in India, but are still only half of those recorded in Western Europe and North America (FAOSTAT, 2015). These differences show the importance of comparing GHG emissions on a *per*-kg-production-basis, as GHG emissions will be greater for low yielding crops than for higher yielding ones using a *per*-kg metric. For instance, according to FAOSTAT (2015) rice yields in China are around twice those in India.

The highest GHG emissions among crops are associated with paddy rice production. Emissions of CH_4 from rice production are recognised as a significant source of GHG emissions globally, and many studies show that changes in water management can substantially reduce CH_4 emissions (Liu *et al.*, 2010; Nayak *et al.*, 2015 ; Yan *et al.*, 2005). It is possible to reduce CH_4 emissions and increase yield through optimising drainage and manure management (Banerjee *et al.*, 2002; Malla *et al.*, 2005 ; Thu *et al.*, 2016). Specifically, changing a continuously-flooded system to intermittent irrigation shows potential to greatly reduce CH_4 emissions. Although some studies show that N_2O emissions may increase under intermittent irrigation the decrease in CH_4 emissions more than compensates this effect (Liu *et al.*, 2010 ; Nayak *et al.*, 2015). Nayak et al. (2015) summarised management opportunities to mitigate GHG emissions from agriculture in China, and these can largely be adapted to Indian agriculture. In rice management, key elements are fertiliser management by reducing synthetic fertiliser inputs, increasing organic manure, and improved water management, as discussed above.

4.2. GHG emissions from livestock products

As expected, GHG emissions from livestock products are generally higher than those from crop production (Figure 2, Table 1). This reflects the inefficiencies of conversion of plant protein to animal protein in herbivores (Ripple et al., 2014), and is also impacted by additional sources of emissions resulting from manure management and enteric fermentation in ruminants. GHG emissions associated with livestock products depend largely on feed inputs, and in other studies have been shown to range between 0.8–2.4 kg CO_2eq kg^{-1} milk, 1.7–6.6 kg CO_2eq kg^{-1} eggs, 2.5–6.9 kg CO_2eq kg^{-1} poultry meat and 10–20 kg CO_2eq kg^{-1} mutton and lamb (Bellarby et al., 2013). These values are based on different studies, mainly from model exercises which focus on Europe. Our milk and poultry results for India are within the range of these studies. The calculated emissions for mutton are higher than in the above discussed studies, resulting from embedded emissions in feed, which is 50–75% of the total GHG emissions *per*-animal-*per*-year.

Ruminants produce CH_4 through enteric fermentation, and options to mitigate this source are somewhat limited (Beauchemin et al., 2011). Other sources of emissions from livestock production are manure management and changed feed rations. To reduce GHG emissions from manure management options include (i) changes to manure storage, e.g. decreased storage time, manure storage cover with straw, or mechanical intermittent aeration during manure storage (Hristov et al., 2013), (ii) manure acidification (Ndegwa *et al.*, 2011 ; Petersen *et al.*, 2012), (iii) feeding of livestock with nitrate supplements (Van Zijderveld et al., 2011) and (iv) stacking of poultry litter (Gerber et al., 2013). To reduce GHG emissions from feed, all mitigation measures previously discussed for crops could be considered, as well as using the residues from crop production as feed.

4.3. GHG emissions associated with Indian diets

Overall, national GHG emissions associated with diets are greatest for rice and livestock products like milk and eggs (Figure 3), because these are widely consumed products with high GHG emissions *per* unit of product. Although there is limited consumption of ruminant meat in India, its high GHG intensity means that it is the third greatest contributor to GHG emissions.

The mitigation potential in livestock production therefore needs to be further explored. In addition to the mitigation options in on-farm management, dietary change could help to decrease GHG emissions considerably, but advice to change dietary intakes to reduce GHG emissions would need to consider the nutritional implications, so as not to compromise health (Aleksandrowicz *et al.*, 2016; Bajzelj *et al.*, 2014; Smith *et al.*, 2013 ; Smith, 2015). However, in the event of a nutritional transition in India, toward the consumption of a greater volume of livestock products, there is likely to be an increase in GHG emissions unless *per*-product emissions are reduced through more efficient production and targeted mitigation measures, especially in the livestock sector and for rice production.

5. CONCLUSION

This study constructed a national dataset of GHG emissions associated with the production of major food items in India, and incorporates variability in emissions from a range of production systems. We used comprehensive agricultural activity data at the farm-level, and a state-of-the-art greenhouse gas accounting tool. We highlight the risk of a likely increase in GHG emissions if diets transition towards increased consumption of animal-based products, and also observe a wide range of emissions from cereal production. In addition to general measures to improve efficient use of nutrients and organic matter stocks, there is also likely to be benefit in developing support mechanisms to target those products with the highest emissions *per* unit of production. We hypothesise that in such cases, mitigation of GHG emissions will be a co-benefit of improved and more efficient agronomic practice, but these options would have to consider the nutritional and health implication for the Indian diet.

ACKNOWLEDGEMENTS

The study is part of the *Sustainable and Healthy Diets in India* (SAHDI) project funded by the Wellcome Trust under the 'Our Planet, Our Health' programme (Grant number 103932) and the *India Greenhouse Gas Mitigation Study* led by the International Maize and Wheat Improvement Center (CIMMYT) and funded by the Climate Change, Agricultural and Food Security (CCAFS) programme of the Consultative Group on International Agricultural Research (CGIAR). The work also contributes to the Belmont Forum/FACCE-JPI-funded DEVIL project (*via* UK NERC project (NE/M021327/1). We thank Drs Paresh B. Shirsath and Hanuman S. Jat of CIMMYT for their expert opinion and help during data collection.

REFERENCES

1. Aleksandrowicz, L., Green, R., Joy, E.J.M., Smith, P., Haines, A., 2016. The impacts of adopting environmentally sustainable and healthy diets on greenhouse gas emissions, land use, and water use: a systematic review. PLoS One in press.
2. Bajzelj, B., Richards, K.S., Allwood, J.M., Smith, P., Dennis, J.S., Curmi, E., Gilligan, C.A., 2014. Importance of food-demand management for climate mitigation. Nat. Clim. Change 10, 924–929.
3. Baker, P., Friel, S., 2014. Processed foods and the nutrition transition: evidence from Asia. Obes. Rev. 15, 564–567.
4. Banerjee, B., Pathak, H., Aggarwal, P., 2002. Effects of dicyandiamide, farmyard manure and irrigation on crop yields and ammonia volatilization from an alluvial

soil under a rice (Oryza sativa L.)-wheat (Triticum aestivum L.) cropping system. Biol. Fertil. Soils 36, 207–214.

5. Beauchemin, K.A., Janzen, H.H., Little, S.M., McAllister, T.A., McGinn, S.M., 2011. Mitigation of greenhouse gas emissions from beef production in western Canada—evaluation using farm-based life cycle assessment. Anim. Feed Sci. Technol. 166–167, 663–677.
6. Bellarby, J., Tirado, R., Leip, A., Weiss, F., Lesschen, J.P., Smith, P., 2013. Livestock greenhouse gas emissions and mitigation potential in Europe. Glob. Change Biol. 19, 3–18.
7. Bowen, L., Ebrahim, S., De Stavola, B., Ness, A., Kinra, S., Bharathi, A.V., Prabhakaran, D., Reddy, K.S., 2011. Dietary intake and rural-urban migration in india: a crosssectional study. PLoS One 6, e14822.
8. CSO, 2002. Manual on Cost of Cultivation SurveysCentral Statistical Organisation, Government of India. . Accessed 01 June 2015 http://eands.dacnet.nic.in/Cost_of_Cultivation.htm.
9. CTA-CCAFS, 2011. Farming's Climate-smart Future: Placing Agriculture at the Heart of Climate-change Policy. Technical Centre for Agricultural and Rural Cooperation ACP-EU (CTA) and CGIAR Research Program on Climate Change, Agriculture and Food Security (CCFAS).
10. Carlton, R., West, J., Smith, P., Fitt, B.L., 2012. A comparison of GHG emissions from UK field crop production under selected arable systems with reference to disease control. Eur. J. Plant Pathol. 133, 333–351.
11. Drewnowski, A., Popkin, B.M., 1997. The nutrition transition: new trends in the global diet. Nutr. Rev. 55, 31–43.
12. FAO, 2001. FAO Food Balance Sheets—A Handbook. Food and Agriculture Organization of the United Nations (FAO), Rome.
13. FAOSTAT, 2015. FAOSTAT Common Database. . Accessed 15 June 2015 http://faostat3.fao.org/home/E.
14. GOI, 2012. Government of India: 19th Livestock Census—2012. Department of Animal Husbandry, Dairying and Fisheries, Ministry of Agriculture, Krishi Bhawan, New Delhi, India.
15. Gadde, B., Menke, C., Wassmann, R., 2009. Rice straw as a renewable energy source in India, Thailand, and the Philippines: overall potential and limitations for energy contribution and greenhouse gas mitigation. Biomass Bioenergy 33, 1532–1546.
16. Gan, Y., Liang, C., Campbell, C.A., Zentner, R.P., Lemke, R.L., Wang, H., Yang, C., 2012. Carbon footprint of spring wheat in response to fallow frequency and soil carbon changes over 25 years on the semiarid Canadian prairie. Eur. J. Agron. 43, 175–184.
17. Gerber, P.J., Steinfeld, H., Henderson, B., Mottet, A., Opio, C., Dijkman, J., Falcucci, A., Tempio, G., 2013. Tackling Climate Change Through Livestock—A Global Assessment of Emissions and Mitigation Opportunities. Food and Agriculture Organization of the United Nations (FAO), Rome.
18. Gupta, P.K., Gupta, V., Sharma, C., Das, S.N., Purkait, N., Adhya, T.K., Pathak, H., Ramesh, R., Baruah, K.K., Venkatratnam, L., Singh, G., Iyer, C.S.P., 2009. Development of methane emission factors for Indian paddy fields and estimation of national methane budget. Chemosphere 74, 590–598.
19. Herrero, M., Havlík, P., Valin, H., Notenbaert, A., Rufino, M.C., Thornton, P.K., Blümmel, M., Weiss, F., Grace, D., Obersteiner, M., 2013. Biomass use, production, feed efficiencies, and greenhouse gas emissions from global livestock systems. Proc. Natl. Acad. Sci. 52, 20888–20893.
20. Hillier, J., Walter, C., Malin, D., Garcia-Suarez, T., Mila-i-Canals, L., Smith, P., 2011. A farm-focused calculator for emissions from crop and livestock production. Environ. Modell. Softw. 9, 1070–1078.
21. Hristov, A.N., Lee, C., Cassidy, T., Heyler, K., Tekippe, J.A., Varga, G.A., Corl, B., Brandt, R.C., 2013. Effect of Origanum vulgare L. leaves on production and milk fatty acid composition in lactating dairy cows. J. Dairy Sci. 96, 1189–1202.

22. ICAR, 2015. VISION 2020—Indian Council of Agricultural Research. ICAR, New Delhi, India. INCCA, 2010. India : Greenhouse Gas Emissions 2007. Indian Network for Climate Change Assessment, Ministry of Environment and Forests, Government of India.
23. IPCC, 1997. An introduction to simple climate models. In: Houghton, J.T., Filho, L.G. M., Griggs, D.J., Maskell, K. (Eds.), IPCC Second Assessment Report. Intergovernmental Panel on Climate Change.
24. IPCC, 2006. IPCC guidelines for national greenhouse gas inventories, prepared by the National Greenhouse Gas Inventories Programme. In: Eggleston, H.S., Buendia, L., Miwa, K., Ngara, T., Tanabe, K. (Eds.), Vol. 4: Agriculture, Forestry and Other Land Use.
25. IGES, Japan. Jain, N., Bhatia, A., Pathak, H., 2014. Emission of air pollutants from crop residue burning in India. Aerosol Air Qual. Res. 14, 422–430.
26. Li, C., Salas, W., DeAngelo, B., Rose, S., 2006. Assessing alternatives for mitigating net Greenhouse Gas Emissions and increasing yields from rice production in China over the next twenty years. J. Environ. Qual. 35, 1554–1565.
27. Liu, S., Qin, Y., Zou, J., Liu, Q., 2010. Effects of water regime during rice-growing season on annual direct N2O emission in a paddy rice–winter wheat rotation system in southeast China. Sci. Total Environ. 408, 906–913.
28. Malla, G., Bhatia, A., Pathak, H., Prasad, S., Jain, N., Singh, J., 2005. Mitigating nitrous oxide and methane emissions from soil in rice-wheat system of the IndoGangetic plain with nitrification and urease inhibitors. Chemosphere 58, 141– 147.
29. Misra, A., Singhal, N., Sivakumar, B., Bhagat, N., Jaiswal, A., Khurana, L., 2011. Nutrition transition in India: secular trends in dietary intake and their relationship to diet-related non-communicable diseases. J. Diabetes 3, 278–292.
30. National Sample Survey Organisation, 2007. Nutritional Intake in India 2004–2005.
31. NSS 61st Round. Government of India, New Delhi, India. Nayak, D., Saetnan, E., Cheng, K., Wang, W., Koslowski, F., Cheng, Y., Zhu, W.Y., Wang, J., Liu, J., Moran, D., Yan, X., Cardenas, L., Newbold, J., Pan, G., Lu, Y., Smith, P., 2015. Management opportunities to mitigate greenhouse gas emissions from Chinese agriculture. Agric. Ecosyst. Environ. 209, 108–124.
32. Ndegwa, P.M., Hristov, A.N., Ogejo, J.A., 2011. Ammonia emission from animal manure: mechanisms and mitigation techniques. In: He, Z. (Ed.), Environmental Chemistry of Animal Manure. Nova Science Publishers, Hauppauge, NY, pp. 107–151.
33. Ogle, S.M., Buendia Butterbach-Bahl, L.K., Breidt, F.J., Hartman, M., Yagi, K., Nayamuth, R., Spencer, S., Wirth, T., Smith, P., 2013. Advancing national greenhouse gas inventories for agriculture in developing countries: improving activity data, emission factors and software technology. Environ. Res. Lett. 8, 015030.
34. Pathak, H., Jain, N., Bhatia, A., Patel, J., Aggarwal, P.K., 2010. Carbon footprints of Indian food items. Agric. Ecosyst. Environ. 1–2, 66–73.
35. Petersen, S.O., Andersen, A.J., Eriksen, J., 2012. Effects of cattle slurry acidification on ammonia and methane evolution during storage. J. Environ. Qual. 41, 88–94.
36. Popkin, B.M., Adair, L.S., Ng, S.W., 2012. Global nutrition transition and the pandemic of obesity in developing countries. Nutr. Rev. 70, 3–21.
37. Powlson, D.S., Stirling, C.M., Jat, M.L., Gerard, B.G., Palm, C.A., Sanchez, P.A., Cassman, K.G., 2014. Limited potential of no-till agriculture for climate change mitigation. Nat. Clim. Change 4, 678–683.
38. Röös, E., Sundberg, C., Hansson, P., 2011. Uncertainties in the carbon footprint of refined wheat products: a case study on Swedish pasta. Int. J. Life Cycle Assess. 16, 338–350.
39. Ripple, W.J., Smith, P., Haberl, H., Montzka, S.A., McAlpine, C., Boucher, D.H., 2014. Ruminants, climate change and climate policy. Nat. Clim. Change 4, 2–5.
40. Saharawat, Y.S., Singh, B., Malik, R.K., Ladha, J.K., Gathala, M., Jat, M.L., Kumar, V., 2010. Evaluation of alternative tillage and crop establishment methods in a ricewheat rotation in North Western IGP. Field Crops Res. 116, 260–267.

41. Shangguan, W., Dai, Y., Duan, Q., Liu, B., Yuan, H., 2014. A global soil data set for earth system modeling. J. Adv. Model. Earth Syst. 6, 249–263.
42. Singhal, K.K., Mohini, M., 2002. Uncertainty Reduction in Methane and Nitrous Oxide Gases Emission from Livestock in India. Project report, Dairy Cattle Nutrition Division, National Dairy Research Institute, Karnal, India, pp. 62.
43. Smith, P., Martino, D., Cai, Z., Gwary, D., Janzen, H., Kumar, P., McCarl, B., Ogle, S., O'Mara, F., Rice, C., Scholes, B., Sirotenko, O., Howden, M., McAllister, T., Pan, G., Romanenkov, V., Schneider, U., Towprayoon, S., Wattenbach, M., Smith, J., 2008. Greenhouse gas mitigation in agriculture. Philos. Trans. R. Soc. B Biol. Sci. 1492, 789–813.
44. Smith, P., Haberl, H., Popp, A., Erb, K.-H., Lauk, C., Harper, R., Tubiello, F., de Siqueira Pinto, A., Jafari, M., Sohi, S., Masera, O., Böttcher, H., Berndes, G., Bustamante, M., Ahammad, H., Clark, H., Dong, H.M., Elsiddig, E.A., Mbow, C., Ravindranath, N.H., Rice, C.W., Robledo Abad, C., Romanovskaya, A., Sperling, F., Herrero, M., House, J.I., Rose, S., 2013. How much land based greenhouse gas mitigation can be achieved without compromising food security and environmental goals? GCB Bioenergy 19, 2285–2302.
45. Smith, P., Bustamante, M., Ahammad, H., Clark, H., Dong, H., Elsiddig, E.A., Haberl, H., Harper, R., House, J., Jafari, M., Masera, O., Mbow, C., Ravindranath, N.H., Rice, C. W., Robledo Abad, C., Romanovskaya, A., Sperling, F., Tubiello, F., 2014. Agriculture, Forestry and Other Land Use (AFOLU). In: Edenhofer, O., PichsMadruga, R., Sokona, Y., Farahani, E., Kadner, S., Seyboth, K., Adler, A., Baum, I., Brunner, S., Eickemeier, P., Kriemann, B., Savolainen, J., Schlömer, S., von Stechow, C., Zwickel, T., Minx, J. (Eds.), Climate Change 2014: Mitigation of Climate Change. Contribution of Working Group III to the Fifth Assessment Report of the Intergovernmental Panel on Climate Change. Cambridge University Press, Cambridge, United Kingdom and New York, NY, USA, pp. 811– 922.
46. Smith, P., 2015. Malthus is still wrong: we can feed a world of 9–10 billion, but only by reducing food demand. Proc. Nutr. Soc. 3, 187–190.
47. Thu, T.N., Phuong, L.B.T., Van, T.M., Hong, S.N., 2016. Effect of water regimes and organic matter strategies on mitigating Greenhouse Gas Emission from rice cultivation and co-benefits in agriculture in Vietnam. Int. J. Environ. Sci. Dev. 7, 85–90.
48. Van Zijderveld, S.M., Gerrits, W.J.J., Dijkstra, J., Newbold, J.R., Hulshof, R.B.A., Perdok, H.B., 2011. Persistency of methane mitigation by dietary nitrate supplementation in dairy cows. J. Dairy Sci. 94, 4028–4038.
49. Whittaker, C., McManus, M.C., Smith, P., 2013. A comparison of carbon accounting tools for arable crops in the United Kingdom. Environ. Modell. Softw. 46, 228– 239.
50. Yan, X., Yagi, K., Akiyama, H., Akimoto, H., 2005. Statistical analysis of the major variables controlling methane emission from rice fields. Glob. Change Biol. 7, 1131–1141.

CHAPTER 7

Plant Breeding for Harmony between Modern Agriculture Production and the Environment

João Carlos da Silva Dias

Instituto Superior de Agronomia, University of Lisbon, Lisbon, Portugal

ABSTRACT

The world population is estimated to be 9.2 billion in 2050. To sufficiently feed these people, the total food production will have to increase 60% - 70%. Climate models predict that warmer temperatures and increases in the frequency and duration of drought during the present century will have negative impact on agricultural productivity. These new global challenges require a more complex integrated agricultural and breeding agenda that focuses on livelihood improvement coupled with agro-ecosystem resilience, eco-efficiency and sustainability rather than just on crop productivity gains. Intensifying sustainability agro-ecosystems by producing more food with lower inputs, adapting agriculture to climate change, conserving agro-biodiversity through its use, and making markets to work for the small farmers are needed to address the main issues of our time. Plant breeding has played a vital role in the successful development of modern agriculture. Development of new cultivars will be required while reducing the impact of agriculture on the environment and maintaining sufficient production.

Conventional plant breeding will remain the backbone of crop improvement strategies. Genetic engineering has the potential to address some of the most challenging biotic constraints faced by farmers, which are not easily addressed through conventional plant breeding alone. Protective measures and laws, especially patenting, must be moderated to eliminate coverage so broad that it stifles innovation. They must be made less restrictive to encourage research and free flow of materials and information. Small farmers have an important role in conserving and using crop biodiversity. Public sector breeding must remain vigorous, especially in areas where the private sector does not function. This will often require benevolent public/private partnerships as well as government support. Active and positive connections between the private and public breeding sectors and large-scale gene banks are required to avoid a possible conflict involving breeders' rights, gene preservation and erosion. Plant breeding can be a powerful tool to bring "harmony" between agriculture and the environment, but partnerships and cooperation are needed to make this a reality.

Keywords: Breeding, Modern Agriculture, Climate Changes, Biodiversity, Environment, Ecosystems, Transgenic Crops, Small Farmers

1. INTRODUCTION

There are now almost 7.25 billion human beings inhabiting this planet, and it has been projected that world population growth may exceed 70 million annually over the next 40 years. The world population will be approximately 9.2 billion in 2050, when the concentration of carbon dioxide and ozone will be 550 ppm and 60 ppm, respectively and the climate will be warmer by 2°C [1] . At that time it is expected that approximately 90% of this global population will reside in Asia, Africa, and Latin American countries [2] [3] . Currently, about 1 billion human beings suffer from hunger; 3 billion malnourished people suffer one or more micronutrient deficiencies (especially vitamin A, iodine and iron) and live with less than 2 US dollars per day; and anthropogenic climate change continues to affect food output and quality [4] [5] . By 2050, to sufficiently feed all these people, the total food production will have to increase 60% to 70% to meet a net demand of 1 billion tonnes of cereal for food and to feed, and 200 million tonnes of meat [6] -[8] , depending on assumptions of population growth, income growth and dietary changes. This projected increase of global crop demand is partly due to a growing global population, but a larger driver is increasing global affluence and associated changes in diet due to higher incomes [4] [8] . As global incomes increase, diets typically shift from those comprised of mostly grains, to diets that contain more vegetables and fruits and a greater proportion of meat, dairy products, and eggs [4] [8] -[10] .

This shift from plant-based diets to more intensive demand for animal products is termed as the "Livestock Revolution" [11] and it is estimated that approximately 40% of the world's population will undergo this revolution to more animal consumption by the year 2050 [8] . In order to meet these demands, global livestock production systems are shifting from using mostly waste products, crop residues, and marginal lands to more industrial systems which require less land and use of higher value feed crops [11] [12] . In developing countries with high rates of increasing animal product demands, a greater proportion of cereals are being directed towards animals [13] . Increasing demand for meat and dairy is also of importance to the global environment because their production requires more land and other resources than plant-based foods [14] -[16] . In fact, livestock production is the single largest anthropogenic use of land. According to a 2011 analysis, 75% of all agricultural land (including crop and pasture land) is dedicated to animal production [17] . Livestock production is also responsible for other environmental impacts. Livestock production is estimated to be responsible for 18% of total greenhouse gas emissions [18] , and animal products generally have a much higher water footprint than plant-based foods [19] .

In addition to growing meat and dairy demands, affluent nations are also directing a growing proportion of high-value feedstock to bio-fuel production. A great majority of bio-fuel feedstock are human-edible, especially from maize in the United States and sugarcane in Brazil. In 2010 global bio-fuel production represented 2.7% of global fuel for road transportation (at 107 billion liters produced), which is more than a 450% increase from the year 2000 [20] . To produce these fuels the United States and Brazil combined dedicated over 460 million tonnes of maize and sugarcane, respectively, to bio-fuel production in 2010, which is 6% of global crop production [21] .

In 2008, the world's arable land amounted to 1386 M ha, out of a total 4883 M ha land used for agriculture [22] . Each year, arable and agricultural land is lost due to deforestation, overgrazing, agricultural activities, gathering and overexploitation for fuel-wood, urbanization and industrialization. The most direct negative impact of agriculture on biodiversity is due to the considerable loss of natural habitats, which is caused by the conversion of natural ecosystems into agricultural land. The arable land is limited. Increases in arable land can only be done by deforestation. Agricultural production should be increased without further deforestation. This requires innovation and better technologies, as well as substantial investment, to increase yields on existing agricultural land.

Climate models predict that warmer temperatures and increases in the frequency and duration of drought during the twenty-first century will have negative impact on agricultural productivity [23] -[28] . For example, maize production in Africa could be at risk of significant yield losses as researchers predict that each degree-day that the crop spends above 30°C reduces yields by 1% if the plants receive sufficient water [27] . These predictions are similar to those reported for maize yield in the United States [29] . Lobell et al. [27] further showed that maize yields in Africa decreased by 1.7% for each degree-day the crop spent at temperatures of over 30°C under drought. Wheat production in Russia decreased by almost one-third in 2010, largely due to the summer heat wave. Similarly, wheat production declined significantly in China and India in 2010, largely due to drought and sudden rise in temperature respectively, thereby causing forced maturity [30] . Warming at +2°C is predicted to reduce yield losses by 50% in Australia and India [31] [32] . Likewise, the global maize and wheat production, as a result of warming temperatures during the period of 1980 to 2008, declined by 3.8% and 5.5%, respectively [28] . So climate change poses a serious threat to species fitness [33] [34] , and to agro-ecosystems essential to food production [35] .

Climatic variation and change are already influencing the distribution and virulence of crop pest and diseases, but the interactions between the crops, pests and pathogens are complex and poorly understood in the context of climate change [36] . We will need to integrate plant biology into the current paradigm with respect to climate change to succeed in defeating emerging pests and pathogens posing a new threat to agriculture due to climate change [37] -[39] .

In this context we can ask: can we feed and clothe the growing world population while simultaneously preserving or improving ecosystems and the natural environment?

History shows that modern agriculture has the potential to feed the world population but also to be worst and even catastrophically with the natural environment. Some examples are deforestation, overgrazing and erosion, in many parts of the world, which contributed to the outright collapse of ecosystems. One classical example is Madagascar's central highland plateau that has become virtually totally barren (about ten percent of the country), as a result of slash-and-burn deforestation, an element of shifting cultivation practiced by many natives. Intensification of production systems have also led to reduction in crop and livestock biodiversity, and increased genetic vulnerability and erosion. In contrast, the "Green Revolution", which began providing high-yielding crop cultivars and high-input management techniques to developing countries in the 1960s, has prevented mass starvation and improved living standards throughout the world [40] . Dwarfing, photoperiod insensitive genes and host plant resistance genes to pathogens and pests were bred for various crops during the "Green Revolution" [41] . Production in agriculture was increased in many nations of Asia and Latin America by the "Green Revolution". Crop yields in the developing world would have been at least 20% less and food prices about 19% higher than they were in 2000 without the innovations of the "Green Revolution" [42] . Calorie consumption would have dropped by about 5% and the number of malnourished children would have increasing by at least 2%; i.e., the "Green Revolution" helped to improve the health status of 32 to 42 million pre-school children. Since the beginning of the "Green Revolution" in 1960, land devoted to crops increased some 10%, land under irrigation has doubled, pesticide use by agriculture has tripled, fertilizer use is up 23-fold, pesticide use is up by a factor of 53. Nowadays, forty per cent of crop production comes from the 16% of agricultural land that is irrigated. Irrigated lands account for a substantial portion of increased yields obtained during the "Green Revolution". However, the global rate of increase in the irrigated area is declining, per capita irrigated area has declined by 5% since 1978, and new dam construction may allow only a 10% increase in water for irrigation over the next 30 years [43] . The enhancement of yield achieved in the "Green Revolution" (29% in food supplies per capita since 1960) may have been associated with an increased level of greenhouse gas emissions associated with higher fertilizer production and application, but, overall, its net effect has been calculated to have reduced CO_2 emission by some 161 gigatons of carbon (GtC) over the period 1961-2005 [44] , implying that gains in crop productivity can make a positive contribution to reducing greenhouse gas emissions.

Developing sustainable agriculture in environmentally sensitive systems is the great challenge of the coming decades. More food, animal feed, fiber, fuel, and forest products must be produced with less available land, water, and nutrients, to meet basic human needs and improve the sustainability of production [45] . In addition, pressure from an increasing global human population will necessitate more efficient and diversified land use. High crop yields of main staples will still be needed for freeing land to cash or more nutritious crops such as fruits and vegetables, as well as to prevent biodiversity losses and protect the environment by avoiding the use of today's forests, woodlands, pastures, rangelands and mountain sides for agriculture. Furthermore, recent research shows that investment in crop yield gains compares favorably with other commonly proposed climate change mitigation strategies, and should be a priority target to reduce greenhouse gas emissions.

Identifying the most appropriate technologies and practices to achieve these objectives are critical. This requires the building of a knowledge base to support such tasks. Fortunately, the World already possesses the know-how and research capability required to achieve the main objective to transform agriculture into a fully natural resource-based system which manages eco-efficiently its surrounding environment and associated biodiversity. Agro-ecological approaches are known to improve farming system productivity, reduce pollution through sound methods of nutrient and pest management, maintain biodiversity reserves, and enhance habitat quality through careful management of soil, water, and natural vegetation. Infrastructure (particularly roads and irrigation), incentives, institutions and other innovations beyond agriculture are also needed to implement such a global change of rural landscapes in the short- to mid-terms. The agenda for a new "Green Revolution" needs to consider new approaches to promote innovations in plant science, agricultural and management practices and benefits to farmers and consumers.

Modern production agriculture in the developed world is highly industrialized. There is considerable discussion about the inadequacy of the dominant model of agricultural intensification and growth, which relies on increased use of capital inputs, such as fertilizer and pesticides [43] . Technology and purchased inputs, e.g. fertilizer, pesticides and water are required to maintain high levels of production, and use of these inputs continues to increase in the developing world. Despite the critical need for agricultural production and continued improvements in management practices, current systems are still not in "harmony" with the environment because they can create many problems for ecosystems and human communities. The generation of unacceptable levels of environmental damage and problems of economic feasibility are cited as key problems with this model of industrial agriculture [45] [46] . Specific external costs of industrial agriculture which should be improved include soil deterioration, erosion, declining surface water and groundwater quality, limited recycling of nutrients, excessive use of off-farm fertilizers and pesticides, diminished biodiversity within the agricultural system (both in terms of the variety of crops sown and coexisting species), lapses in food safety, and the loss of rural employment. By developing new field crops, and trees that meet societal needs, plant breeding plays a distinctive and crucial role in addressing these challenges, which must be dealt with immediately to develop sustainable agronomic systems for the future.

In this article two general ways are described in which plant breeders can engage in environmental issues: 1) by breeding plants that are better adapted to environment and environmental stresses, producing more with less and where productivity can be maintained in the face of increasingly variable weather patterns and sub-optimal conditions, as well as pest and disease pressures; and 2) by breeding plants that can alter and "improve" environments, as breeding alternative crops and crops for new uses or breeding for local adaptation and sustainable solutions. It is also presented the impacts of transgenic crops, commonly referred to as genetically modified crops on biodiversity.

2. PLANT BREEDING, AGRICULTURE AND ENVIRONMENT

2.1. Introduction

Farming and plant breeding have been closely associated since the early days when crops were first domesticated. The domestication of staple crops, for example, rice and soybean in eastern Asia; wheat in the Middle East; sorghum in Africa; and maize, beans, and potatoes in the Americas [47] , began independently, in multiple locales, 5000 - 12,000 years ago [47] . For thousands of years, these crops were grown and morphologically altered by farmers, who selected the most desirable and adaptable cultivars to plant in the next growing season. Without understanding the science behind it, early farmers saved the seed from the best portion of their crop each season. Over the years, they selected the traits which they liked best, transforming and domesticating the crops they grew.

After the discoveries of Darwin and Mendel, scientific knowledge was applied to plant breeding in the late 1800s [40] . Commercial hybridization of crop species began in the United States in the middle of the 1920s with sweet corn and followed by onions in the 1940s [4] . With the implementation of hybrid crop breeding, yield per unit land area rapidly increased in the United States [48] and since that time, public and private breeding companies have been placing more and more emphasis on the development of hybrids, and many species have been bred as hybrid cultivars for the marketplace. Besides heterosis, hybrids also allow breeders to combine the best traits and multiple disease and stress resistances. Furthermore, if the parents are homozygous, the hybrids will be uniform, an increasingly important trait in commercial market production. The creation of hybrid cultivars requires homozygous inbred parental lines, which provide a natural protection of plant breeders' rights without legal recourse and ensure a market for seed companies.

In the 1970's breeders' rights protection has been provided through International Union for the Protection of New Varieties of Plants (UPOV), which coordinates an international common legal regime for plant variety protection. Protection was granted for those who develop or discover cultivars that are new, distinct, uniform, and stable [49] . Cultivars may be either sexually or asexually propagated. Coverage for herbaceous species is 20 years. Protective ownership was extended by UPOV in 1991 to include essentially derived cultivars [49] . At the same time, the farmer's exemption (which permitted farmers to save seed for their own use) was restricted; giving member states the option to allow farmers to save seed. Additionally, in Europe after 1998 and the United States after 2001, plant breeding companies can take advantages of patent laws to protect not only the cultivar itself but all of the plant's parts (pollen, seeds), the progeny of the cultivar, the genes or genetic sequences involved, and the method by which the cultivar was developed [50] . The seed can only be used for research that does not include development of a commercial product i.e., another cultivar, unless licensed by the older patent. The patents are considered the ultimate protective device allowing neither a farmer's exemption nor a breeder's exemption (that permitted the protected cultivar to be used by others in further breeding to create new cultivars) [51] . The use of patents for transgenic crops introduces additional problems according to the IAASTD report [46] developed with the contribution from 400 scientists around the world, and adopted by 58 governments. In developing countries, especially instruments such as patents may boost up costs and restrict experimentation by individual farmers whereas potentially undermining local practices for securing food and economic sustainability. Thus, there is particular concern regarding present intellectual property rights instruments, which may inhibit seed-saving, exchange, sale, and access to proprietary materials of vital importance to the independent research community, specifically in view of the need for analyses and long term experimentation on climate change impacts [49] [50] .

Research and development (hereafter R & D) for improved seed development is expensive. Such product protection has presented a business incentive to corporations to invest in the seed industry, which supported an enormous increase in private R & D leading to strong competition in the marketplace between the major seed companies. The majority of current crop cultivars sold nowadays are proprietary products developed by private R & D. A significant consequence of this increase in R & D has been a reduction in public breeding programs. As a result, the cost for R & D to develop new crop cultivars is shifting from the publicly supported research programs to the customers of the major seed companies [4] [52] .

One of the main factors to determine success in plant breeding is crop biodiversity and genetic capacity. Access to genetic variation, biodiversity, is required to achieve crop cultivar improvement. No practical breeding program can succeed without large numbers of lines (genotypes) to evaluate, select, recombine and inbreed (fix genetically). This effort must be organized in order for valid conclusions to be reached and decisions to be made. Scientists, breeders, support people and facilities, budgets, and good management are requirements to assure success in the seed business.

Science must be state-of-the-art to maximize success in a competitive business environment. The continued need for fundamental breeding research is critical to support development of new technology and expansion of the knowledge base which supports cultivar development, competition among proprietary cultivar results in owner-companies striving to do the best possible research to develop their own products and to compete on genetic and physiological quality of crop seed in the marketplace. Reasonable profit margins are essential to pay back the R & D costs to the owner and to fund future research on developing even better crop cultivars to stay competitive. There is considerable genetic variation within the numerous crop species, which can be exploited in the development of superior proprietary cultivars. The consequences of this dynamic situation will mean relatively short-lived cultivars replaced by either the owner of the cultivar or a competitor seed company. This intense competition means constantly improved and more sophisticated cultivars. Seed companies are in the business of manipulating genes to improve plant cultivar performance for a profit. The success of the research is judged by the success of the product in making a reasonable profit. The research must improve economic performance starting with the seed production costs and including the farmer-ship- per/processor and the end user. If any link in this sequence of events is weak or broken, the new cultivar will likely fail [4] [53] .

Modern plant breeding is the science of improving plants to achieve farmer needs and better fit production environments, but it is a long-term proposition. Each released cultivar represents a culmination of a decade or more of work, from initial crosses through final testing. The rate of improvement is a function of the amount of heritable genetic variation present in a population, the time it takes to complete a breeding cycle (from seed production through selection to seed production again), which can range from multiple generations per year (e.g. maize on field sites in both hemispheres) to decades (some trees require 8 years of growth before flowering). In hybrid crops, several years (multiple breeding cycles) are necessary to develop inbred lines that must then be tested in hybrid combinations. Many years of testing under various environmental conditions must be conducted to ensure that the new cultivar (inbred, hybrid, or population) will perform well for the farmer, consumer, or end-user before any substantial additional investment is made to increase production and distribution of the cultivar.

Biotechnology is a new and potentially powerful tool that has been added by all the major seed corporations to their crop breeding research programs, and is part of an ongoing public research for developing genetic engineered crop projects. It can augment and/or accelerate conventional cultivar development programs through time saved, better products, and more genetic uniformity, or achieve results not possible by conventional breeding [54] . Genetic engineering provides innovative methods for modern plant breeding to adapt crops to agricultural systems facing new challenges brought by the changing climate. New breeding methods, relying on genetic engineering, can accelerate the pace to improve crops, or be more precise in transferring desired genes into plant germplasm. Some limited target traits already available in transgenic cultivars include those adapting agriculture to climate change and reducing their emissions of greenhouse gases.

Plant breeding may benefit from recent advances in genotyping and precise phenotyping, and by increasing the available agro-biodiversity through the use of genomics-led approaches. Today marker-assisted breeding is applied to a broad range of crops and could facilitate domesticating entirely new crops. Marker-assisted selection is particularly important for improving complex, quantitatively inherited traits that alter yield, and for speeding up the breeding process [55] . Crop genomics has also been improving in the last decade and today there are faster and cheaper systems being increasingly used in gene banks, genetic research and plant breeding, e.g. for studying interactions between loci and alleles such as heterosis, epistasis and pleiotropy, or analyzing genetic pathways. Advances in crop genomics are providing useful data and information for identifying DNA markers, which can be further used for both germplasm characterization and marker-assisted breeding. Genomics-assisted breeding approaches along with bioinformatics capacity and metabolomics resources are becoming essential components of crop improvement programs worldwide [49] [56] .

Progress in crop genome sequencing, high resolution genetic mapping and precise phenotyping will accelerate the discovery of functional alleles and allelic variation associated with traits of interest for plant breeding. Genome sequencing and annotation include an increasing range of species such as banana/plantain, cassava, citrus, grape, model legumes, maize, potato, rice, sorghum, sugarcane, soybean, among other species. Perhaps, one day further research on the genome of a plant species from a drought-prone environment may assist in breeding more hardy and water efficient related crops due to gene synteny.

Transgenic breeding involves the introduction of foreign DNA. The use of transgenic crops remains controversial worldwide after more than 1.5 decades of introducing them into the agro-ecosystems using specific frameworks to regulate their release and commercialization. While conventional plant breeding utilizing non- transgenic approaches will remain the backbone of crop improvement strategies, transgenic crop cultivars should not be excluded as products capable of contributing to development goals. Breeding of transgenic crops so far seems to have been responsible and regulatory agencies have proceeded with caution in releasing transgenic crops. Available commercial transgenic crops and products are at least as safe in terms of food safety as those ensuing from conventional plant breeding [54] [57] -[59] . Farm level profitability will ultimately determine whether farmers adopt and retain new transgenic crop technology, which may also depend on much more than technical performance particularly in the developing world; e.g. national research capacity, environmental and food safety regulations, intellectual property rights and agricultural input markets.

Decisions, policies and procedures about monitoring transgenic crops should be science-based, and this approach requires education. There will be continuing assessment on the need for, and type of monitoring as new (and unique) products are developed and released to agro-ecosystems. For example, when deploying transgenic crops with host plant resistance to an insect pest (e.g. expressing Cry insecticidal proteins derived from Bacillus thurigiensis) numerous experiments are conducted to determine effective insect resistance management strategies for farmers. Collecting baseline data is indeed essential for effective monitoring and guiding research on transgenic crops. Environmental, food and feed safety aspects should also be investigated before releasing trans- genic crops. Another research area should focus on the unintentional spread of transgenic traits into conventionally-bred crop or landrace gene pools of the same species, particularly in the centers of crop diversity or origin. Transgene flow raises a new set of ecological and economic issues for scientists and policymakers to consider for transgene containment. Appropriate measurements should also be taken when transgenic and conventional crops of the same species coexist in the same locations if some farmers wish to grow crops for transgenic-free markets. The global spread of transgenic crops also has significant implications for organizations involved in germplasm conservation and genetic enhancement. Protocols, which are most likely based on polymerase chain reaction (PCR) markers for detecting specific recombinant DNA sequences in bulked samples collected from sentinel plots, are therefore needed to monitor unintentional transgene flow in gene banks and breeding plots. Using plant-derived genes to introduce useful traits and plant-derived promoters, will also overcome some concerns about the development of genetically engineered crops. In this regard, cisgenesis addresses some negative views regarding the use of genes from non-crossable species for breeding crops. Cisgenesis involves only genes from the plant itself or from a crossable close relative, and these genes could also be transferred by conventional breeding methods. Crop wild relatives are therefore a valuable source of traits for cisgenesis. The impact of transgenic crops on biodiversity is discussed in chapter 3.

Plant breeders need to understand the various valuation strategies very early in the breeding process if they are to direct long-term selection toward reducing agriculture's negative environmental impacts and achieving greater sustainability while maintaining productivity. Regardless of method, breeding objectives can be broadened to include traits which reduce the environmental footprint of traditional production systems (e.g. nutrient and water use efficiencies that reduce off-farm inputs), to adapt crops to new climates, to host plant resistance to tackle old and emerging pathogen epidemics, or new cultivars for new production systems (e.g. perennial polycultures that mimic the biodiversity of natural systems), albeit with some reduction in rate of gain for the traditional agronomic traits of interest. Interdisciplinary crop improvement strategies accounting for ecological, socio-economic and stakeholder considerations will help identify traits leading to plant cultivars using fewer inputs, less land, and less energy, thereby resulting in a more sustainable agricultural ecosystem.

The impact of breeding on crop production is dependent upon the complex relationships involving the farmers, the cultivars available to them, and the developers of those cultivars. Farmers consist of commercial producers with varying size land holdings ranging from moderately small farms to very large ones, and subsistence farmers with small farms often on marginal lands. The subsistence farmers are usually poor. Several types of cultivars are available. The least sophisticated in terms of methods of development are landraces, also known as local cultivars. Modern cultivars consist of development by crossing and selection alone, those developed by crossing and selection with specific important improvements are often obtained from crosses with wild species or by transgenic methods, and F_1 hybrids between desirable inbred lines. The developers of landraces are usually farmers themselves, and are obtained by repeated simple selection procedures of generation after generation. Improved cultivars and hybrids are created either by public sector breeders or seed companies.

Nearly 70% of the world's farmers, from 570 million world exploitations, are small/subsistence and poor farmers. They feed 1.5 billion of the world's population. So they are also a key for biodiversity and for improving the sustainability. For these farmers improved cultivars, hybrids or transgenic seeds tend to be riskier than landraces, since the higher costs associated with seeds and production impose a greater income risk. The lack of capital available denies them the opportunity to invest in production inputs. Small farmers may have lower production costs with landraces because they achieve adequate yields with fewer inputs. In addition, profits from improved hybrid or transgenic cultivars tend to be more variable. Yields are often higher but market prices tend to be inconsistent. For example in India states of Andhra Pradesh and Maharashtra, farmers have been promised higher yields and lower pesticide costs when using Bt cotton, thus they acquired loans to afford the costly seeds (Monsanto has control over 95% of the Indian Bt cotton seed market and this near monopoly has resulted in great increased prices). When, in many cases, the farmers found the yields failed to meet their expected result, the consequences were usually very serious and many farmers died by committing suicide over the past 15 years, perhaps due to this reason. This situation of using Bt cotton seeds was explained by the absence of irrigation systems combined with specialization in high-cost crops, and played low market prices. Without collateral help these farmers are usually unable to secure a loan from a bank or money lender [53] [60] . Rates are often unmanageably high for those able to get a loan, with strict penalties for late payments. Similarly, a lack of education, resources, skill training and support prevent these farmers from using improved cultivars and then to generate a stable income from their production. In addition, governments do not usually regulate the price of crops or even provide market information. Improving market information systems for crops and facilitating farmers' access to credit are then essential components for a strategy to enable poor farmers to grow improved cultivars. A major obstacle to success in crop production using improved cultivars is the shortage of affordable credit. Desperate for cash, subsistence farmers are forced to sell their crops immediately after the harvest to middlemen or their creditors at unfavorable prices. Low cost quality seeds are essential for these poor farmers to improve their life [60] .

2.2. Breeding to Adapt Plants to the Environment

2.2.1. Producing More with Less

In the coming decades we will need to produce more with less. Fresh water suitable for irrigation is expected to become increasingly scarce and the costs of fertilizer and other agricultural inputs will increase as fossil-fuel costs rise. Nevertheless, continuing gains in production per hectare must be realized to offset the loss of premium agricultural lands (e.g. from urbanization and industrialization), while supplying a growing population. By developing resource efficient plants, plant breeders can continue to improve the sustainability of agricultural ecosystems. Plants requiring fewer off-farm input applications (specifically water, pesticides, nitrogen, phosphorus, and other nutrients) decrease the cost of production, lower fossil energy use, and reduce contamination of water systems, which help to improve public health and stabilize rural economies [61] [62] .

Although modern plant breeding efforts initially focused on improving uptake of inputs, recent efficiency gains have been made in physiologically increasing yield and biomass production without further increasing inputs. Many crops already have genetic variation in nutrient use efficiency, utilization, and uptake [63] -[65] and plant breeding will further improve these traits. Intensive agro-ecosystems, central to food security and reducing rural poverty, should emphasize improvements in system productivity, host plant resistance and enhance use- efficiency of inputs such as water and fertilizers.

Water use-efficiency and water productivity are therefore being sought by agricultural researchers worldwide to address water scarcity in drought-prone environments across the world. Under water scarcity, yields, of some crops, are a function of the amount of water used by the crop, how efficiently the crop uses this water for biomass-growth (i.e., water use efficiency or above-ground biomass/water use), and the harvest index; i.e., the proportion of grain yield to above-ground biomass. Water use efficiency is the ratio of total dry matter accumulation to evapotranspiration and other water losses; i.e., water entering and being lost from the system which is not transpired through the plant. An increase in transpiration efficiency or reduction in soil evaporation will increase water use efficiency. More recently, water productivity was re-defined at the crop level as the ratio of biomass with economic value produced (for example grain yield of cereals) compared to the amount of water transpired. This water productivity has been labeled as "productive" because transpiration is the only water flow in a field actually passing through the crop. Both water use efficiency and water productivity may be improved through plant breeding, as biomass accumulation and harvest index. Farooq et al. [66] discuss the advances in transgenic breeding for drought-prone environments. In their review, they noted the testing of 10 transgenic rice events [unique DNA recombination taking place in one plant cell and thereafter to be used for generating entire transgenic plant(s)] under water scarcity. It seems the transgenic expression of some stress-regulated genes leads to increased water use efficiency.

Agriculture contributes significantly to greenhouse gas emissions. Nitrous oxide and dioxide are potent greenhouse gases released by manure or nitrogen (N) fertilizer, particularly in intensive cropping systems. Nitrous oxide (N_2O), which is a potent greenhouse gas susceptible to denitrification, is generated through use of manure or N fertilizer. In many intensive cropping systems common N fertilizer practices lead to high fluxes of N_2O and nitric oxide (NO). Several groups of heterotrophic bacteria use NO_3 as a source of energy by converting it to the gaseous forms N_2, NO, and NO_2 (nitrous dioxide). N_2O is therefore often unavailable for crop uptake or utilization. Proper amounts and timing of N applications can reduce NO_2 emissions (50% less) in intensive irrigated agro-ecosystems without significantly affecting crop yields. An optical, hand held sensor which calculates the normalized differential vegetative index, thereby assessing yield potential as plants grow, can reduce unneeded N fertilizer inputs, saving farmers' money and protecting the environment by reducing trace gas emissions. Genetic enhancement of crops shows great potential for reducing N_2O emissions from soils into the atmosphere. Some plants possess the capacity to modify nitrification in situ because they produce chemicals which inhibit nitrification in soil. This release of chemical compounds from plant roots suppressing soil nitrification has been called biological nitrification inhibition, which seems to vary widely among and within species, and appears to be a widespread phenomenon in some tropical pasture grasses, e.g. Brachiaria humidicola. Nitrification inhibition also enhances agro-ecosystem fertility in a sustainable way especially under high nitrate leaching and denitrification fluxes, which may account for the ecological advantage of African grasses over indigenous grasses of other tropical pastures. Biological nitrification inhibition may be an interesting target trait of crop genetic engineering for mitigating climate change.

Almost one-fifth of global methane emissions are from enteric fermentation in ruminant animals. Apart from various rumen manipulation and emission control strategies, genetic engineering is a promising tool to reduce these emissions. The amount of methane produced varies substantially across individual animals of the same ruminant species. Efforts are ongoing to develop low methane-emitting ruminants without impacting reproductive capacity and wool and meat quality. A recent study by Shi et al. [67] , to understand why some sheep produce less methane than others, deployed high-throughput DNA sequencing and specialized analysis techniques to explore the contents of the rumens of sheep. The study showed that the microbiota present in sheep rumen was solely responsible for the differences among high and low methane emitting sheep. It was further observed that the expression levels of genes involved in methane production varied more substantially across sheep, suggesting differential gene regulation. There is an exciting prospect that low-methane traits can be slowly introduced into sheep.

Crops are bred for N use efficiency (hereafter NUE) because this trait is a key factor for reducing N fertilizer pollution, improving yields in N limited environments, and reducing fertilizer costs. The use of plant species or genotypes of same species efficient in absorption and utilization of N is an important strategy in improving NUE in sustainable agricultural systems. Whole plant physiology, quantitative genetics, and forward and reverse genetic approaches are providing a better understanding of the physiological and molecular controls of N assimilation in crops under varying environments. Crops are being bred for NUE because this trait will be a key factor for reducing N fertilizer pollution as well as for improving yields in N limiting environments. There are various genetic engineering activities for improving NUE in crops [64] [68] . The gene Alanine aminotransferase from barley, which catalyzes a reversible transamination reaction in the N assimilation pathway, seems to be a promising candidate for accomplishing this plant breeding target. Transgenic plants over-expressing this enzyme can increase N uptake especially at early stages of growth. This gene technology was licensed to a private biotechnology company, which was founded in the last decade with the aim of promoting sustainable agriculture by running a profitable company [69] . A patent issued a few years ago gave this company the rights to use the NUE gene technology in major cereals (wheat, rice, maize, sorghum, and barley), as well as, in sugarcane. Today they are testing the technology with rice in China, and researching further with rice and wheat in India, and planning to assess its value for maize and rice in sub-Saharan Africa through private-public partnerships.

Keeping N in ammonium form will affect how N remains available for crop uptake and will improve N recovery, thus reducing losses of N to streams, groundwater and the atmosphere. There are genes in tropical grasses such as B. humidicola and in the wheat wild relative Leymus racemosus that inhibit or reduce soil nitrification by releasing inhibitory compounds from roots and suppressing Nitrosomonas bacteria [70] . Their value for genetic engineering crops for reducing nitrification needs to be further investigated.

2.2.2. Adapting to Global Climate Change and for Abiotic and Biotic Stress Tolerance

Extreme weather events are expected to increase in both number and severity in coming years [71] . Climate change impacts agro-ecosystems through changes over the long-term in key variables affecting plant growth (e.g. rising temperatures) and through increasing the variability (frequency and intensity) of weather conditions (rainfall, drought, waterlogging and elevated temperatures). These changes affect both crop productivity and quality. In addition to physically destroying crops, climate change has altered host-pathogen relationships and resulted in increased disease incidence, in insect-pest borne stress in crop plants, and in invasive pests which feed and damage them.

There are two ways to adapt crops to new environments: developing new crops (long-term endeavor starting with domestication) and introducing target traits into existing crops through plant breeding, which includes genetic engineering. However, the job of crop improvement is becoming increasingly difficult. Cultivars which are not only high yielding but are also efficient in use of inputs are needed, tailored to ever more stringent market demands, able to maintain stability under increasing climate variability, and potentially contribute to climate mitigation. These multi-trait demands for new cultivars provide significant challenges for crop breeders, and standard selection approaches struggle under such complexity. To maintain productivity in the face of increased climatic variability, both the population and the plant cultivars will need to be continually developed to withstand "new" climate extremes and the stresses which these will entail [72] .

Many breeding programs are already developing plants which tolerate extreme weather conditions, including drought, heat, and frost [73] [74] . Plant breeders are also beginning to address expected changes due to increased climate variability, by increasing genetic diversity sources and by adjusting selection and testing procedures [75] .

More frequent weather extremes will likely affect the existing ranges of not only agronomic cultivars but also local native plant species [76] . Because some genetic variation useful for climate change adaptation will be found only in wild plant relatives of cultivated crops, preserving genetic biodiversity is essential in order for breeders to select plants that will be well-suited for future environmental conditions [77] .

Global climate change notwithstanding, additional stress tolerances in crop species are needed to maintain productivity and survival. In the near future, tolerance to various soil conditions including acidic, aluminum-rich soils (particularly in the tropics) and saline soils (especially those resulting from irrigation), will be increasingly important for production on marginal agricultural lands or as the salt content of irrigated lands increases [78] . Bhatnagar-Mathur et al. [79] suggested that genetic engineering could accelerate plant breeding to adapt crops to stressful environments. They further underline that engineering the regulatory machinery involving transcription factors (TF; a protein binding specific DNA sequences and thereby governing the flow of genetic information from DNA to messenger RNA) provides the means to control the expression of many stress-responsive genes. There are various target traits for adapting crops, through genetic engineering, to high CO_2 and high O_3 environments of the changing climate [80] . Ortiz [81] , Jewell et al. [82] , and Dwivedi et al. [83] [84] provide the most recent overviews on research advances in genetic engineering for improved adaptation to drought, salinity or extreme temperatures in crops. The most cited include TF, and genes involved in: 1) signal sensing, perception, and transduction; 2) stress-responsive mechanisms for adaptation; and 3) abscisic acid biosynthesis for enhanced adaptation to drought. Transporter, detoxifying and signal transduction genes as well as TF are cited for tolerance to salinity. Genes related to reactive oxygen species, membrane and chaperoning modifications, late abundance embryogenesis proteins, osmoprotectants/compatible solutes and TF are pursued in crop genetic engineering for temperature extremes.

Transgenic crops provide the means to adapt crops to climate change, particularly in terms of drought and salinity. Duration and intensity of drought has increased in recent years, consistent with expected changes of the hydrologic cycle under global warming. Drought dramatically reduces crop yields. Genetic engineering may be one of the biotechnology tools for developing crop cultivars with enhanced adaptation to drought [85] . It should be seen as complementary to conventional plant breeding rather than as an alternative to it. The function of a TF such as the Dehydration-Responsive Element Binding (DREB) gene in water stress-responsive gene expression has been extensively investigated [86] . The main research goal was to gain a deep understanding of TF in developing transgenic crops targeting drought-prone environments [87] . For example, the DREB1A gene was placed under the control of a stress-inducible promoter from the rd29A gene and inserted via biolistic transformation into wheat bread [88] . Plants expressing this transgene demonstrated significant adaptation to water stress when compared to controls under experimental greenhouse conditions as manifested by a 10-day delay in wilting when water was held. Saint Pierre et al. [89] indicated, however, that these transgenic lines did not generally out-yield the controls under water deficit in confined field trials. Nonetheless, they were able to identify wheat lines combining acceptable or high yield under enough irrigation which also showed stable performance across the water deficit treatments used in their experiments; i.e., severe stress, stress starting at anthesis, and terminal stress.

Soils affected by salinity are found in more than 100 countries, and about 1/5 of irrigated agriculture is adversely affected by soil salinity. Therefore, breeding salt-tolerant crops should be a priority because salinity will most likely increase under climate change. Mumms [90] lists some candidate genes for salinity tolerance, indicating the putative functions of these genes in the specific tissues in which they may operate. Genes involved in tolerance to salinity in plants, limit the rate of salt uptake from the soil and the transport of salt throughout the plant, adjust the ionic and osmotic balance of cells in roots and shoots, and regulate leaf development and the onset of plant senescence. The most promising genes for the genetic engineering of salinity tolerance in crops, as noted by Chinnusamy et al. [91] , are those related to ion transporters and their regulators, as well as the C-re- peat-binding factor. The recent genome sequencing of Thellungiella salsuginea, a close relative of Arabidopsis thriving in salty soils, will provide more resources and evidence about the nature of defense mechanisms constituting the genetic basis underlying salt tolerance in plants [92] .

In the quest for breeding transgenic rice and tomato, advances showing salt tolerance have occurred. Plett et al. [93] were able to show an improved salinity tolerance in rice by targeting changes in mineral transport. They initially observed that cell type-specific expression of AtHKT1 (a sodium transporter) improved sodium (Na^+) exclusion and salinity tolerance in Arabidopsis. Further research explored the GAL4-GFP enhancer trap (transgenic construction inserted in a chromosome and used for identifying tissue-specific enhancers in the genome) to drive expression of AtHKT1 in the root cortex in transgenic rice plants. The transgenic rice plants had a higher fresh weight under salinity stress due to a lower concentration of Na^+ in the shoots. They also noted that root-to- shoot transport of $^{22}Na^+$ decreased and was correlated with an up-regulation of OsHKT1, the native transporter responsible for Na^+ retrieval from the transpiration stream. Moghaieb et al. [94] bred transgenic tomato plants producing ectoine (a common compatible solute in bacteria living in high salt concentrations). Ectoine synthesis was promoted in the roots of transgenic tomato plants under saline conditions, which led to increased concentration of photosynthesis in improving water uptake. Likewise, the photosynthetic rate of ectoine-transgenic tomato plants increased through enhancing cell membrane stability in oxidative conditions under salt stress.

Transgenic crops can also contribute to climate change mitigation efforts by reducing input use intensity [95] . The integration of genetic engineering with conventional plant breeding, within an interdisciplinary approach, will likely accelerate the development and adoption of crop cultivars with enhanced adaptation to climate change related stresses [96] . Global warming will reduce yields in many crops about 6% and 5% average yield loss per 1°C in C_3 and C_4 crops, respectively, whose optimum temperature ranges are 15°C - 20°C and 25°C - 30°C [97] . The extent of yield loss depends on crop, cultivar, planting date, agronomy and growing area. For instance, an increase of 1°C in the night time maximum temperature translates into a 10% decrease in grain yield of rice, whereas a rise of 1°C above 25°C shortens the reproductive phase and the grain-filling duration in wheat by at least 5%, thereby reducing grain yield proportionally. Heat stress will exacerbate climate change impacts in the tropics, while it may put agriculture at risk in high latitudes where heat-sensitive cultivars are grown today. Hence, new cultivars must be bred to address heat stress. Ainsworth and Ort [98] suggested giving priority to traits improving photosynthesis for adapting to heat stress. However, plants have various mechanisms to cope with high temperatures, e.g. by maintaining membrane stability, or by ion transporters, proteins, osmoprotectants, antioxidants, and other factors involved in signaling cascades and transcriptional control [99] [100] . Furthermore, Gao et al. [101] noted that bZIP28 gene (a gene encoding a membrane-tethered TF) up-regulated in response to heat in Arabidopsis. Some of these genes can be used in crop genetic engineering to enhance plant adaptation to heat stress. For example, some stress-associated genes such as ROB5, a stress inducible gene isolated from bromegrass, enhanced performance of transgenic canola and potato at high temperatures [102] . Likewise, Katiyar-Agarwal et al. [103] introduced hsp101 gene (a heat shock protein gene from Arabidopsis) in basmati rice.

This transgenic rice had a better growth in the recovery phase after suffering heat stress.

Globalization has, among other consequences, led to the rapid spread of plant disease and invasive pests. Being immobile, plants are unable to escape pathogens causing plant disease and pests which feed and damage them. Plant disease is mainly caused by fungi, bacteria, viruses, and nematodes. Approximately 70,000 species of pests exist in the world, but of these, only 10% are considered serious [104] . Synthetic pesticides have been applied to crops since 1945 and have been highly successful in reducing crop losses to some pest insects, plant pathogens, weeds and in increasing crop yields [104] . One estimate suggests that without pesticides, crop losses to pests might increase by 30%. Despite pesticide use, insects, pathogens and weeds continue to exact a heavy toll on world crop production, approaching 40% [104] [105] . Pre-harvest losses are globally estimated at 15% for insect pests, 13% for damage by pathogens, and about 12% for weeds [104] . Developing resistant cultivars reduces the need for expensive and environmentally damaging pesticides to be applied. For example, a recent outbreak of Xanthomonas campestris pv. musacearum led to the devastating Xanthomonas wilt of banana in the Great Lakes Region of Africa, thereby threatening the food security and income of millions of East and Central African people who depend on this crop. Transgenic banana plants with the hypersensitivity response-assisting protein (Hrap) gene from sweet pepper did not show any infection symptoms after artificial inoculation of potted plants with Xanthomonas wilt in the screen house [106] . Selected transgenic banana plants with putative host plant resistance to Xanthomonas wilt are ongoing confined field-testing in East Africa, where elevated temperatures, due to the changing climate, will likely favor banana production.

Weather influences how pathogens and pests affect and interact with crops and their host plant resistance, and thus climate change can also have wide-ranging impacts on pests and diseases [84] . Late blight, which is caused by Phytophthora infestans, ranks as the most damaging potato pest. Late blight accounts for 20% of potato harvest failures worldwide, translating into 14 million tonnes valued at 7.6 billion US dollars. Global warming will increase late blight spread, e.g. expanding its range above 3000 meters in the Andes [107] . Chemical control may lead to more aggressive strains of the pathogen and chemical control is often regarded as being environmentally damaging. Cisgenic potato cultivars with late blight resistance are becoming available and will impact growers, consumers and the environment favorably [108] . Related wild Solanum species can be a source of alleles to enhance host plant late blight resistance in potato. For example, S. bulbocastum (a wild relative with high resistance to late blight from Mexico) was used to breed the cultivar "Fortuna" using genetic engineering.

Cisgenesis allows inserting several host plant resistance genes from wild crop species in one step without linkage drag (reduction in cultivar fitness).

2.3. Breeding Plants to Improve the Environment

In general, plants are bred for their most obvious end products, including grain, fiber, sugar, biomass yield, fruit quality, or ornamental qualities. However, plants deployed across the landscape in agricultural or forestry settings affect the environment in measurable ways. Perennial crops have environmentally beneficial properties not present in annual crops, such as helping to prevent erosion in agricultural systems, providing wildlife habitat, and acting as sinks for carbon and nutrients. Traditionally, perennial crops have not been a major focus of breed- ing programs because they generally take more time and scientific knowledge to improve, and therefore, products such as new cultivars are often not produced within the timeframe of funding cycles. Current tree breeding programs are developing elms (Ulmus spp.), chestnuts (Castanea dentata), hemlocks (Tsuga spp.), and other species which are resistant to introduced diseases and insects [109] [110] . As compared with natural selection, artificial selection via plant breeding has overcome these stresses more effectively by rapidly incorporating diverse exotic genetic sources of resistance, hybridizing to include multiple, different genetic resistances into the same plant, and making use of off-season locations or artificial conditions to shorten generation cycles. A more complex example which may be feasible in the future is tree breeding for larger and improved root systems to decrease soil erosion, sequester carbon, and improve soil quality by increasing soil organic matter.

New crop cultivars developed by plant breeders must help improve soil health, reduce soil erosion, prevent nutrient and chemical runoff, and maintain biodiversity. The goal to breed projects for forages, which include several species, is to produce a high yield of leaf and stem biomass, as opposed to grain, for ruminant animals. In the tropics many forages are perennial, providing year-round erosion control, improving water infiltration as compared with that, from annual cropping systems, and in some cases, sequestering carbon. The forage breeding program at the University of Georgia (UG) has developed cultivars in several species and has been proactive in developing agreements with private-sector commercial partners to oversee seed production and marketing of new cultivars. Among the cultivars developed at UG is "Jesup MaxQ" tall fescue, a cultivar carrying a non-toxic endophytic fungus that was both highly persistent under grazing and greatly improved animal weight gain and feed efficiency over standard cultivars. In addition, this program developed the first true dual purpose, grazing and hay, alfalfa cultivar "Alfagraze", followed by several further improved alfalfa cultivars like "Buldog 805" which persist through summer under cattle grazing [111] .

Cover crops are annual species planted in rotation with crops to specifically improve soil conditions and to control weeds, soil-borne diseases, and pests [112] -[114] . Continuous cover crops can reduce on-farm erosion, nutrient leaching, and grain losses due to pest attacks and build soil organic matter as well as improve the water balance, leading to higher yields [115] [116] . For instance, Kaumbutho and Kienzle [117] showed that maize yield increased from 1.2 to 1.8 - 2.0 t/ha in Kenya with the use of a mucuna (velvet bean) cover crop using case studies conducted from 2004 to 2007; and Pretty and Hine [118] found that farmers who adopted mucuna cover cropping benefited from higher yields of maize with less labor input for weeding (maize following mucuna yields 3 - 4 t/ha without application of nitrogen fertilizer, similar to yields normally obtained with recommended levels of fertilization of 130 Kg·N/ha).

Many current perennial and cover crop cultivars are essentially wild species bred from germplasm collections and developed to increase success in managed agro-ecosystems. As compared with non-native vegetation, plant species native to a particular region are generally thought to survive on less water, use fewer nutrients, require minimal pesticide applications, and be non-invasive; however, counter examples for both native and non-native species are plentiful [119] . As potentially valuable species are identified, breeding to improve them for traits of consumer importance will be needed to broaden available biodiversity in cultivated landscapes. With a changing climate, species considered critical to the landscape may require human-assisted hybridization with distant relatives to better ensure survival from threats posed by novel pests or diseases.

Alternative crops are also being bred for new uses, such as removing toxic chemicals and excess nutrients and improving degraded soils, including mine spoils [120] . Phytoremediation is a biotechnology to clean the contaminated sites of toxic elements (e.g. Cd, Cu, Zn, As, Se, Fe) via plant breeding, plant extracting, and plant volatilizing [121] . Phytoremediation of land contaminated with inorganic and/or organic pollutants has attracted much attention and research over the last decade [122] -[124] . Among the various approaches comprising phytoremediation, phytoextraction of metals and metalloids is probably the most challenging task. Where soils are impacted by industrial or mining activities, the degree of pollution is usually severe, making phytoextraction unfeasible within a reasonable time frame because of the high quantity of the pollutants present in the soil. Simple mass-balance calculations show that phytoextraction is potentially feasible only in low or moderately contaminated soils. For more heavily contaminated soils, phytostabilization with tolerant plants may be used to stabilize the contaminated sites and reduce the risk of erosion and leaching of pollutants to water bodies. Hyper-accumulation of metals or metalloids is important for the phytoextraction strategy [125] [126] . The last few years have seen a steady expansion in the list of hyper-accumulator species, which could be valuable plant resources for phytoremediation. A word of caution however, as the hyper-accumulation ability of some reported "hyper-accumulators" have yet to be confirmed in studies using field contaminated soils. Chaney et al. [125] discussed situations where phytoextraction may be applicable, such as paddy soils contaminated with moderate levels of cadmium (Cd) giving rise to Cd concentrations in rice grain exceeding the safe limit. An Indica-Japonica hybrid cultivar of rice was found to be an effective Cd phytoextractor, removing 7% - 14% of soil Cd; this had the effect of decreasing subsequent Cd accumulation in soybean seeds in a pot study by 24% - 46% [127] . The efficient translocation of Cd from roots to shoots appears to be the main reason for the efficient Cd accumulation in this rice cultivar, although the underlying mechanisms remain unclear. A major QTL responsible for the root to shoot translocation of Cd has been identified in an F_2 population constructed from the parental lines of rice differing in Cd accumulation in shoots by 13-fold [128] . While the trait for high Cd accumulation in shoots is useful for phytoextraction, the opposite is true for the strategy of breeding crops low in Cd accumulation for the benefit of food safety [129] . Many of the previous studies on phytoextraction have focused on hyper-accumulators. Small-scale field trials have shown that an ecotype of the Zn/Cd hyper-accumulator Thlaspi caerulescens from southern France was able to phytoextract Cd efficiently through the different seasons with good growth of biomass [130] [131] . This ecotype possesses a high-affinity Cd uptake system which is not suppressed by Zn [132] . The Chinese brake fern Pteris vittata has a strong ability to hyper-accumulate arsenic (As) and shows promising potential for phytoextraction of and from contaminated soils under field conditions [133] [134] , but the plant thrives only in the humid tropic/subtropical climates. Phytoextraction using high biomass plants such as willow (Salix sp.) and poplar (Populus sp.) has also been proposed [135] . Some Salix species are good accumulators of Cd and Zn, and up to 20% of soil Cd was removed by cropping three S. asmithiana in a lysimeter study [136] . A large proportion of the metals are stored in leaves, so plants either have to be harvested before leaves fall or the fallen leaves are to be collected. Such biomass plants may be grown on contaminated areas not suitable for food production, allowing gradual phytoextraction of metals while the biomass may be used to generate energy by pyrolysis [137] .

New perennial crops and tree species [e.g. switchgrass (Panicum virgatum), poplar (Populus spp.), Miscanthus, Arundo, etc.] are being developed as improved cellulosic feedstock for biofuels that will have a higher yield and energy content than was previously available [138] [139] . Cellulosic biofuels provide one approach for mitigating the impacts of global warming associated with fossil-fuel combustion, but concerns over appropriate implementation and environmental impacts remain [140] . Frontier approaches should be applied to study the possible advantages of perennial biofuel crops more photosynthetically productive, entail lower input costs, and improve soil nutrient input and retention. Through alliances with the bio-energy industry, research should also adapt industrial processes to biomass sources and sources to promising processes. Biofuels should form therefore part of a global, cross-cutting agenda of agricultural research, involving partners in the farming and energy sectors.

Simply developing more productive feedstock does not necessarily lead to enhanced environmental health. Without crop rotation, further monocultures of grain maize or increased palm oil production could have net negative environmental effects in the long term, but such efforts may be a necessary transition to facilitate infrastructure development for cellulosic feedstock. There is a concern however, that energy crops might inadvertently compete for land currently allocated for food crop production, and thereby raise food prices, must be considered carefully. Breeding alternative crops needs to be undertaken in close consultation with agronomists, economists, ecologists, and the commercial sector or industry, to ensure that new cultivars have the proper traits making them both profitable and sustainable.

The increasing demand, particularly by the industrialized world, for biofuels should take into account the agro-ecosystems and its biodiversity to ensure their healthy management. The agricultural systems required for producing biofuels need to be sustainable for an efficient use of biomass, and partitioning it among energy, feed, food and CO_2 fixation demands. They should be more eco-efficient for using existing farmland or marginal (dry, waterlogged, saline) tracts. Although some advocate that bio-energy can play a role for mitigating climate change by reducing greenhouse emissions, appropriate life-cycle analysis will be needed on a case by case basis to determine the use of land resources and estimate net carbon emissions of each suggested renewable energy technology. The agenda for plant breeding may include increasing plant grain and biomass productivity, optimizing the chemical and physical attributes of biofuel sources, and improving specific traits in first- and second-generation biofuel crops, within a framework of sustainable agriculture.

A major goal of harmonizing agriculture with the environment is to "tailor" crops to individual landscapes. Plant breeding has always maximized production by selecting for adaptation in the target environments of interest, using local environmental forces for plant selection [75] . By selecting breeding germplasm growing under local environmental conditions, individual cultivars can be optimized for small regional areas of production that fit prevailing environmental and weather patterns. Likewise, plants could be tailored to provide specific ecosystem services to local environments, to address local needs. One cost-effective way to achieve this is through participatory plant breeding, which involves local farmers in the breeding process.

Alternative crop rotations, planting densities, and tillage systems may make production more environmentally benign but will require altering breeding targets and an understanding that systems biology is complex and rarely has simple solutions. For example, no-tillage systems used for soil conservation can lead to colder soils in spring and change the prevalence and onset of various soilborne diseases, thus requiring the addition of specific disease resistances in the breeding objectives [141] . Breeders must select from conditions prevailing under new management practices to ensure cultivars will be optimally productive.

3. IMPACT OF TRANSGENIC CROPS ON CROP BIODIVERSITY AND AGRO-BIODIVERSITY

The potential impact of transgenic crops on biodiversity has been a topic of interest both in general as well as specifically in the context of the Convention on Biological Diversity. In a recent review, Carpenter [142] took a biodiversity lens to the substantial body of literature existing on the potential impacts of genetically modified crops on the environment, considering the impacts at the crop, farm, and landscape scales. Overall, the review finds that currently commercialized transgenic crops have reduced the impacts of agriculture on biodiversity, through enhanced adoption of conservation tillage practices, reduction of insecticide use, and use of more environmentally benign herbicides. Increasing yields also alleviate pressure to convert additional land into agricultural use.

Respecting crop biodiversity is widely accepted, that greater varietal and species diversity would enable agricultural systems to maintain productivity over a wide range of conditions. With the introduction of transgenic crops, concern has been raised that crop genetic biodiversity will decrease since breeding programs will concentrate on a smaller number of high value cultivars. Three studies (two in the United States on cotton and soybean and one in India on cotton) have analyzed the impact on the introduction of transgenic crops within-crop genetic biodiversity [142] . Studies in the United States of genetic diversity on cotton and soybean both concluded that the introduction of transgenic cultivars was found to have little or no impact on biodiversity. In contrast, the introduction of Bt cotton in India initially resulted in a reduction of on-farm varietal biodiversity due to the introduction of technology in only a small number of cultivars, which has since been offset by more Bt cultivars becoming available over time [142] . Carpenter [142] concluded that from a broader perspective, transgenic crops may actually increase crop biodiversity by enhancing underutilized alternative crops, making them more suitable for widespread domestication.

Respecting agro-biodiversity plants have a major influence on soil communities of micro- and other organisms fundamental to many functions of soil systems, such as nitrogen cycling, decomposition of wastes, and mobilization of nutrients. The potential impact of Bt crops on soil organisms is well studied. A comprehensive review of the available literature, by Icoz and Stotzky [143] , on the effects of Bt crops on soil ecosystems included the results of 70 scientific articles. The review found that, in general, few or no toxic effects of Cry proteins on woodlice, collembolans, mites, earthworms, nematodes, protozoa, and the activity of various enzymes in soil have been reported. Although some effects ranging from no effect to minor and significant effects, of Bt plants on microbial communities in soil have been reported, they were mostly the result of differences in geography, temperature, plant cultivar, and soil type and, in general, were transient and not related to the presence of the Cry proteins. Studies published since the Icoz and Stotzky review [143] have reached similar conclusions, including novel studies on snails.

In a study conducted in the northeastern part of the United States, Hoheisel and Fleischer [144] investigated the seasonal dynamics of coccinellids and their food (aphids and pollen) in a vegetable farm system containing plantings of Bt sweet corn, Bt potato, and transgenic insect-resistant squash [145] . The results indicated the transgenic vegetable crops provided conservation of coccinellids and resulted in a 25% reduction in insecticide use. In a similar study with these same crops, Leslie et al. [145] compared the soil surface-dwelling communities of Coleoptera and Formicidae in the transgenic crops and their isolines and found no differences in species richness and species composition but found that the transgenic vegetables required fewer insecticide applications. Such results are clear that genetically modified technology can be introduced within vegetable integrated pest management (IPM) systems and transgenic vegetables can offer novel and effective ways of controlling insects and the pathogens they transmit [54] [59] .

Another main concern of transgenic crops is the unintentional spread of transgenic traits into weedy species [146] . There are examples of transgene escape and some evidence for selective advantage of herbicide resistance picked up by weeds [147] [148] . The risk of herbicide resistant genes from a transgenic crop cultivar being transferred to weed relatives has been demonstrated in field crops such as canola/oilseed rape [149] [150] and sugar beet [151] . Rose et al. [152] has demonstrated that a "transgenic mitigation strategy" may impart a negative genetic load to hybrids ensuing from crosses between the weed field mustard and oilseed rape crop. The transgenic mitigation measure was a fitness-mitigating dwarfing gene beneficial for crops but deleterious for weeds (i.e., the hybrid weed is dwarfed due to this mitigation gene and is therefore outgrown by its non-trans- genic counterparts). This finding challenges the view that a transgenic plant might always endow a wild relative with a so-called fitness gene, making it harder and giving it the potential to become a "super-weed". Furthermore, Palaudelmàs et al. [153] found that transgenic maize volunteers had low plant vigor, rarely had cobs, and produced pollen which cross-fertilized neighbor plants only at low levels (0.16% in the worst-case scenario and was below the Regulation EC 1830/2003 establishing the adventitious threshold of 0.9% for coexistence). None- theless, transgene flow raises a new set of ecological and economic issues for scientists and policy makers to consider for transgene containment.

Crop production practices also have significant effects on the composition of weed communities. Changes in the types of weeds that are important locally are termed weed shifts. Such shifts are particularly relevant for managing weeds in herbicide tolerant crop systems in which tillage practices and herbicide use both play major roles in shaping the weed community. There are reports in literature of fourteen weed species or groups of closely related species that have increased in abundance in glyphosate resistant crops [142] . At the same time in six states of the United States, in a survey of transgenic maize, soybean, and cotton growers, between 36% and 70% of growers indicated that weed pressure had declined after implementing rotations using glyphosate resistant crops. The use of herbicides can also result in changes to weed communities through the development of herbicide tolerant weed populations. Globally, glyphosate resistant weeds have been confirmed for 21 weeds in 15 countries [142] . Most of these cases have been reported where glyphosate resistant crops are commonly grown. The development of weeds resistant to glyphosate will most likely require modification to weed control programs where practices, in addition to applying glyphosate, are needed to control the resistant populations.

The introduction of herbicide tolerant transgenic crops has been associated with the increased adoption of conservation tillage practices, which decreases run-off, increases water infiltration and reduces erosion. Trends in the adoption of conservation tillage have been studied in the United States and Argentina, the largest growers of herbicide tolerant transgenic crops. While conservation tillage was already being adopted by some growers prior to the introduction of genetically modified herbicide tolerant crops in both countries, studies have shown a positive two-way causal relationship between the adoption of conservation tillage and the adoption of genetically modified herbicide tolerant crops.

As stated, the most direct negative impact of agriculture on biodiversity is due to the considerable loss of natural habitats, which is caused by the conversion of natural ecosystems into agricultural land. Increases in crop yields allow less land to be dedicated to agriculture than it would otherwise be necessary. According to Carpenter [142] a large and growing body of literature has shown that the adoption of transgenic crops has increased yields, particularly in developing countries. A review of the results of global farmer surveys made by Carpenter [154] , found the average yield increases for developing countries range from 16% for insect-resistant corn to 30% for insect-resistant cotton, with an 85% yield increase observed in a single study on herbicide-tolerant corn. On average, developed-country farmers report yield increases ranging from no change for herbicide-tolerant cotton to a 7% increase for herbicide-tolerant soybean and insect-resistant cotton. Brookes et al. [155] have estimated the benefit of these yield improvements by reducing conversion of land into agricultural use. They estimate that 2.64 million hectares of land would probably be brought into grain and oilseed production if biotechnology traits were no longer used.

Storer et al. [156] stated that the most direct landscape level effects of growing Bt crops would be expected for target pest species for which the crop is a primary food source and is mobile across the landscape. Area-wide pest suppression not only reduces losses to adopters of technology, but may also benefit non-adopters and growers of other crops by reducing crop losses and/or the need to use pest control measures such as insecticides [142] .

Several studies have investigated the impact on the introduction of Bt corn and cotton on regional outbreaks of pest populations, reporting evidence of regional pest suppression in Bt corn and cotton in various areas of the United States and in Bt cotton growing regions of China [142] [157] -[161] . The effects of transgenic crops on above-ground non-target invertebrates have been the subject of a large number of laboratory and field studies. By the end of 2008, over 360 original research papers had been published on non-target effects of Bt crops [161] . A comprehensive review of the literature by Naranjo [161] included 135 laboratory-based studies on nine Bt crops from 17 countries and 63 field-based studies on five Bt crops from 13 countries, which were analyzed using meta-analysis techniques. In general, laboratory studies identified greater levels of hazard than field studies, at least partially explained by differences in organisms studied, and frequently higher protein exposure in lab studies compared to exposure levels in the field [161] [162] . Field studies demonstrated few harmful non-target effects, with non-target effects of insecticides being much greater than Bt crops[161] [162] . More recent literature on non-target impacts of Bt crops are largely consistent with Naranjo's conclusions [142] . Studies on non- target impacts of herbicide tolerant crops, such as the United Kingdom Farm Scale Evaluations (FSE), have found that the effects on various groups of arthropods followed the effects on the abundance of their resources. Where weed control was more effective, the reduction in weeds and weed seeds led to decreases in insects which live in or on weeds, and vice versa. Other studies on non-target impacts of herbicide tolerant crops, conducted for herbicide tolerant soybean and maize in the United States and herbicide tolerant canola in Canada, have reached similar conclusions. FSE results on the bird survey were in accord with differences in food availability found in the studies. Specifically, a greater abundance of granivores was found on conventional than on genetically engineered herbicide tolerant sugar beet, as well as on genetically engineered herbicide tolerant maize after application of herbicides to the genetically modified herbicide tolerant field. No differences were detected in spring oilseed rape. In the subsequent winter season, granivores were more abundant in fields where conventional sugar beet had been grown than on genetically engineered herbicide tolerant fields. Several bird species were more abundant on maize stubbles following genetically engineered herbicide tolerant treatment [142] .

The pest management traits embodied in currently commercialized transgenic crops have led to changes in the use of pesticides which may have impacts on biodiversity. If the planting of genetically modified pest-resistant crop cultivars eliminates the need for broad-spectrum insecticidal control of primary pests, naturally occurring control agents are more likely to suppress secondary pest populations, maintaining a diversity and abundance of prey for birds, rodents, and amphibians. In addition to the studies on the non-target impacts of transgenic crops compared to conventional practices, many studies have quantified changes in pesticide use since the introduction of genetically modified crops. Reductions ranging from 14% to 75% of total active ingredients have been reported for Bt crops compared to conventional crops in Argentina, Australia, China, India, and the United States [142] . Brookes and Barfoot [163] indicated that transgenic crops grown by farmers were able to lower pesticide spraying by 443 kg (9.1%) of active ingredients, thereby decreasing the environmental impact associated with herbicide and insecticide use by 17.9%. These authors also emphasize that transgenic crops significantly reduced the release of greenhouse gas emissions from their cropping area, which was equivalent, in 2010, to removing 8.6 million cars from the roads. Fewer surveys have captured changes in herbicide use in genetically modified herbicide tolerant crops, perhaps because the impact of genetically modified herbicide tolerant crops has largely been a substitution between herbicides applied at different rates, and therefore, changes in the amount of herbicide used is a poor indicator of environmental impact. Several studies have been done to apply environmental indicators to observed changes in pesticide use related to the adoption of both insect resistant and herbicide tolerant crops, which all show a reduction in the environmental impact of pesticides used on transgenic crops [142] . A life cycle assessment made by Bennet et al. [164] showing that transgenic sugar beet tolerating herbicide would be less harmful to the environment than the conventional sugar beet crop because of the lower emissions from herbicide manufacture, transport, and field operations.

Some benefits of genetically modified crops are expected to decline over time, and potential benefits and risks may become more numerous as technology is applied to more crops [165] . For example the plants from Bt cotton seed have been effective in controlling damage from bollworm (Helicoverpa armigera) in Chinese cotton production since 1999, reducing the need for pesticides and increasing incomes of Chinese farmers. However, field data collected in 2004 indicates that these benefits are being eroded by an increasing use of pesticides aimed at the control of secondary pests [166] . This was confirmed in 2009 by Wang et al. [167] , who claimed that the increased problems with secondary pests were of less importance than the decreased use of insecticides due to growing Bt cotton. In United States maize fields another pest, the rootworm has developed resistance to the toxin from Bt, as reported by the Environmental Protection Agency [168] .

Knowledge gained over the past 15 years that genetically modified crops have been grown commercially indicates the impacts on biodiversity are positive on balance. By increasing yields, decreasing insecticide use, increasing use of more environmentally friendly herbicides, and facilitating adoption of conservation tillage, transgenic crops have contributed to increasing agricultural sustainability. Previous reviews have also reached the general conclusion that genetically modified crops have had little to no negative impact on the environment [142] [161] . Most recently, the United States National Research Council released a comprehensive assessment of the effect of genetically modified crop adoption on farm sustainability in the United States which concluded: "generally, genetically modified crops have had fewer adverse effects on the environment than non-genetically modified crops produced conventionally" [165] . Thus, transgenic crops can continue to decrease pressure on biodiversity as global agricultural systems expand to feed a world population expected to continue to increase for the next 30 to 40 years.

4. CONSERVATION AND USE OF BIODIVERSITY: OPPORTUNITIES FOR COOPERATION AND NEW PARTNERSHIPS

Plant genetic resources for food and agriculture are the quintessential global public good. No nation is self-re- liant. A viable market for their conservation and trade does not exist. The conservation of plant genetic resources is a prerequisite for addressing climate change, as well as water and energy constraints, which will grow in importance in the next decades. The Svalbard Global Seed Vault is an International Treaty which establishes a multilateral System to facilitate access and benefit sharing of plant genetic resources. The Treaty has an insurance policy and provides legal framework for a cooperative and global approach to manage this essential resource. The Svalbard Global Seed Vault has a mechanism for ensuring the permanent conservation of unique crop biodiversity, the Global Crop Diversity Trust, which is structured as an endowment fund [169] .

Plant breeding is vital to protect the yield gains made to date, and to further increase the genetic yield potential of all crops. As a result of the Green Revolution, global productivity of the main food staples steadily rose since the 1960s. Such achievements ensued from crop genetic enhancement partnerships. They are models illustrating partnering for exchange, evaluation, release and use of plant genetic resources worldwide. These partnerships include national agricultural research institutes and international agricultural research centers. For many decades the global wheat yield increased due to an effective International Wheat Improvement Network (IWIN) officially founded as an international organization in 1966 [170] . This wheat network deployed cutting-edge science alongside practical multi-disciplinary applications, resulting in the development of genetically enhanced wheat germplasm, which has improved food security and the livelihoods of farmers in the developing world [171] . The spring wheat germplasm bred in Mexico under the leadership of Nobel Peace Laureate Norman Borlaug was further used for launching the Green Revolution in India, Pakistan and Turkey [172] . The network was broadened during the 1970s to include Brazil, China and other major developing country wheat producers. It resulted in wheat cultivars with broader host plant resistance (especially to rusts), better adaptation to marginal environments, and tolerance to acid soils. Nowadays IWIN, an international "alliance", operates field evaluation trials in more than 250 locations, in roughly 100 countries it tests improved breeding lines of wheat in different environments. The number of wheat cultivars released annually in the developing world doubled to more than 100 cultivars by early 1990s due to this networking and the strengthening of national capacity [173] . The widespread adoption of newly bred wheat cultivars, especially in South Asia and Latin America, due to yield increases, led to 50% average annual rates of investment returns [174] . The urban poor also benefited significantly because grain harvest increases drove wheat prices down. Every year, nursery sets and trials are sent to various researchers worldwide, who share their data from these trials to catalogue and analyze. The returned data are used to identify parents for subsequent crosses and to incorporate new genetic variability into advanced wheat lines that are consequently able to cope with the dynamics of abiotic and biotic stresses affecting wheat farming systems. The full pedigree and selection histories are known and phenotypic data cover yield, agronomic, patho- logical and quality data [171] .

The International Network for Genetic Evaluation of Rice (INGER) is one more example of world cooperation. It was established in 1975 as a consortium of national agricultural research systems of rice-growing countries and Centers of today's CGIAR Consortium. INGER was initially founded as an International Rice Testing Program, but soon became an integral component of world national rice breeding program. INGER partners can share rice breeding lines. Every year partners provide about 1000 genetically diverse breeding lines, which have been grown in about 600 experiment stations from 80 countries. This network facilitated the release of 667 cultivars worldwide, which translated into 1.5 billion US dollars of economic benefits. It was estimated that ending INGER could lead to a reduction of 20 rice cultivars per year and to an economic loss of 1.9 billion US dollars [175] . Further analysis by Jackson and Huggan [176] has shown how genetic conservation of landraces can lead to significant gains in rice breeding.

Two other examples of cooperation and partnership are the Latin American Maize Project (LAMP) and the Germplasm Enhancement of Maize (GEM). The LAMP was established as a partnership between Latin America and the United States to assess national germplasm and facilitate the exchange of maize genetic resources across the American continent [177] . The United States Department of Agriculture, the participating national agricultural research systems and a multinational seed corporation provided the funding. The aim of LAMP was to obtain information about the performance of maize germplasm and to share it with plant breeders for developing genetically enhanced open pollinated and hybrid cultivars. The maize germplasm was tested for agronomic characteristics from sea level to 3300 m, and from 41°N to 34°S across 32 locations in the first stage and in 64 locations (two per region) in the second stage. These locations were clustered according to five homologous areas: lowland tropics, temperate and three altitudes.

There were a total five LAMP breeding stages [177] . In the first stage, 14,847 maize accessions belonging to a region were planted for evaluation in trials using a randomized complete block design with two replications of 10 m^2 plots at a single location, which was environmentally similar to that from where these landraces were originally collected. The next step included the assessment of the upper quintile (20%) of those accessions evaluated for agronomic performance in the previous stage. These accessions were planted in two locations with two replications, and the upper 5% were further selected according to their performance. These best selected accessions of each country were interchanged among regions belonging to the same homologous area in the third stage. They were tested in two locations with two replications in each region. The selected maize accessions from the same homologous area were mated with the best tested accession of the region in an isolated field within each region. In the fourth stage, combining ability tests of 268 selected maize accessions were carried out with a local tester using two replications at two locations within each region. The elite maize germplasm was integrated into breeding programs in the fifth stage, which was the last. The best cross combinations and heterotic pools were also determined by LAMP. Maize breeders obtained access to the most promising accessions identified by LAMP to widen the crop genetic base. A LAMP core subset has been made available for encouraging further use in broadening of maize genetic diversity [178] .

The GEM was set up to introgress useful genetic diversity from Latin American maize landraces and other tropical maize donor sources (lines and hybrids) into United States' maize germplasm, to broaden the genetic base of the "corn-belt" hybrids [179] [180] . GEM owes its existence to LAMP because it has used the Latin American landrace maize accessions selected by LAMP in crosses with elite temperate maize lines from the private seed companies in North America [177] . GEM used a pedigree breeding system to develop S_3 lines. The GEM breeders arranged their crosses into non-Stiff Stalk and Stiff Stalk heterotic groups [181] .

LAMP provided the first step through the sharing of information needed to select gene bank maize accessions for further germplasm enhancement. GEM completed the process by returning to genetically enhanced breeding materials derived from gene bank accessions. This improved germplasm can be further used in maize breeding in the United States and elsewhere. LAMP and GEM are very nice examples of international and national public-private partnerships in crop germplasm enhancement.

Agricultural plant breeding is a typical commodity- or species-oriented and solves problems within a species, rather than making breeding choices based on system wide needs. For example, maize breeders currently maximize the area in which maize can be grown, and maximize the amount of maize produced throughout that area. If environmental harmony is to be a key breeding objective, then a change in agricultural thinking to appropriately value whole cropping systems will be required. Achieving these goals will require collaboration among the private, public, and non-profit sectors, and with society as a whole. Programs within the private sector excel at breeding major, profitable crops, and have economies of scale to increase the efficiency of production and ultimately provide farmers with seed. As a valuable complement to commercial breeding programs, public and non-profit breeding programs may focus on developing alternative crops, breeding for small target regions, tackling long-term and high-risk problems, evaluating diverse genetic resources, and, importantly, conducting basic research on breeding methodology to enhance efficiency. Only publicly funded breeding programs, and in particular those based at universities, can provide the necessary education and training in plant breeding and in specialized fields such as ecology. Without trained students from public programs, private commercial breeding programs suffer from an erosion of intellectual capital. Conversely, without the private sector to commercialize public-sector-derived products, beneficial traits and new cultivars cannot easily and quickly be put in the hands of farmers, as has been seen in developing countries without a developed seed industry [182] . As stated, seed production is high technology and a cost intensive venture and only well organized seed companies with good scientific manpower and well equipped research facilities can afford seed production.

Although due to globalization, most breeding research and cultivar development in the world is presently conducted and funded in the private sector, mainly by huge multinational seed companies. Public breeders, cultivar development activities and research are disappearing worldwide. In general, this means there are fewer decision-making centers for breeding and cultivar development. This has also resulted in the focus on relatively few major crops produced worldwide, to the detriment of all the other cultivated crops. It is imperative that national governments and policymakers, as part of a social duty, invest in breeding research and cultivar development of traditional open-pollinated cultivars and in the minor crops. More investments in this area will mean less expensive seed for growers to choose from, and an increased preservation of crop biodiversity. To accomplish these goals new approaches may be required to crop breeding research and development by both the public and private sector. Until recently, breeding research and development which targets small-scale and poor farmers has largely been undertaken by public sector institutions and national agricultural research institutes. However, the capacity to undertake the work was mainly dependent on national or international funding and expertise. The work has been limited by the capacity of these institutions to pay for it. As a result, crop breeding advancement has varied enormously among countries and even within regions in developed and still developing countries. In the area of plant breeding, the process to produce improved cultivars is slow, and it requires long-term sustained commitment that may not fit the continuing changes in the national and international politics to fund research. The application of biotechnology promises acceleration in some aspects of plant breeding, but the adoption of more advanced technology raises the cost of research significantly at a time when investment funding has diminished. Public plant breeding remains a key component of crop breeding research systems worldwide, especially in developing countries. However, the increasing presence of private sector breeding and a decrease in national and international support makes it difficult for the public sector to continue operating in the traditional manner. Declining funding for public crop breeding coupled with the rapid increase of crop production and an urbanizing population has created a difficult situation. Public sector breeding must be strengthened. More public sector crop breeders are needed worldwide to select and to produce non-hybrid cultivars of the minor crops. Breeding of major crops and other minor crops must continue as a viable endeavor. This will benefit small farmers, and will safeguard biodiversity and food security in developing countries.

While the maintenance of vigorous public sector breeding programs in areas where private companies are not interested in providing low cost cultivars is highly desirable, an additional approach to maximize crop and agricultural research input would be the development of global programs with public-private partnerships. The public sector may support portions of crop and agricultural R & D, unattractive to the private sector, and feed improved breeding lines and systems to the private sector for exploitation in regions where the private sector is active, and nurture private sector development in regions where it is lacking. In recent years, private plant breeding programs have increased in number and size. Financial investment also increased, as well as interest in intellectual property protection. The spirit of original attempts to protect plant breeders' rights was that granting a certificate of protection should not inhibit the flow of information and products through continued research by the entire plant breeding community [49] [50] . In a classic sense, the patent is a defensive tool to prevent competitors from reaping benefits which rightfully belong to the inventor. In the modern context, it is an offensive weapon, to stifle competition, prevent further innovation by others and maximize income [49] [51] . The United States utility patent, it is a way to slow down the flow of progress in plant breeding research, unless the research is within the company holding the patent. While obviously benefiting that company, it is a big step backwards for the plant breeding community and by far, for agriculture itself. The intellectual property protection must encourage research and free flow of materials and information [49] [51] . Protection should be for the cultivar only. There should be no constraint against other breeders using that cultivar in further research, including further breeding. Another breeder should be free to use the protected cultivar in a cross, followed by further development through pedigree breeding. Another breeder should also be free to transfer genes controlling economic traits into the protected cultivar by the backcross method or by genetic transformation procedures [49] [50] .

5. CONCLUSIONS

The growing demand for food in the next decades poses major challenges to humanity. We have to safeguard both biodiversity and arable land for future agricultural food production, and we need to protect genetic biodiversity to safeguard ecosystem resilience. We must produce more food with less input, while deploying every effort to minimize risk. Agricultural sustainability is no longer an option, it is mandatory!

Plant breeding is the science of improving plants to further improve the human condition. Plant breeding has played a vital role in the successful development of modern agriculture via "new" cultivars. Plant breeders are continually improving the ability of cultivars to withstand various environmental conditions. By reducing the impact of agriculture on the environment while maintaining sufficient production will require the development of new cultivars.

Climate change is altering the availability of resources and the conditions crucial to plant performance. Plants respond to these changes through environmentally induced shift in phenotype. Understanding these responses is essential to predict and manage the effects of climate change on crop plants.

In the foreseeable future and an increase in population will need significant production. Breeding and modern agricultural technologies can increase yield on existing agricultural land. As a result, they can make a significant contribution to biodiversity conservation by limiting the need to expand agricultural land and by allowing nature to be maintained for conservation purposes and harmony between agriculture and the environment.

There is still an on-going debate among researchers and in the media on the best strategy to keep pace with global population growth and increasing food demand. One strategy favors the use of transgenic crops, while another strategy focuses on agricultural biodiversity. There are short research funds for agro-biodiversity solutions in comparison with funding for research in genetic modification of crops. Favoring biodiversity does not exclude any future biotechnological contributions, but favoring biotechnology threatens future biodiversity resources. The future breeding programs should encompass not only knowledge of existing practices but also conservation of a wide pool of genetic resources of existing crops and breeds, including their wild relatives, to provide the genes necessary to cope with changes in agricultural production. Therefore, agro-biodiversity should be a central element of future sustainable agricultural development, instead of just a source of traits which can be used in current breeding programs. The concept of sustainability rests on the principle that the present needs must be addressed without compromising the ability of future generations to meet their own needs. Sustainable agriculture is an alternative to solve future fundamental and applied issues related to food production in an ecological way.

Farmers in developing countries, especially small farmers, have problems specific to their cultural, economic and environmental conditions, such as limited purchasing power to access improved cultivars and proprietary technologies. These farmers have an important role in conserving and using crop biodiversity. The future of the world food security depends on stored crop genes as well as on farmers who use and maintain crop genetic diversity on a daily basis. In the long run, the conservation of plant genetic diversity depends not only on a small number of institutional plant breeders and seed banks, but also on the vast number of farmers who select, improve, and use crop diversity, especially in marginal farming environments. Their extensive farming systems using landraces or open-pollinated cultivars increase sustainability and less impact from stresses caused by drought, insect and diseases, due to long-term in situ selection of these crops cultivated as opposed to the fertilizer, herbicide, and pesticide demands in an intensive crop based system with improved, hybrid, or transgenic cultivars. That is why we should also be alerted and particularly alarmed by the current trend to exclusively use improved, hybrid, and transgenic crop cultivars. Farmers do not just save seeds; they are plant breeders who constantly adapt their crops to specific farming conditions and needs. This genetic biodiversity is the key to maintain and improve the world's food security, and agriculture sustainability.

The introduction of genetically modified technology has been hailed as a gene revolution similar to the "Green Revolution" of the 1960s. The "Green Revolution" had an explicit strategy for technology development and diffusion, targeting farmers in developing countries, in which improved germplasm was made freely available as a public good, a particular success in Asia. In contrast to the "Green Revolution", the push for genetically modified crops is based largely on private agricultural research, with cultivars provided to farmers on market terms. To date efforts on genetically modified crops have been focused on crops considered to be profitable enough by large plant breeding companies, not on solutions to problems confronted by the world's small farmers. Existing biodiversity in combination with plant breeding has much more to offer the many world's farmers and consumers, while genetically modified crops have more to offer the agro-industry and some large- scale farms, and this explains why they have received so much attention and research funding. Genetically modified crops and their creation may attract investment in agriculture, but it can also concentrate ownership of agricultural resources. In developing countries, patents may drive up costs; restrict experimentation by the public researcher or individual farmer, while undermining local practices that enhance food security and economic sustainability. There is particular concern that present intellectual property rights instruments, including genetically modified organisms, will inhibit sowing of own seeds, seed exchange, and sale.

Transgenic crops can continue to decrease pressure on biodiversity as global agricultural systems expand to feed a world population expected to continue to increase for the next 30 to 40 years. Due to higher income elasticity of demand and population growth, these pressures will be greater in developing countries. Both current and pipeline technology hold great potential in this regard. The potential of currently commercialized genetically modified crops to increase yields, decrease pesticide use, and facilitate the adoption of conservation tillage has yet to be realized, as there continue to be countries where there is a good technological fit, but they have not yet approved these technologies for commercialization. In addition to the potential benefits on expanded adoption of current technology, several pipeline technologies offer additional promises of alleviating the impacts of agriculture on biodiversity. Continued yield improvements in crops such as rice and wheat are expected with insect resistant and herbicide tolerant traits that are already commercialized in other crops. Technologies such as drought tolerance and salinity tolerance would alleviate the pressure to convert high biodiversity areas into agricultural use by enabling crop production on sub-optimal soils. Drought tolerance technology, which allows crops to withstand prolonged periods of low soil moisture, is anticipated to be commercialized within less than five years. This technology has particular relevance for areas like sub-Saharan Africa, where drought is a common occurrence and access to irrigation is limited. Salt tolerance addresses the increasing problem of saltwater encroachment on freshwater resources. Nitrogen use efficient technology is also under development, which can reduce run-off of nitrogen fertilizer into surface waters. Technology promises to decrease the use of fertilizers while maintaining yields, or increase yields achievable with reduced fertilizer rates where access to fertilizer input is limited. This technology is slated to be commercialized within the next 10 years.

One of the major arguments for genetic modified technology is that new cultivars can be developed more quickly than in traditional plant breeding, but like new cultivars derived from conventional breeding methods, transgenic cultivars require several years of field trials to ensure that the inserted traits will actually become expressed and have the desired effects in local environments. When genes coding for certain traits are transferred, typically from one plant species to another, the desired traits are not always expressed unless the environment interacts with the genes in the anticipated way triggering the desired response, which depends on the regulating sequences inserted with the gene. This means that new transgenic cultivars, developed under laboratory conditions in a controlled climate, have to be tested under field conditions, as in more traditional breeding methods, so currently there is little difference in the speed with which either method will result in the release of new cultivars.

The knowledge gained from basic plant research will underpin future crop improvements, but effective mechanisms for the rapid and effective translation of research discoveries into public good agriculture remain to be developed. Maximum benefit will be derived if robust plant breeding and crop management programs have ready access to all the modern crop biotechnological techniques, both transgenic and non-transgenic, to address food security issues. This will require additional investments in capacity building for research and development, in developing countries. Technology implementation alone is not sufficient to address such complex questions as food security. Biotechnologies will make new options available but are not a global solution. We must ensure that society will continue to benefit from the vital contribution that plant breeding offers, using both conventional and biotechnological tools. Genetic engineering has the potential to address some of the most challenging biotic constraints faced by farmers, which are not easily addressed through conventional plant breeding alone. Besides other promising traits seems to be host plant resistance to insects and pathogens. However, transgenic cultivars will have one or a few exogenous genes whereas the background genotype will still be the product of non-transgenic (or conventional) crop breeding. One should follow a pragmatic approach when deciding whether to engage in transgenic plant breeding. Biotechnology products will be successful if clear advantages and safety are demonstrated to both farmers and consumers.

João Carlos da Silva Dias There is a need of investment in research breeding and cultivar development in traditionally open-pollinated cultivars and in the minor crops. More investments in this area will mean cheaper cultivars for growers to choose from and more preservation of crop biodiversity. In recent years, private plant breeding programs have increased in number and size. Financial investment also increased, as well as interest in intellectual property protection. Protective measures, especially patenting, must be moderated to eliminate coverage so broad that it stifles innovation. The intellectual property protection laws for plants must be made less restrictive to encourage research and free flow of materials and information. Public sector breeding must remain vigorous, especially in areas where the private sector does not function. This will often require benevolent public/private partnerships as well as government support. Intellectual property rights laws for plants must be made less restrictive to encourage freer flow of materials. Active and positive connections between the private and public breeding sectors and large-scale gene banks are required to avoid a possible conflict involving breeders' rights, gene preservation and erosion. Partnerships between policy makers with public and private plant breeders will be essential to address future challenges. Many current breeding efforts remain under-funded and disorganized. There is a great need for a more focused, coordinated approach to efficiently utilize funding, share expertise, and continue progress in technologies and programs.

REFERENCES

1. Jaggard, K.W., Qi, A.M. and Ober, E.S. (2010) Possible Changes to Arable Crop Yields by 2050. Philosophical Transactions of the Royal Society B: Biological Sciences, 365, 2835-2851.
2. FAO (2012) The State of the World Population Report. By Choice, Not by Chance: Family Planning, Human Rights and Development. United Nations Population Fund, New York.
3. da Silva Dias, J.C. (2014) Guiding Strategies for Breeding Vegetable Cultivars. Agricultural Sciences, 5, 9-32.
4. Dias, J.S. and Ryder, E.J. (2011) World Vegetable Industry: Production, Breeding, Trends. Horticultural Reviews, 38, 299-356.
5. Dias, J.S. (2012) Chapter 1. Vegetable Breeding for Nutritional Quality and Health Benefits. In: Carbone, K., Ed., Cultivars: Chemical Properties, Antioxidant Activities and Health Benefits, Nova Science Publishers, Inc., Hauppauge, 1-81.
6. Tilman, D., Balzer, C., Hill, J. and Befort, B.L. (2011) Global Food Demand and the Sustainable Intensification of Agriculture. Proceedings of the National Academy of Sciences of the United States of America, 108, 20260-20264.
7. Alexandratos, N. and Bruinsma, J. (2012) World Agriculture towards 2030/2050: The 2012 Revision. Paper No. 12-03, Food and Agriculture Organization (FAO), Rome.
8. FAO (2012) How to Feed the World in 2050. FAO, Rome.
9. Delgado, C.L. (2003) Rising Consumption of Meat and Milk in Developing Countries Has Created a New Food Revolution. Journal of Nutrition, 133, 3907S-3910S.
10. Kastner, T., Rivas, M.J.I., Koch, W. and Nonhebel, S. (2012) Global Changes in Diets and the Consequences for Land Requirements for Food. Proceedings of the National Academy of Sciences of the United States of America, 109, 6868- 6872.
11. Delgado, C.L. (1999) Livestock to 2020: The Next Food Revolution. Food, Agriculture, and the Environment Discussion Paper No. 28, International Food Policy Research Institute, Washington DC.

12. The Royal Society (2009) Reaping the Benefits: Science and the Sustainable Intensification of Global Agriculture. The Royal Society Policy Document 11/09, The Royal Society, London.
13. Naylor, R., Steinfeld, H., Falcon, W., Galloway, J., Smil, V., Bradford, E., Alder, J. and Mooney, H. (2005) Losing the Links between Livestock and Land. Science, 310, 1621-1622.
14. Gerbens-Leenes, P. and Nonhebel, S. (2002) Consumption Patterns and Their Effects on Land Required for Food. Ecological Economics, 42, 185-199.
15. Wirsenius, S., Azar, C. and Berndes, G. (2010) How Much Land Is Needed for Global Food Production under Scenarios of Dietary Changes and Livestock Productivity Increases in 2030? Agricultural Systems, 103, 621-638.
16. Pimentel, D. and Pimentel, M. (2003) Sustainability of Meat-Based and Plant-Based Diets and the Environment. American Journal of Clinical Nutrition, 78, 660S-663S.
17. Foley, J.A., Ramankutty, N., Brauman, K.A., Cassidy, E.S., Gerber, J.S., Johnston, M.,Mueller, N.D., O'Connell, C., Ray, D.K., West, P.C., Balzer, C., Bennett, E.M., Carpenter, S.R., Hill, J., Monfreda, C., Polasky, S., Rockström, J., Sheehan, J., Siebert, S., Tilman, D. and Zaks, D.P. (2011) Solutions for a Cultivated Planet. Nature, 478, 337-342.
18. Steinfeld, H., Gerber, P., Wassenaar, T., Castel, V., Rosales, M. and De Haan, C. (2006) Livestock's Long Shadow: Environmental Issues and Options. FAO, Rome.
19. Mekonnen, M.M. and Hoekstra, A.Y. (2012) A Global Assessment of the Water Footprint of Farm Animal Products. Ecosystems, 15, 401-415.
20. World Watch Institute (2010) Biofuel Production Up Despite Economic Downturn Vital Signs. World Watch Institute, New York.
21. Food and Agricultural Policy Research Institute (FAPRI) (2011) World Biofuels: FAPRI-ISU 2011 Agricultural Outlook. FAPRI, Ames.
22. FAO (2013) FAO Statistical Yearbook—Land Use. FAOSTAT, FAO, Rome, PA4.
23. Kucharik, C.J. and Serbin, S.P. (2008) Impact of Recent Climate Change on Wisconsin Corn and Soybean Yield Trends. Environmental Research Letters, 3, Article ID: 034003, 10 p.
24. Battisti, D.S. and Naylor, R.L. (2009) Historical Warnings of Future Food Insecurity with Unprecedented Seasonal Heat. Science, 323, 240-244.
25. Schlenker, W. and Lobell, D.B. (2010) Robust Negative Impacts of Climate Change on African Agriculture. Environmental Research Letters, 5, Article ID: 014010, 8 p.
26. Roudier, P., Sultan, B., Quirion, P. and Berg, A. (2011) The Impact of Future Climate Change on West African Crop Yields: What Does the Recent Literature Say? Global Environmental Change, 21, 1073-1083.
27. Lobell, D.B., Schlenker, W. and Costa-Roberts, J. (2011) Climate Trends and Global Crop Production since 1980. Science, 333, 616-620.
28. Lobell, D.B., Bänziger, M., Magorokosho, C. and Vivek, B. (2011) Nonlinear Heat Effects on African Maize as Evidenced by Historical Yield Trials. Nature Climate Change, 1, 42-45.
29. Schlenker, W. and Roberts, M.J. (2009) Nonlinear Temperature Effects Indicate Severe Damages to US Crop Yields under Climate Change. Proceedings of the National Academy of Sciences of the United States of America, 106, 15594- 15598.
30. Gupta, R., Gopal, R., Jat, M.L., Jat, R.K., Sidhu, H.S., Minhas, P.S. and Malik, R.K. (2010) Wheat Productivity in Indo-Gangetic Plains of India during 2010: Terminal Heat Effects and Mitigation Strategies. PACA Newsletter, 14, 1-11.
31. Asseng, S., Foster, I. and Turner, N.C. (2011) The Impact of Temperature Variability on Wheat Yields. Global Change Biology, 17, 997-1012.
32. Lobell, D.B., Sibley, A. and Ortiz-Monasterio, J.I. (2012) Extreme Heat Effects on Wheat Senescence in India. Nature Climate Change, 2, 186-189.
33. Bell, G. and Collins, S. (2008) Adaptation, Extinction and Global Change. Evolutionary Applications, 1, 3-16.
34. Kelly, A.E. and Goulden, M.L. (2008) Rapid Shifts in Plant Distribution with Recent Climate Change. Proceedings of the National Academy of Sciences of the United States of America, 105, 11823-11826.

35. Shanthi-Prabha, V., Sreekanth, N.P., Babu, P.K. and Thomas, A.P. (2011) The Trilemma of Soil Carbon Degradation, Climate Change and Food Insecurity. Disaster Risk and Vulnerability Conference 2011, The Applied Geoinformatics for Society and Environment, Germany, 107-112.
36. Gregory, P.J., Johnson, S.N., Newton, A.C. and Ingram, J.S.I. (2009) Integrating Pests and Pathogens into the Climate Change/Food Security Debate. Journal of Experimental Botany, 60, 2827-2838.
37. Patz, J.A. and Kovats, R.S. (2002) Hot Spots in Climate Change and Human Health: Present and Future Risks. Lancet, 368, 859-869.
38. Mcmichael, A., Woodruff, R.E. and Hales, S. (2006) Climate Change and Human Health: Present and Future Risks. Lancet, 367, 859-869.
39. Ziska, L.H., Epstein, P.R. and Schlesinger, W.H. (2009) Rising CO_2, Climate Change, and Public Health: Exploring the Links to Plant Biology. Environmental Health Perspectives, 117, 155-158.
40. Borlaug, N. (1983) Contributions of Conventional Plant Breeding to Food Production. Science, 219, 689-693.
41. Trethowan, R.M., Reynolds, M.P., Ortiz-Monasterio, I. and Ortiz, R. (2007) The Genetic Basis of the Green Revolution in Wheat Production. Plant Breeding Reviews, 28, 39-58.
42. Evenson, R.E. and Gollin, D. (2003) Assessing the Impact of the Green Revolution, 1960 to 2000. Science, 300, 758- 762.
43. Tilman, D., Cassman, K.G., Matson, P.A., Naylor, R. and Polasky, S. (2002) Agricultural Sustainability and Intensive Production Practices. Nature, 418, 671-677.
44. Burney, J.A., Davis, S.J. and Lobell, D.B. (2010) Greenhouse Gas Mitigation by Agricultural Intensification. Proceedings of the National Academy of Sciences of the United States of America, 107, 12052-12057.
45. Edgerton, M.D. (2009) Increasing Crop Productivity to Meet Global Needs for Feed, Food, and Fuel. Plant Physiology, 149, 7-13.
46. IAASTD (2009) Agriculture at the Crossroads. International Assessment of Agricultural Knowledge, Science and Technology for Development (IAASTD). Island Press, Washington DC.
47. Harlan, J.R. (1992) Crops and Man. American Society of Agronomy and Crop Science Society of America, Madison.
48. Pratt, R.C. (2004) An Historical Examination of the Development and Adoption of Hybrid Corn: A Case Study in Ohio. Maydica, 49, 155-172.
49. Dias, J.S. (2011) Biodiversity and Vegetable Breeding in the Light of Developments in Intellectual Property Rights. In: Grillo, O. and Verona, G., Eds., Ecosystems Biodiversity, Chapter 17, INTECH publisher, Rijeka, 389-428.
50. Dias, J.S. (2012) Impact of the Vegetable Breeding Industry and Intellectual Property Rights in Biodiversity and Food Security. In: Jones, A.M. and Hernandez, F.E., Eds., Food Security: Quality, Management, Issues and Economic Implications, Nova Science Publishers Inc., Hauppauge, 57-86.
51. Dias, J.S. (2013) Impact of Vegetable Breeding Industry and Intellectual Property Rights in Food Security. In: Nath, P., Ed., The Basics of Human Civilization-Food, Agriculture and Humanity, Vol. I. Present Scenario, Prem Nath Agricultural Science Foundation (PNASF), Bangalore & New India Publishing Agency (NIPA), New Delhi, 173-198.
52. Dias, J.S. and Ryder, E. (2012) Impact of Plant Breeding on the World Vegetable Industry. Acta Horticulturae, 935, 13-22.
53. Dias, J.S. (2010) Impact of Improved Vegetable Cultivars in Overcoming Food Insecurity. In: Nath, P. and Gaddagimath, P.B., Eds., Horticulture and Livelihood Security, Scientific Publishers, New Dehli, 303-339.
54. Dias, J.S. and Ortiz, R. (2012) Transgenic Vegetable Crops: Progress, Potentials and Prospects. Plant Breeding Reviews, 35, 151-246.
55. Dias, J.S. (1989) The Use of Molecular Markers in Selection of Vegetables. SECH, Actas de Horticultura, 3, 175-181.

56. Dias, J.S. (1991) The Use of Computers in Plant Breeding. SECH, Actas de Horticultura, 8, 367-371.
57. Dias, J.S. and Ortiz, R. (2012) Transgenic Vegetable Breeding for Nutritional Quality and Health Benefits. Food and Nutrition Sciences, 3, 1209-1219.
58. Dias, J.S. and Ortiz, R. (2013) Transgenic Vegetables for Southeast Asia. In: Holmer, R., Linwattana, G., Nath, P. and Keatinge, J.D.H., Eds., Proceedings. Regional Symposium on High Value Vegetables in Southeast Asia: Production, Supply and Demand (SEAVEG 2012), Chiang Mai, 24-26 January 2012, 361-369.
59. Dias, J.S. and Ortiz, R. (2013) Transgenic Vegetables for 21st Century Horticulture. Acta Horticulturae, 974, 15-30.
60. Dias, J.C. (2010) Impact of Improved Vegetable Cultivars in Overcoming Food Insecurity. Euphytica, 176, 125-136.
61. Tilman, D. (1999) Global Environmental Impacts of Agricultural Expansion: The Need for Sustainable and Efficient Practices. Proceedings of the National Academy of Sciences of the United States of America, 96, 5995-6000.
62. Robertson, G.P. and Swinton, S.M. (2005) Reconciling Agricultural Productivity and Environmental Integrity: A Grand Challenge for Agriculture. Frontiers in Ecology and the Environment, 3, 38-46.
63. Hirel, B., Le Gouis, J., Ney, B. and Gallais, A. (2007) The Challenge of Improving Nitrogen Use Efficiency in Crop Plants: Towards a More Central Role for Genetic Variability and Quantitative Genetics within Integrated Approaches. Journal of Experimental Botany, 58, 2369-2387.
64. Foulkes, M.J., Hawkesford, M.J., Barraclough, P.B., Holdsworth, M.J., Kerr, S., Kightley, S. and Shewry, P.R. (2009) Identifying Traits to Improve the Nitrogen Economy of Wheat: Recent Advances and Future Prospects. Field Crops Research, 114, 329-342.
65. Korkmaz, K., Ibrikci, H., Karnez, E., Buyuk, G., Ryan, J., Ulger, A.C. and Oguz, H. (2009) Phosphorus Use Efficiency of Wheat Genotypes Grown in Calcareous Soils. Journal of Plant Nutrition, 32, 2094-2106.
66. Farooq, M., Kobayashi, N., Wahid, A., Ito, O. and Basra, S.M.A. (2009) Strategies for Producing More Rice with Less Water. Advances in Agronomy, 101, 351-388.
67. Shi, W., Moon, C.D., Leahy, S.C., Kang, D., Froula, J., Kittelmann, S., Fan, C., Deutsch, S., Gagic, D., Seedorf, H., Kelly, W.J., Atua, R., Sang, C., Soni, P., Li, D., Pinares-Patiño, C.S., Mcewan, J.C., Janssen, P.H., Chen, F., Visel, A., Wang, Z., Attwood, G.T. and Rubin, E.M. (2014) Methane Yield Phenotypes Linked to Differential Gene Expression in the Sheep Rumen Microbiome. Genome Research, 24, 1517-1525.
68. Shrawat, A.K. and Good, A.G. (2008) Genetic Engineering Approaches to Improving Nitrogen Use Efficiency. Plant Research News. ISB Report, May 2008. Information Systems for Biotechnology (ISB) News Report, Blackburg. http://www.isb.vt.edu/news/2008/news08.may.htm#may0801
69. Daemrich, A., Reinhardt, F. and Shelman, M. (2008) Arcadia Biosciences: Seeds of Change. Harvard Business School, Boston.
70. Subbarao, G.V., Tomohiro, B., Masahiro, K., Osamu, I., Samejima, H., Wang, H.Y., Pearse, S.J., Gopalakrishnan, S., Nakahara, K., Zakir Hossain, A.K.M., Tsujimoto, H. and Berry, W.L. (2007) Can Biological Nitrification Inhibition (BNI) Genes from Perennial Leymus racemosus (Triticeae) Combat Nitrification in Wheat Farming? Plant and Soil, 299, 55-64.
71. IPCC (Intergovernmental Panel on Climate Change) (2009) The Physical Science Basis. In: Solomon, S., Qin, D., Manning, M., Chen, Z., Marquis, M., Averyt, K.B., Tignor, M. and Miller, H.L., Eds., Contribution of Working Group I to the Fourth Assessment Report of the Intergovernmental Panel on Climate Change, Cambridge University Press, Cambridge.
72. Ortiz, R., Sayre, K.D., Govaerts, B., Gupta, R., Subbarao, G.V., Ban, T., Hodson, D., Dixon, J.M., Ortiz-Monasterio, J.I. and Reinolds, M. (2008) Climate Change: Can Wheat Beat the Heat? Agriculture, Ecosystems & Environment, 126, 46-58.
73. Araus, J., Slafer, G., Royo, C. and Serret, M.D. (2008) Breeding for Yield Potential and Stress Adaptation in Cereals. Critical Reviews in Plant Sciences, 27, 377-412.

74. Cattivelli, L., Rizza, F., Badeck, F.W., Mazzucoteli, E., Mastrangelo, A.M., Francia, E., MarÈ, C., Tondelli, A. and Stanca, A.M. (2008) Drought Tolerance Improvement in Crop Plants: An Integrated View from Breeding to Genomics. Field Crops Research, 105, 1-14.
75. Ceccarelli, S. and Grando, S. (2007) Decentralized-Participatory Plant Breeding: An Example of Demand Driven Research. Euphytica, 155, 349-360.
76. Burke, M.B., Lobell, D.B. and Guarino, L. (2009) Shifts in African Crop Climates by 2050, and the Implications for Crop Improvement and Genetic Resources Conservation. Global Environmental Change, 19, 317-325.
77. Jarvis, D.I., Brown, A.H.D., Cuong, P.H., Collado-Panduro, L., Latoumerie-Moreno, L., Gyawali, S., Tanto, T., Sawadogo, M., Mar, I., Sadiki, M., Hue, N.T., Arias-Reyes, L., Balma, D., Bajracharya, J., Castillo, F., Rijal, D., Belqadi, L., Rana, R., Saidi, S., Quedraogo, J., Zangre, R., Rhrib, K., Chavez, J.L., Schoen, D., Shapit, B., Santis, P.D., Fadda, C. and Hodgkin, T. (2008) A Global Perspective of the Richness and Evenness of Traditional Crop-Variety Diversity Maintained by Farming Communities. Proceedings of the National Academy of Sciences of the United States of America, 105, 5326-5331.
78. Witcombe, J.R., Hollington, P.A., Howarth, C.J., Reader, S. and Steele, K.A. (2008) Breeding for Abiotic Stresses for Sustainable Agriculture. Philosophical Transactions of the Royal Society B: Biological Sciences, 363, 703-716.
79. Bhatnagar-Mathur, P., Vadez, V. and Sharma, K.K. (2007) Transgenic Approaches for Abiotic Stress Tolerance in Plants: Retrospect and Prospects. Plant Cell Reports, 27, 411-424.
80. Ainsworth, E., Rogers, A. and Leakey, A.D.B. (2008) Targets for Crop Biotechnology in a Future High-CO_2 and High-O_3 World. Plant Physiology, 147, 13-19.
81. Ortiz, R. (2008) Crop Genetic Engineering under Global Climate Change. Annals of Arid Zone, 47, 343-354.
82. Jewell, M.C., Campbell, B.C. and Godwin, I.D. (2010) Transgenic Plants for Abiotic Stress Resistance. In: Kole, C., Michler, C.H., Abbott, A.G. and Hall, T.C., Eds., Transgenic Crop Plants, Springer-Verlag, Berlin-Heidelberg, 67- 132.
83. Dwivedi, S.L., Upadhyaya, H., Subudhi, P., Gehring, C., Bajic, V. and Ortiz, R. (2010) Enhancing Abiotic Stress Tolerance in Cereals through Breeding and Transgenic Interventions. Plant Breeding Reviews, 33, 31-114.
84. Dwivedi, S.L., Sahrawat, K., Upadhyaya, H. and Ortiz, R. (2013) Food, Nutrition and Agrobiodiversity under Global Climate Change. Advances in Agronomy, 120, 1-128.
85. Ruane, J., Sonnino, A., Steduto, P. and Deane, C. (2008) Coping with Water Scarcity: What Role for Biotechnologies? Land and Water Discussion Paper 7, Food and Agriculture Organization of the United Nations, Rome.
86. Sakuma, Y., Maruyama, K., Qin, F., Osakabe, Y., Shinozaki, K. and Yamaguchi-Shinozaki, K. (2006) Dual Function of an Arabidopsis Transcription Factor DREB2A in Water-Stress-Responsive and Heat-Stress-Responsive Gene Expression. Proceedings of the National Academy of Sciences of the United States of America, 103, 18822-18827.
87. Ortiz, R., Iwanaga, M., Reynolds, M.P., Wu, X. and Crouch, J.H. (2007) Overview on Crop Genetic Engineering for Drought-Prone Environments. Journal of SAT Agricultural Research, 4, 1-30.
88. Pellegrineschi, A., Reynolds, M., Pacheco, M., Brito, R.M., Almeraya, R., Yamaguchi-Shinozaki, K. and Hoisington, D. (2004) Stress-Induced Expression in Wheat of the Arabidopsis thaliana DREB1A Gene Delays Water Stress Symptoms under Greenhouse Conditions. Genome, 47, 493-500.
89. Saint Pierre, C.S., Crossa, J.L., Bonnett, D., Yamaguchi-Shinozaki, K. and Reynolds, M.P. (2012) Phenotyping Transgenic Wheat for Drought Resistance. Journal of Experimental Botany, 63, 1799-1808.
90. Mumms, R. (2005) Genes and Salt Tolerance: Bringing Them Together. New Phytologist, 167, 645-663.

91. Chinnusamy, V., Jagendorf, A. and Zhu, J.K. (2005) Understanding and Improving Salt Tolerance in Plants. Crop Science, 45, 437-448.
92. Wu, H.J., Zhang, Z., Wang, J.Y., Oh, D.H., Dassanayake, M., Liu, B., Huang, Q., Sun, H.X., Xia, R., Wu, Y., Wang, Y.N., Yang, Z., Liu, Y., Zhang, W., Zhang, H., Chu, J., Yan, C., Fang, S., Zhang, J., Wang, Y., Zhang, F., Wang, G., Yeol Lee, S., Cheeseman, J.M., Yang, B., Li, B., Min, J., Yang, L., Wang, J., Chu, C., Chen, S.Y., Bohnert, H.J., Zhu, J.K., Wang, X.J. and Xie, Q. (2012) Insights into Salt Tolerance from the Genome of Thellungiella salsuginea. Proceedings of the National Academy of Sciences of the United States of America, 109, 12219-12224.
93. Plett, D., Safwat, G., Gilliham, M., Skrumsager-Møller, I., Roy, S., Shirley, N., Jacobs, A., Johnson, A. and Tester, M. (2010) Improved Salinity Tolerance of Rice through Cell Type-Specific Expression of Athkt1;1. PLoS ONE, 5, e12571.
94. Moghaieb, R.E., Nakamura, A., Saneoka, H. and Fujita, K. (2011) Evaluation of Salt Tolerance in Ectoine-Transgenic Tomato Plants (Lycopersicon esculentum) in Terms of Photosynthesis, Osmotic Adjustment, and Carbon Partitioning. GM Crops, 2, 58-65.
95. Lybbert, T. and Sumner, D. (2011) Agricultural Technologies for Climate Change Mitigation and Adaptation in Developing Countries: Policy Options for Innovation and Technology Diffusion. ICTSD-IPC Platform on Climate Change. Agriculture and Trade Issues Brief 6. International Centre for Trade and Sustainable Development, Geneva, Switzerland.
96. Varshney, R.K., Bansal, K.C., Aggarwal, P.K., Datta, S.K. and Craufurd, P.Q. (2011) Agricultural Biotechnology for Crop Improvement in a Variable Climate: Hope or Hype? Trends in Plant Science, 16, 363-371.
97. Yamori, W., Hikosaka, K. and Way, D.A. (2013) Temperature Response of Photosynthesis in C_3, C_4, and CAM Plants: Temperature Acclimation and Temperature Adaptation. Photosynthesis Research, 119, 101-117.
98. Ainsworth, E.A. and Ort, D.R. (2010) How Do We Improve Crop Production in a Warming World? Plant Physiology, 154, 526-530.
99. Wahid, A., Gelani, S., Ashraf, M. and Foolad, M.R. (2007) Heat Tolerance in Plants: An Overview. Environmental and Experimental Botany, 61, 199-223.
100. Hasanuzzaman, M., Nahar, K., Alam, Md., Roychowdhury, R. and Fujita, M. (2013) Physiological, Biochemical, and Molecular Mechanisms of Heat Stress Tolerance in Plants. International Journal of Molecular Sciences, 14, 9643-9684.
101. Gao, H., Brandizzi, F., Benning, C. and Larkin, R.M. (2008) A Membrane-Tethered Transcription Factor Defines a Branch of the Heat Stress Response in Arabidopsis thaliana. Proceedings of the National Academy of Sciences of the United States of America, 105, 16398-16403.
102. Gusta, L. (2012) Abiotic Stresses and Agricultural Sustainability. Journal of Crop Improvement, 26, 415-427.
103. Katiyar-Agarwal, S., Agarwal, M. and Grover, A. (2003) Heat-Tolerant Basmati Rice Engineered by Over-Expression of hsp101. Plant Molecular Biology, 51, 677-686.
104. Pimentel, D. (1997) Techniques for Reducing Pesticide Use: Economic and Environmental Benefits. Wiley, New York.
105. Oerke, E.C., Dehne, H.W., Schonbeck, F. and Weber, A. (1994) Crop Production and Crop Protection: Estimated Losses in Major Food and Cash Crops. Elsevier, Amsterdam.
106. Tripathi, L., Mwaka, H., Tripathi, J.N. and Tushemereirwe, W. (2010) Expression of Sweet Pepper Hrap Gene in Banana Enhances Resistance to Xanthomonas campestris pv. Musacearum. Molecular Plant Pathology, 11, 721-731.
107. Ortiz, R., Jarvis, A., Aggarwal, P.K. and Campbell, B.M. (2014) Plant Genetic Engineering, Climate Change and Food Security. CCAFS Working Paper No. 72. CGIAR Research Program on Climate Change, Agriculture and Food Security (CCAFS), Copenhagen.
108. Haverkort, A.J., Boonekamp, P.M., Hutten, R., Jacobsen, E., Lotz, L.A.P., Kessel, G.J.T., Visser, R.F.G. and Van Der Vossen, E.A.G. (2008) Societal Costs of Late Blight in

Potato and Prospects of Durable Resistance through Cisgenic Modification. Potato Research, 51, 47-57.

109. Jacobs, D.F. (2007) Toward Development of Silvical Strategies for Forest Restoration of American Chestnut (Castanea dentata) Using Blight Resistant Hybrids. Biological Conservation, 137, 497-506.
110. Santini, A., La Porta, N., Ghelardini, L. and Mittempergher, L. (2007) Breeding against Dutch Elm Disease Adapted to the Mediterranean Climate. Euphytica, 163, 45-56.
111. Bouton, J. (2007) The Economic Benefits of Forage Improvement in the United States. Euphytica, 154, 263-270.
112. Pimentel, D., Allen, J., Beers, A., Guinand, L., Linder, R., Mclaughlin, P., Meer, B., Musonda, D., Perdue, D., Poisson, S., Siebert, S., Stoner, K., Salazar, R. and Hawkinset, A. (1987) World Agriculture and Soil Erosion. Erosion Threatens World Food Production. Bioscience, 37, 277-283.
113. Glover, J.D., Cox, C.M. and Reganold, J.P. (2007) Future Farming: A Return to Roots? Scientific American, 297, 82-89.
114. Jackson, W., Cox, S., Dehaan, L., Glover, J., Van Tassel, D. and Cox, C. (2009) The Necessity and Possibility of an Agriculture Where Nature Is the Measure. In: Bohlen, P.J. and House, G., Eds., Sustainable Agroecosystem Management, CRC Press, Boca Raton, 61-71.
115. Blanco, H. and Lal, R., Eds. (2008) Principles of Soil Conservation and Management. Springer, New York.
116. Olson, K.R., Ebelhar, S.A. and Lang, J.M. (2010) Cover Crops Effects on Crop Yields and Soil Organic Content. Soil Science, 175, 89-98.
117. Kaumbutho, P. and Kienzle, J. (2008) Conservation Agriculture as Practiced in Kenya: Two Case Studies. Food and Agriculture Organization of the United Nations, Rome.
118. Pretty, J.N. and Hine, R. (2001) Reducing Food Poverty with Sustainable Agriculture: A Summary of New Evidence. Final Report of the "SAFE-World" (The Potential of Sustainable Agriculture to Feed the World) Research Project. Centre for Environment and Society, University of Essex, Colchester.
119. Kendle, A.D. and Rose, J.E. (2000) The Aliens Have Landed! What Are the Justifications for "Native Only" Policies in Landscape Plantings? Landscape and Urban Planning, 47, 19-31.
120. Zhao, F.J. and Mcgrath, S.P. (2009) Biofortification and Phytoremediation. Current Opinion in Plant Biology, 12, 373- 380.
121. Yin, X., Yuan, L., Liu, Y. and Lin, Z. (2012) Phytoremediation and Biofortification: Two Sides of One Coin. In: Yin, X. and Yuan, L., Eds., Phytoremediation and Biofortification, Springer Briefs in Green Chemistry for Sustainable, Springer, New York, 1-6.
122. Pilon-Smits, E. (2005) Phytoremediation. Annual Review of Plant Biology, 56, 15-39.
123. Kramer, U. (2005) Phytoremediation: Novel Approaches to Cleaning up Polluted Soils. Current Opinion in Biotechnology, 16, 133-141.
124. Doty, S.L. (2008) Enhancing Phytoremediation through the Use of Transgenics and Endophytes. New Phytologist, 179, 318-333.
125. Chaney, R.L., Angle, J.S., Broadhurst, C.L., Peters, C.A., Tappero, R.V. and Parks, D.L. (2007) Improved Understanding of Hyperaccumulation Yields Commercial Phytoextraction and Phytomining Technologies. Journal of Environmental Quality, 36, 1429-1443.
126. Mcgrath, S.P. and Zhao, F.J. (2003) Phytoextraction of Metals and Metalloids from Contaminated Soils. Current Opinion in Biotechnology, 14, 277-282.
127. Murakami, M., Ae, N., Ishikawa, S., Ibaraki, T. and Ito, M. (2008) Phytoextraction by a High-Cd-Accumulating Rice: Reduction of Cd Content of Soybean Seeds. Environmental Science & Technology, 42, 6167-6172.
128. Ueno, D., Kono, I., Yokosho, K., Ando, T., Yano, M. and Ma, J.F. (2009) A Major Quantitative Trait Locus Controlling Cadmium Translocation in Rice (Oryza sativa). New Phytologist, 182, 644-653.

129. Grant, C.A., Clarke, J.M., Duguid, S. and Chaney, R.L. (2008) Selection and Breeding of Plant Cultivars to Minimize Cadmium Accumulation. Science of the Total Environment, 390, 301-310.
130. Mcgrath, S.P., Lombi, E., Gray, C.W., Caille, N., Dunham, S.J. and Zhao, F.J. (2006) Field Evaluation of Cd and Zn Phytoextraction Potential by the Hyperaccumulators Thlaspi caerulescens and Arabidopsis halleri. Environmental Pollution, 141, 115-125.
131. Maxted, A.P., Black, C.R., West, H.M., Crout, N.M.J., Mcgrath, S.P. and Young, S.D. (2007) Phytoextraction of Cadmium and Zinc from Arable Soils Amended with Sewage Sludge Using Thlaspi caerulescens: Development of a Predictive Model. Environmental Pollution, 150, 363-372.
132. Zhao, F.J., Hamon, R.E., Lombi, E., Mclaughlin, M.J. and Mcgrath, S.P. (2002) Characteristics of Cadmium Uptake in Two Contrasting Ecotypes of the Hyperaccumulator Thlaspi caerulescens. Journal of Experimental Botany, 53, 535-543.
133. Kertulis-Tartar, G.M., Ma, L.Q., Tu, C. and Chirenje, T. (2006) Phytoremediation of an Arsenic-Contaminated Site Using Pteris vittata L.: Two-Year Study. International Journal of Phytoremediation, 8, 311-322.
134. Salido, A.L., Hasty, K.L., Lim, J.M. and Butcher, D.J. (2003) Phytoremediation of Arsenic and Lead in Contaminated Soil Using Chinese Brake Ferns (Pteris vittata) and Indian Mustard (Brassica juncea). International Journal of Phytoremediation, 5, 89-103.
135. Dickinson, N.M. and Pulford, I.D. (2005) Cadmium Phytoextraction Using Short-Rotation Coppice Salix: The Evidence Trail. Environment International, 31, 609-613.
136. Wieshammer, G., Unterbrunner, R., Garcia, T.B., Zivkovic, M.F., Puschenreiter, M. and Wenzel, W.W. (2007) Phytoextraction of Cd and Zn from Agricultural Soils by Salix ssp. and Intercropping of Salix caprea and Arabidopsis halleri. Plant and Soil, 298, 255-264.
137. Lievens, C., Yperman, J., Cornelissen, T. and Carleer, R. (2008) Study of the Potential Valorisation of Heavy Metal Contaminated Biomass via Phytoremediation by Fast Pyrolysis. Part II. Characterisation of the Liquid and Gaseous Fraction as a Function of the Temperature. Fuel, 87, 1906-1916.
138. Rooney, W., Blumenthal, J., Bean, B. and Mullet, J. (2007) Designing Sorghum as a Dedicated Bioenergy Feedstock. Biofuels, Bioproducts and Biorefining, 1, 147-157.
139. Jessup, R.W. (2009) Development and Status of Dedicated Energy Crops in the United States. In Vitro Cellular & Developmental Biology-Plant, 45, 282-290.
140. Robertson, G.P., Dale, V.H., Doering, O.C., Hamburg, S.P., Melillo, J.M., Wander, M.M., Parton, W.J., Adler, P.R., Barney, J.N., Cruse, R.M., Duke, C.F., Fearnside, P.M., Follett, R.F., Gibbs, H.K., Goldemberg, J., Mladenoff, D.J., Ojima, D., Palmer, M.W., Sharpley, A., Wallace, L., Weathers, K.C., Wiens, J.A. and Wilhelm, W.W. (2008) Sustainable Biofuels Redux. Science, 322, 49-50.
141. Cook, R.J. (2006) Toward Cropping Systems that Enhance Productivity and Sustainability. Proceedings of the National Academy of Sciences of the United States of America, 103, 18389-18394.
142. Carpenter, J. (2011) Impacts of GM Crops on Biodiversity. GM Crops, 2, 7-23.
143. Icoz, I. and Stotzky, G. (2008) Fate and Effects of Insect-Resistant Bt Crops in Soil Ecosystems. Soil Biology and Biochemistry, 40, 559-586.
144. Hoheisel, G.A. and Fleischer, S.J. (2007) Coccinelids, Aphids, and Pollen in Diversified Vegetable Fields with Transgenic and Isoline Cultivars. Journal of Insect Science, 7, 1-12.
145. Leslie, T.W., Hoheisel, G.A., Biddinger, D.J., Rohr, J.R. and Fleisher, S.J. (2007) Transgenes Sustain Epigeal Insect Biodiversity in Diversified Vegetable Farm Systems. Environmental Entomology, 36, 234-244.
146. Jørgensen, R.B. and Andersen, B. (1994) Spontaneous Hybridization between Oilseed Rape (Brassica napus) and Weedy B. campestris (Brassicaceae): A Risk of

Growing Genetically Modified Oilseed Rape. American Journal of Botany, 81, 1620-1626.
147. Hansen, L.B., Siegismund, H.R. and Jørgensen, R.B. (2003) Progressive Introgression between Brassica napus (Oilseed Rape) and B. rapa. Heredity, 91, 276-283.
148. Stewart Jr., C.N., Halfhill, M.D. and Warwick, S.I. (2003) Transgene Introgression from Genetically Modified Crops to Their Wild Relatives. Nature Reviews Genetics, 4, 806-817.
149. Brown, J. and Brown, A.P. (1996) Gene Transfer between Canola (Brassica napus and B. campestris) and Related Weed Species. Annals of Applied Biology, 129, 513-522.
150. Mikkelsen, T.R., Anderson, B. and Jorgensen, R.B. (1996) The Risk of Crop Transgene Spread. Nature, 380, 31.
151. Boudry, P., Broomberg, K., Saumitou-Laprade, P., Morchen, M., Cuegen, J. and Van Dijk, H. (1994) Gene Escape in Transgenic Sugar Beet: What Can Be Learned from Molecular Studies of Weed Beet Populations? Proceedings of the 3rd International Symposium on the Biosafety, Results of Field Tests of Genetically-Modified Plants and Microorganisms, University of California, Division of Agriculture and Natural Resources, Oakland, 75-83.
152. Rose, C.W., Millwood, R.J., Moon, H.S., Rao, M.R., Halfhill, M.D., Raymer, P.L., Warwick, S.I., Al-Ahmad, H., Gressel, J. and Stewart Jr., C.N. (2009) Genetic Load and Transgenic Mitigating Genes in Transgenic Brassica rapa (Field Mustard) × Brassica napus (Oilseed Rape) Hybrid Populations. BMC Biotechnology, 9, 93.
153. Palaudelmàs, M., Peñas, G., Mele, E., Serra, J., Salvia, J., Pla, M., Nadal, A. and Messeguer, J. (2009) Effect of Volunteers on Maize Gene Flow. Transgenic Research, 18, 583-594.
154. Carpenter, J.E. (2010) Peer-Reviewed Surveys Indicate Positive Impact of Commercialized GM Crops. Nature Biotechnology, 28, 319-321.
155. Brookes, G., Yu, T.H., Tokgoz, S. and Elobeid, A. (2010) The Production and Price Impact of Biotech Corn, Canola, and Soybean Rops. AgBioForum, 13, 25-52.
156. Storer, N.P., Dively, G.P. and Herman, R.A. (2008) Landscape Effects of Insect-Resistant Genetically Modified Crops. In: Romeis, J., Shelton, A.M. and Kennedy, G.G., Eds., Integration of Insect-Resistant Genetically Modified Crops within IPM Programs, Springer, New York, 273-302.
157. Naranjo, S.E. (2005) Field Studies Assessing Arthropod Non-Target Effects of Bt Transgenic Crops: Introduction. Environmental Entomology, 34, 1178-1180.
158. Naranjo, S.E. (2005) Long-Term Assessment of the Effects of Transgenic Bt Cotton on the Abundance of Non-Target Arthropod Natural Enemies. Environmental Entomology, 34, 1193-1210.
159. Naranjo, S.E. (2005) Long-Term Assessment of the Effects of Transgenic Bt Cotton on the Function of the Natural Enemy Community. Environmental Entomology, 34, 1211-1223.
160. Wolfenbarger, L.L., Naranjo, S.E., Lundgren, J.G., Bitzer, R.J. and Watrud, L.S. (2008) Bt Crop Effects on Functional Guilds of Non-Target Arthropods: A Meta-Analysis. PLoS ONE, 3, e2118.
161. Naranjo, S.E. (2009) Impacts of Bt Crops on Non-Target Invertebrates and Insecticide Use Patterns. CAB Reviews: Perspectives in Agriculture, Veterinary Science, Nutrition and Natural Resources, 4, 1-23.
162. Duan, J.J., Lundgren, J.G., Naranjo, S. and Marvier, M. (2009) Extrapolating Non-Target Risk of Bt Crops from Laboratory to Field. Biology Letters, 6, 74-77.
163. Brookes, G. and Barfoot, P. (2012) Global Economic and Environmental Benefits of GM Crops Continue to Rise. PG Economics 2012. http://www.pgeconomics.co.uk/page/33/global-impact-2012
164. Bennet, R., Phipps, R., Strange, A. and Grey, P. (2004) Environmental and Human Health Impacts of Growing Genetically Modified Herbicide-Tolerant Sugar Beet: A Life-Cycle Assessment. Plant Biotechnology Journal, 2, 273-278.

165. National Research Council (2010) The Impact of Genetically Engineered Crops on Farm Sustainability in the United States. National Academies, Washington DC.
166. Wang, S., Just, D.R. and Pinstrup-Andersen, P. (2008) Bt-Cotton and Secondary Pests. International Journal of Biotechnology, 10, 113-120.
167. Wang, Z.J., Lin, H., Huang, J., Hu, R., Rozelle, S. and Pray, C. (2009) Bt Cotton in China: Are Secondary Insect Infestations Offsetting the Benefits in Farmer Fields? Agricultural Sciences in China, 8, 83-90.
168. Gassmann, A.J., Petzold-Maxwell, J.L., Keweshan, R. and Dunbar, M.W. (2011) Field-Evolved Resistance to Bt Maize by Western Corn Rootworm. PLoS ONE, 6, e22629.
169. Fowler, C. (2011) Conserving Diversity: The Challenge of Cooperation. Acta Horticulturae, 916, 19-24.
170. Ortiz, R., Mowbray, D., Dowswell, C. and Rajaram, S. (2007) Norman E. Borlaug: The Humanitarian Plant Scientist Who Changed the World. Plant Breeding Reviews, 28, 1-37.
171. Ortiz, R., Braun, H.J., Crossa, J., Crouch, J.H., Davenport, G., Dixon, J., Dreisigacker, S., Duveiller, E., He, Z., Huerta, J., Joshi, A.K., Kishii, M., Kosina, P., Manes, Y., Mezzalama, M., Morgounov, A., Murakami, J., Nicol, J., Ortiz-Ferrara, G., Ortiz-Monasterio, J.I., Payne, T.S., Peña, R.J., Reynolds, M.P., Sayre, K.D., Sharma, R.C., Singh, R.P., Wang, J., Warburton, M., Wu, H. and Iwanaga, M. (2008) Wheat Genetic Resources Enhancement by the International Maize and Wheat Improvement Center (CIMMYT). Genetic Resources and Crop Evolution, 55, 1095-1140.
172. Reynolds, M.P. and Borlaug, N.E. (2006) International Collaborative Wheat Improvement: Impacts and Future Prospects. Journal of Agricultural Science, 144, 3-17.
173. Lantican, M.A., Dubin, M.J. and Morris, M.L. (2005) Impacts of International Wheat Breeding Research in the Developing World, 1988-2002. Centro Internacional de Mejoramiento de Maíz y Trigo, México D.F.
174. Alston, J.M., Marra, M.C., Pardey, P.G. and Wyatt, T.J. (2000) Research Returns Redux: A Meta-Analysis of the Returns to Agricultural R&D. Australian Journal of Agricultural and Resource Economics, 44, 185-215.
175. Evenson, R.E. and Gollin, D. (1997) Genetic Resources, International Organizations, and Improvement in Rice Varieties. Economic Development and Cultural Change, 45, 471-500.
176. Jackson, M.T. and Huggan, R.D. (1993) Sharing the Diversity of Rice to Feed the World. Diversity, 9, 22-25.
177. Salhuana, W. and Pollak, L. (2006) Latin American Maize Project (LAMP) and Germplasm Enhancement of Maize (GEM) Project: Generating Useful Breeding Germplasm. Maydica, 51, 339-355.
178. Taba, S., Díaz, J., Franco, J., Crossa, J. and Eberhart, S.A. (1999) A Core Subset of LAMP from the Latin American Maize Project. CD-Rom. Centro Internacional de Mejoramiento de Maíz y Trigo, México D.F.
179. Balint-Kurti, P., Blanco, M., Milard, M., Duvick, S., Holland, J., Clements, M., Holley, R., Carson, M.L. and Goodman, M. (2006) Registration of 20 GEM Maize Breeding Germplasm Lines Adapted to the Southern USA. Crop Science, 46, 996-998.
180. Goodman, M.M. (2005) Broadening the U.S. Maize Germplasm Base. Maydica, 50, 203-214.
181. Ortiz, R., Taba, S., Chávez-Tovar, V.H., Mezzalama, M., Xu, Y., Yan, J. and Crouch, J.H. (2010) Conserving and Enhancing Maize Genetic Resources as Global Public Goods—A Perspective from CIMMYT. Crop Science, 50, 13-28.
182. Delmer, D.P. (2005) Agriculture in the Developing World: Connecting Innovations in Plant Research to Downstream Applications. Proceedings of the National Academy of Sciences of the United States of America, 102, 15739-15746.

CHAPTER 8

Climate Change Impacts and Adaptation in Rainfed Farming Systems: A Modeling Framework for Scaling-Out Climate Smart Agriculture in Sub-Saharan Africa

Berhanu F. Alemaw[1], Timothy Simalenga[2]

[1]*Department of Geology, University of Botswana, Gaborone, Botswana*

[2]*Centre for Coordination of Agricultural Research & Development in Southern Africa (CCARDESA), Gaborone, Botswana*

ABSTRACT

Improving agricultural water productivity, under rainfed or irrigated conditions, holds significant scope for addressing climate change vulnerability. It also offers adaptation capacity needs as well as water and food security in the southern African region. In this study, evidence for climate change impacts and adaptation strategies in rainfed agricultural systems is explored through modeling predictions of crop yield, soil moisture and excess water for potential harvesting. The study specifically presents the results of climate change impacts under rainfed conditions for maize, sorghum and sunflower using soil-water-crop model simulations, integrated based on daily inputs of rainfall and evapotranspiration disaggregated from GCM scenarios. The research targets a vast farming region dominated by heavy clay soils where rainfed agriculture is a dominant practice. The potential for improving soil water productivity and improved water harvesting have been explored as ways of climate change mitigation and adaptation measures. This can be utilized to explore and design appropriate conservation agriculture and adaptation practices in similar agro-ecological environments, and create opportunities for outscaling for much wider areas. The results of this study can suggest the need for possible policy refinements towards reducing vulnerability and adaptation to climate change in rainfed farming systems.

Keywords: Climate Change, Adaptation, Rainfed Farming Systems, A Modeling, Climate Smart Agriculture, Southern Africa

1. INTRODUCTION

Climate change and other global drivers of socio-economic, energy, global trade, resources and demographic changes are set to affect present and future human development including the vulnerable areas such as southern Africa region (SADC) [1] . Faced with imperatives of increased food production and poverty alleviation, present day conditions call for high vigilance in developing and preserving the raw materials for food production, specifically through land and water management [2] . Faced with imperatives of increased food production and poverty alleviation, present day conditions call for high vigilance in developing and preserving the raw materials for food production: land and water. Agricultural water (under rainfed or irrigation settings) holds significant scope for addressing climate change vulnerability and adaptation needs as well as water and food insecurity in the region [1] -[3] .

The observed human-induced changes to climate pose a threat to food security the world over [3] and Southern Africa is no exception. Climate model studies show that the average temperature of earth's surface is expected to increase by 3°C over the next century, if greenhouse gas emissions continue to rise at the current rates [3] . This has a negative impact on crop yields and food security in the Southern African region where 60% - 80% of the population is directly dependent on agriculture for their livelihoods [2] .

Crop and livestock production systems will have to change in response to the changing agro-ecological conditions. This manuscript aims at developing a set of practical approaches to agriculture in order for farmers to be resilient and adapt to the predicted climate changes. Here, a regional framework for the implementation of climate smart agriculture concepts is also illustrated.

The specific objective of this study is to inform adaptation policy making processes. Also it will support investment decisions in climate change adaptation. This paper seeks to inform scientists and experts in the fields of agriculture, climate change and socio-economics to collectively build a strong base of evidence on climate change and variability impacts on rainfed cropping systems. It also tries to devise agricultural interventions and practices that enhance general resilience in the quest to overcome climatic shocks and develop adaptation strategies.

The main objectives of this manuscript are to report, for a southern African study region, on agricultural impacts assessment under climate variability and change scenarios for rainfed systems to illustrate the regional challenges of climate change and variability in southern Africa. The research uses a selected study area, known as the Pandamatenga Plains, which is located in northern Botswana, and it considers cropping of maize, sorghum and sunflower under rainfed conditions.

The specific objectives are: 1) to show real world climate change through the scientific understanding of downscaled climate scenarios; 2) to integrate downscaled climate scenarios with a crop model and adaptation option models, and with agricultural production information; 3) to assess the impact of climate change on crop yield, soil moisture stress, excess runoff etc. in rainfed agricultural systems; 4) to determine the potential for excess moisture enhancement and water harvesting through a modeling study; and finally 5) to recommend adaptation strategies based on modeling evidences.

2. CLIMATE CHANGE AND RAINFED CROP PRODUCTION

The southern African region is vulnerable to climate change that causes multiple biophysical, political, and socioeconomic stresses. The stresses remain a major threat to the region's susceptibility to vulnerability; they restrain the region's populace adaptive capacity to climate changes and variability [2] -[4] . Besides increases in temperature, climate change in sub-Saharan Africa is expected to cause, increases in the incidence of extreme events such as droughts and floods [5] -[9] , changes in rainfall intensity [10] , increases in desertification and increased in drought frequencies [4] [11] [12] .

Present research confirms that while crops would respond positively to elevated CO_2 [13] , the associated impacts of elevated temperatures, altered patterns of precipitation and possibly increased frequency of extreme events, droughts and floods. Taken all else as equal, these events will probably combine to depress crop yields and increase production risks [2] . Expected impacts include shortened or disrupted growing seasons, reductions in the area suitable for agriculture, and declines in agricultural yields in many regions of sub-Saharan Africa [9] [14] . Several studies have already revealed that a combination of increased rainfall variability and increasing ambient air temperatures will cause a significant decline in yields of major staple crops, particularly for maize [15] .

In a recent work, Lobel et al. [16] used a data set of more than 20,000 historical maize trials in combination with daily weather data. It showed that for each degree day spent above 30°C maize yield was reduced by 1 percent under optimal rain-fed conditions, and by 1.7 percent under drought conditions. In a similar study, maize yield projections in Malawi found a decline of up to 20 percent in the next 50 years [16] . A similar study projected a decline of 10 to 57 percent by 2080 in Zimbabwe [6] [17] , which is mainly due to increased rainfall variability. It is noted that many other factors contribute, but these projections allow to showcase the framework and if business as usual would prevail.

Climate change is emphasized as one of the major sources of challenge for food security, and livelihoods making the southern African region vulnerable to a variety of stresses. It is estimated that the livelihoods of nearly 70% of the region which depends on rain-fed agriculture, an activity that is characterized by small-scale, subsistence farms is affected [18] [19] . Due to its largely adverse effects on African agriculture and livelihoods, climate change is expected to have a negative impact on food security [9] [18] [20] . In a recent study, most farmers in Zambia are unable to afford certain alternatives, such as those of agro-forestry or conservation; they face difficulties in accessing markets due to poor road infrastructure, fluctuating market prices, high costs and late deliveries of farming inputs [21] . Coupled with the low presence of systematic early warning systems in place against natural hazards and disasters, it shows the adaptation capacity of farmers remains limited.

Most of the research on climate change impacts related to food in Africa, as evident in IPCC assessments, focuses on changes in crop yields and food production [9] [10] reported climate change impacts on the yield of maize by considering regional model across southern Africa. These authors experimented on several climate change scenarios and examined the sensitivity of maximized yields to shifting of sowing dates as a means of developing adaptation decisions by keeping yields as high as possible.

It is reported that with current climate change mitigation policies and related sustainable development practices, global GHG emissions will continue to grow over the next few decades [2] . The projected climate change and emission scenarios are well documented in the latest IPCC reports [2] . The various future storylines of GHG emission scenarios are expressed in terms of SRES scenarios [22] . The SRES scenarios, provided in [22] , are grouped into four scenario families (A1, A2, B1 and B2) that represent alternative development pathways, covering a wide range of demographic, economic and technological driving forces and resulting GHG emissions. The emissions projections are widely used in the assessments of future climate change, as inputs to many recent climate change vulnerability and impact assessments.

For the assessment of climate change impact on agricultural productivity and yield, several models have been used. A number of these agricultural models are developed to analyse and model crop-soil-water interactions under different agro ecological and agronomic practices under rainfed systems as well as under irrigated conditions. Some of these crop models are: 1) Decision Support System for Agro technology Transfer (DSSAT) [23] ; 2) Erosion Productivity Impact Calculator (EPIC) [24] ; 3) Crop Environment Resource Synthesis (CERES) model [25] ; 4) The Agricultural Production Systems Simulator (APSIM) [26] ; 5) CROPWAT [27] ; 6) Soil Moisture Accounting Crop-Specific (SMACS) model [28] ; and 7) the CLICROP model [29] .

Of the several crop models, the FAO CROPWAT and SMACS models can easily be adopted with very little field data demand. The latter has also been tested for the same study area for observed climatology of 42 years covering the period 1961-2002, under no-climate change conditions [28] . FAO's CROPWAT is a monthly crop model which considers daily rainfalls applied on selected days of each decade. SMACS is a daily-moisture accounting model adaptable for different crops, and externally coupled with daily rainfall and temperature generation model. Another advantage of the SMACS model is that it easily enables/allows external coupling with GCM outputs with possible disaggregation to daily values using weather generators.

Beyond impact assessment, the SMACS model can be used as a decision support system as it can also be used to calculate water balances such as actual crop ET, excess surface runoff and actual soil moisture besides yield and crop stress indicators. It is therefore suitable to evaluate and investigate the potential of and promote rainwater harvesting and conservation agriculture practices as potential adaptation measures. The water requirement of the crop at a given time of the growing season is calculated by multiplying the reference (Potential) evapotranspiration with a crop coefficient, whose values are published by FAO [27] . SMACS model is considered due to easy adaptability to simulate soil moisture balances, crop water demand and stress for current climatology as well as post climate change conditions.

In the SMACS model, all precipitation in excess of surface runoff (computed using the SCS curve number model [30] , is assumed to infiltrate the soil surface. Soil water flow was modelled following the approach taken in the CERES models [25] , in which water is immediately transferred downward in the soil profile if the amount of water entering the layer exceeds the layer's saturated water content. Water will then continue to drain from a layer until a drained water upper limit or field capacity is reached. Further water can be removed from the soil only through evaporation and transpiration. Soil evaporation is simulated by assuming a limiting water content to which soil evaporation can dry the soil, and that evaporative potential declines continuously as a function of soil water depth.

The ratio of actual transpiration (soil water uptake) to potential transpiration is used as the environmental indicator of water stress and yield potential [31] -[35] . In the crop models considered in the study (maize, sorghum and sunflower), water stress is assumed in the crops to affect growth by limiting photosynthesis in direct proportion to the ratio of actual to potential transpiration. The common approach for estimating crop yield reduction is based on FAO experience [36] , which addressed the relationship between crop yield and water use by proposing a simple equation where relative yield reduction is linearly related to the corresponding relative reduction in evapotranspiration (ET) [36] .

Future climate changes, as well as differences in climates from one location to another, may involve changes in climatic variability as well as changes in the means. In this study, a synthetic weather generator is used to systematically change the within-year variability of temperature and precipitation (and therefore also the interannual variability), without altering long-term mean values. For precipitation, both the magnitude and the qualitative nature of the variability can get manipulated. The synthetic daily weather series serve as input to the three crop simulation models.

3. STUDY AREA

The study area is commonly known as the Pandamatenga Plains. It is located on the north-eastern edge of Botswana and is contiguous with the borders of Namibia, Zambia and Zimbabwe in the southeast. It also meets the confluence of the Chobe River with the Zambezi River. The area also extends with similar agro ecological settings in Zimbabwe (Figure 1), representing a number of farming communities in a number of districts.

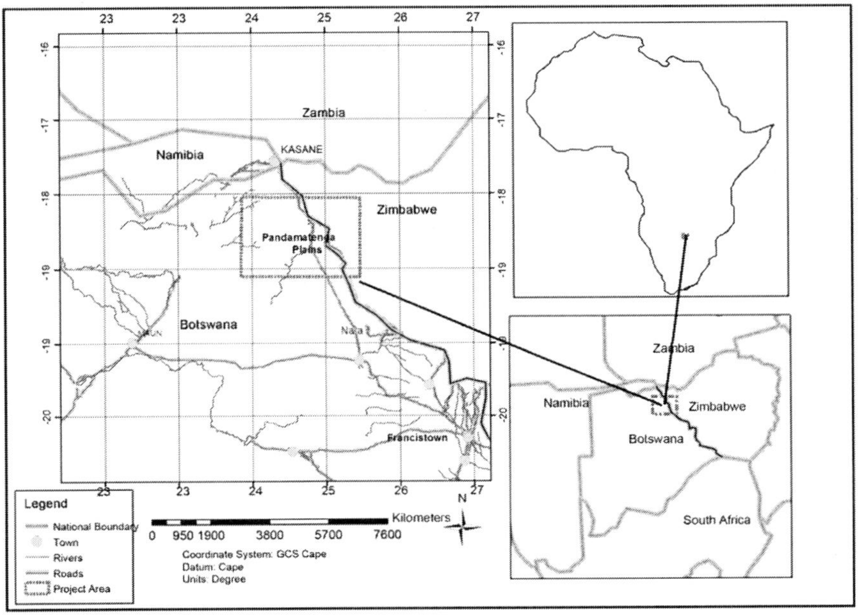

Figure 1. Location map of the study area.

Due to its transboundary linkage and since it exhibits similar agro-ecological characteristics extending and covering Southern Zambia, eastern Namibia, western Zimbabwe and north-eastern Botswana, the site was considered as an ideal site for assessing the climate change impacts of rainfed farming systems. Furthermore, the study site is a vast agricultural area where rainfed agriculture is currently practiced.

The area is characterized by the presence of heavy clay soils. It is understood that during the rainy season they are liable to become sticky, waterlogged and poorly aerated, and they become hard during dry periods. They are reasonably fertile and are capable of retaining both water and nutrients [37] . Successful cultivation and good management in some areas have shown that these soils are able to make a significant contribution to food production. Because of low drainage, harvesting rainwater excesses can be practiced in such soil conditions.

The climate of the project area, like most of the north and south-eastern Botswana, is sub-tropical and semi-arid. The climate records at Kasane and Pandamatenga compiled by the Department of Metrological Services (DMS) have been available for the study. The monthly air temperature variation in the project area computed from 1971-2000 recorded data indicates a mean daily minimum and maximum temperature of 15°C to 29°C, respectively.

The highest temperatures are prevalent during November to March of the year while June and July experience the lowest temperatures.

Figure 2 shows the annual temperature variations and general trend in the daily maximum and minimum temperature data of Francistown, a nearby climatic station for the period 1971 to 2000. Wind direction is normally easterly, north-easterly and south-easterly. The incidence of higher winds is greater around September, October and November with average speeds of above 180km/day. Mean annual rainfall in general varies, with most rain falling during the summer months. Rainfall is characterized by short periods of heavy rain, which cause flush flooding. The men annual rainfall in the study area is generally about 550 mm.

4. MODEL DESIGN AND DATA

4.1. State-of-the-Art in Agricultural Impacts of Climate Change

Assessments of climate change impacts are especially challenging because they are subject to considerable uncertainties of climate predictions and the feedback mechanisms. Several studies highlighted the importance of precipitation, temperatures, soil moisture, and atmospheric CO_2concentrations in crop-soil-atmospheric interactions [13] [38] [39] . These components are projected to change significantly in the coming decades [2] . The knowledge gained in such experimental studies can be formalized in models, helping to structure the complex interactions, which can be purely conceptual, or quantitative [40] .

One such approach is to apply crop models with simulation results of atmospheric general circulation models (GCMs). In the this study, MAGICC/SCENGEN climate predictions were adopted to study the regional and local climate and also analyse the wider variations among various GCM predictions embedded in MAGICC/ SCENGEN [41] and to consider the crop yield sensitivity by the various SRES and GCM scenarios. It was applied in the context of the local climate conditions of the Pandamatenga plains located in northern Botswana considering a square grid with a spatial resolution of 2.5°.

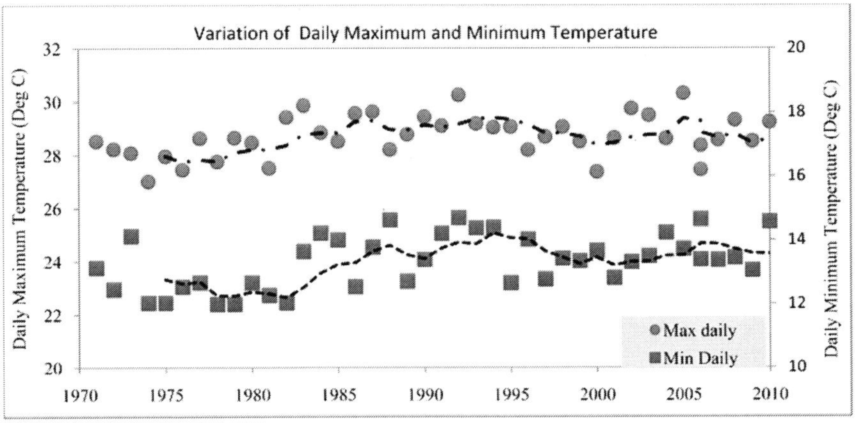

Figure 2. Annual variation of daily maximum and minimum temperature (1971- 2000) at Francistown station. The dashed lines indicate the mean values of 5 years temperature values.

Wider variations among GCM predictions are generally common among climate predictions obtained from the various researches conducted by different organizations and researchers due to possible different in model assumptions, the mathematical model boundaries, climate forcing, etc. However, GCMs provide the most plausible regional climate change scenarios.

It was attempted to determine the climate change scenarios over the study area from various GCM simulations. Figure 3 shows changes in temperature in °C and change in precipitation in percentage form as given by the model for a 2.5° square grid located with center at Latitude 21.25°E, Longitude 28.75°S around the center of the catchments based on the SRES scenarios of A1B-A1M, which is the illustrative scenario adopted [41] .

The coordinates a square area covering the study area were identified and the corresponding model output of simulated monthly climate scenarios were extracted for a set GCM and SRES scenarios. Figure 3 illustrate the variations in the predicted changes of precipitation and GCMS simulated by the climate model. Of which three scenarios are summarized in Table 1 simulated changes and temperatures for the 2050s at a 2.5° square pixel centered at the Pilot area (around the center of the Pandamatenga Plains) from three selected GCM scenarios that were adopted as they typically present dry, moderate and wet conditions. Due to wide mix of the models and their prediction, the study was based on the following three scenarios. These are summarized in Table 1 and described as:

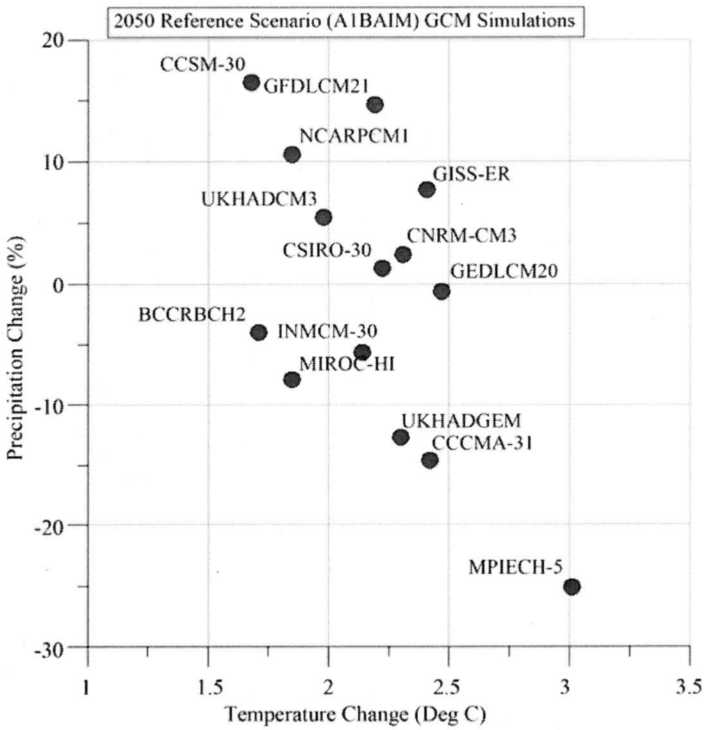

Figure 3. Comparison of GCM projections at a 2.5° square GCM grid centered at Pandamatenga/Mid-Zambezi Basin (between 17.5°S - 20°S and 25°E - 27.5°E).

Table 1. Projected changes in temperature and precipitation during the baseline period in the study area.

GCM	Temperature change (°C)	Precipitation change (%)	Remark
CCCMA-31	2.42	−14.60	Warm/Dry
GFDLCM21	2.19	14.70	Warm/Wet
UKHADCM3	1.98	5.50	Moderate

- Warm and wet conditions (Scenario 1). This is a typical condition represented by GFDLCM21.
- Warm and dry conditions (Scenario 2). This is a typical condition represented by CCCMA-31.

- Moderate conditions (Scenario 3). This is a typical condition represented by UKHADCM3.

4.2. Yield Reduction and Climate Change Impacts on Rainfed Agriculture

The impact of agricultural drought on crop production can be largely expressed by yield reduction. For this, yield reduction due to water deficiency was computed within SMACS model software. Yield reduction was calculated from water balance output combined with an empirical formula developed which recommends a formula for percentage yield reduction,

$$Y_r = K_y\left(1 - E_a/E_p\right)\times 100 \tag{1}$$

where E_a is the actual evapotranspiration, E_p is the total water requirement without water stress, and K_y is a crop dependent stress indicator, which is known as the yield response factor.

Based on the analysis of an extensive amount of the available literature on crop-yield and water relationships and deficit irrigation, K_y values of 1.25 for Maize, 0.9 for Sorghum and 0.95 for Sunflower were adopted from [36] .

4.3. Risk, Resilience and Reliability of Rainfed Agriculture

To understand climate change and related impacts on rainfed agriculture, the investigation involved the study of sustainability of rainfed systems through soil moisture modelling and risk analysis of various crops using three indicators: risk, resilience, vulnerability indices. These factors are used as quantitative measures for assessment of soil moisture reliability and sustainability of rainfed systems, details of which are presented in [42] .

There is clear evidence showing that besides moisture stress and soil fertility constraints often constitute the primary limiting factor to crop growth also in drylands [43] . Soil water stress can be assessed from crop-soil- water simulations of long-term meteorological variables especially using crop specific water accounting (SMACS) model [28] . More recent trend has been the assessment of degree of availability of soil moisture which is an integrating variable for the underlying hydroclimatic and agronomic factors of rainfed agricultural areas. The risk level for sustenance of rainfed systems can be determined as a probability at which the soil moisture (S) drops below a given moisture threshold ($S_{WP} + S_a \cdot p$) during the crop's length of the growing period (LGP), $S_a = S_{FC}$? S_{WP}, which is the readily available moisture content i.e. soil moisture content at field capacity minus that at wilting point, and p is the available soil moisture factor which is expressed as $p = (S ? S_{WP})/S_a$. The risk factor for the entire growing period (simulation period) of crops under rainfed conditions can be calculated as defined as:

$$r[\%] = \frac{n}{T} \times 100\% \tag{2}$$

where n is the number of days in which actual soil moisture S, drops below the critical soil moisture threshold ($S_{WP} + S_a \cdot p$) during the total number of days (T) of the entire cropping period over all the entire years of simulation. In a period of years of analysis considered, T in days becomes the product of the number of the simulated years and the length of the growing period (LGP) in days, which is assumed to be constant for each year. The risk factor here is the same as the probability of failure which refers to the proportion of days to the total number of cropping days in the entire number of years simulated, within which the simulated soil moisture drops below the amount which is set at p times the readily available soil moisture content. If other agricultural conditions such as land management and nutrient availability are not altered, then this risk factor integrates the prevailing hydroclimatology, soil moisture availability and crop-soil-water conditions to assess sustainability of the crops cultivated under rainfed conditions.

Reliability and resilience have been widely used in water resources management to express the state of reservoir system [43] , and these concepts have been adopted in this study to explain the state of soil moisture its availability for sustenance of crop growth in the SMACS model. Reliability is a measure of frequency or probability that a system is in a satisfactory state meeting a given criterion. Resiliency generally indicates a measure of how quickly a system recovers from failure once failure has occurred. The computational scheme for these indices in this study is almost similar to that of water management applications used in [44] -[46] , which are specifically tailored for analysing soil moisture stress and the associated risks and sustainability of rainfed systems.

Defining a criterion (C) as the minimum required soil moisture from a rainfed agricultural system, the daily soil moisture depth (S_t) can be classified as a satisfactory state (A) or a failure state (B), i.e.,

$$\text{If } S_t \geq C \quad \begin{cases} \text{then} & S_t \in A \quad \text{and } Z_t = 1 \\ \text{else} & S_t \in B \quad \text{and } Z_t = 0 \end{cases} \tag{3}$$

where Z_t is a generic indicator variable. The daily available moisture content (S_t) simulated in the SMACS model was evaluated against criterion and, thus, system failure occurs when soil moisture is below the criterion at any given day i.e. satisfactory state and failure if otherwise. The criteria C used is $C = S_{WP} + S_a \cdot p$.

5. RESULTS AND ANALYSIS

5.1. Coupling Disaggregated Climate Data with Crop Model

In order to simulate the impact of changed climate simulated by GCMs on the soil water balances, a simplified procedure was used through external coupling of the interaction between monthly weather generation models, the climate models and the crop-water balance model (SMACS). Soil retention parameters (permanent wilting point, field capacity and available moisture content), as well as crop growth factors and crop calendars, for the crops considered in the study were used as input to the SMACS model. The SMACS model simulates soil moisture, actual evapotranspiration, runoff, and indices such as crop yield, risk, reliability, resilience based on simulations at daily time steps from disaggregated inputs of precipitation and computed potential evapotranspiration. These indices in impact assessment of climate change on agriculture resources.

A weathergenerator was used as a procedure of incorporating natural variability in the analysis involves employing an ensemble of scenarios of climate variables that are formulated by stochastic hydrologic methods [47] -[50] . Weather generators are statistical methods that base on observed historical records of climate variable to generate long-term series of synthetic climatic data by preserving statistical properties of the observed data. In these models, the variance across an ensemble at a time step represents the temporal variability of the hydrologic variables and similar approaches have been applied in hydrological impact studies of climate change [48] -[50] .

The monthly changes in temperature and precipitation obtained from SCENGEN as climate change scenarios of 2100s were disaggregated to reconstruct the time series of monthly temperature and precipitation from the end of the baseline period-extending from 2001 up to 2100. They were then made inputs to the crop water balance model, SMACS [28] . Time series of synthetic daily weather values exhibiting the same means as, but different variability from, the base climates were constructed using the approach presented in [51] [52] . The monthly mean values for the baseline period are used to adjust the mean values of the changed climate values.

Under increase of GHG concentrations, GCMs can predict long-term climatic changes with some degree of certainty. Thus, the perturbed/changed climate provided by GCMs for month j is treated as D(j). Thus, in a changing climate, the future climatic mean S^*(j) say for 2100 can be given by:

$$S^*(j) = (1 + \Delta(j))\, S(j) \tag{4}$$

where S(j) represents the baseline climate for month j, S^*(j) is the future mean value and D(j) is the change relative to the present value of the climatic variable (fraction). The monthly change D(j) is considered as the GCM output which is available for grids covering the globe for various periods say 2100s, N years from the current climatic period.

In order to incorporate the year to year variability of these changes, the mean value of the changes is assumed to increase proportionally up to the number of years of simulation—for example—N years up to the 2100s for which GCM changes are considered. The changes in each month j which are linearly varying for each year i and designated by D(i, j), the corresponding monthly value of the future climate S^*(i, j) is calculated as:

$$S^*(i,j) = \left[1 + \Delta(i,j)\frac{i}{N}\right] S(j) \quad j = 1,12 \tag{5}$$

Equation (5) can be used to generate daily time series for N years with linearly increasing/decreasing monthly means. In order to assess soil-water balances, the possible transient responses with respect to these transient climatic variables should be accommodated. This approach is an ideal condition as it is a linear assumption in which the monthly climatic variable is assumed to increase or decrease in each year i constantly and consistently for N years (Equation (5)).

Therefore, the monthly climate averages/values in the case of the precipitation or adjusted harmonic mean values for the temperature, are used in the weather generator model to obtain modified mean values. Then the variability of historical values is used to determine the daily time series data. For precipitation data, the Markov chain model transition matrices were used to determine the dry-wet day sequences and for the precipitation depths for wet days were determined from probability distribution model, a calibrated Weibull model developed for the region [53] . Details of this procedure used in daily rainfall and temperature weather generation is presented in literature [53] [54] . An in house computer program using FORTRAN was written to develop the weather generator and couple it with the SMACS model.

5.2. Yield Reduction and Climate Change Impacts on Rainfed Agriculture

Yield reductions in common grains grown in the study area have been simulated. These common crops are maize, sorghum and sunflower, which are cultivated widely by farming communities in the study area. The crop coefficients and soil coefficients assuming vertisols throughout two lithological layers have been assumed for the purpose of this study.

A summary of yield reductions for time slices of 2001-2050 simulated by the three GCMs in comparison with the baseline scenario are presented in Figure 4. It appears that the yield reductions of all the cultivars considered, maize, sorghum and sunflower, in general are predicted by CCCMA-31 scenario compared to the other scenarios i.e. the UKHDCM3 and the GFDLCM21 scenarios.

In almost 50 percent of the years considered during simulation period of 2001-2050, the yield reductions are above 50 percent (Figure 4), and in general sorghum appears to exhibit lower reduction compared to maize followed by sunflower.

5.3. Risk and Climate Sensitivity

The study attempted to establish and understand the degree of susceptibility of crop to failures in terms of soil moisture availability to meet the evapo-transpirative demands of crops studied. Using the SMACS model, the following indices, namely risk, resilience and reliability were used to investigate the available moisture failure rates in rainfed conditions, where rainfed agriculture is practiced in the study site. In this study, daily simulations from climatic data of 1971-2000 were made for the study area, considering commonly grown crops: maize, sunflower, and sorghum.

Risk, reliability and resilience of maize simulated from the entire 1971-2000 daily record and taking the current sowing dates and actual cropping length, at available soil moisture factor, p = 30% for Pandamatenga Plains is shown in Table 2. The corresponding indices for the post climate change scenarios are also computed. Percentage pie charts for these sustainability indices with available soil moisture factor, p = 30% are presented for the baseline period as well as the three climate changes scenarios in Figure 5.

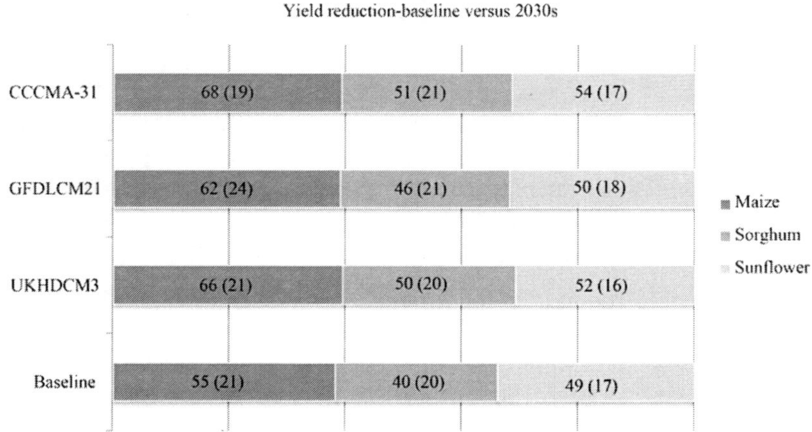

Figure 4. Projected mean percentage yield reductions under baseline and three GCM scenarios. The standard deviations are given in parentheses.

The results of analysis of risk, resilience and reliability of rainfed agriculture can be used as indicators of vulnerability and the extent to which mitigation measures can be taken, such as by adjusting sowing dates for the perceived changes in the climate and weather system. In similar studies [31] observed that maize yields declines could be offset by shifting of sowing dekads/dates as a means of developing adaptation decisions by keeping maize yields as high as possible.

5.4. Mitigating Risk and Potential for Excess Rainwater Harvesting

Another mitigation potential to be explored is harvesting of excess runoff during days where precipitation ex- cesses are prevalent above the infiltration capacity and moisture holding capacity of soils. The potential for har- vesting excess storm runoff in the rainfed agricultural fields of the study site is investigated. Results are summarized in Table 3. The corresponding runoff potential and percentile duration curves are presented for the baseline period as well as the three climate changes scenarios in Figure 6.

Table 2. Comparison of sustainability indicators of risk, reliability and resilience for the baseline climatology.

Indicator	1971-2000 (Baseline Period)		
	Maize	Sunflower	Sorghum
Risk (%)	0.94	4.87	0.85
Reliability (%)	99.1	95.1	99.1
Resilience (%)	38.5	37.1	32.9

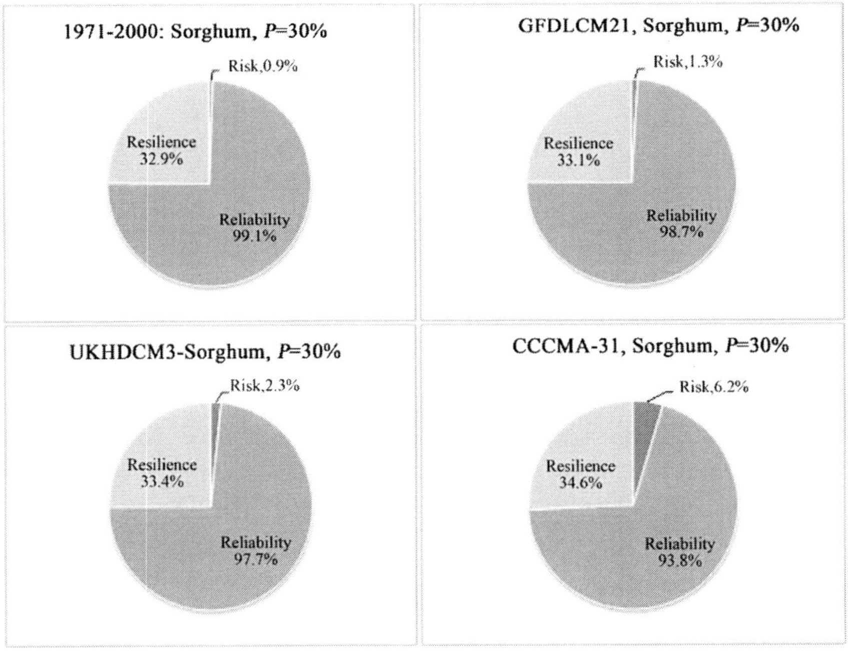

Figure 5. Risk, reliability and resilience of sorghum under rainfed conditions with available soil moisture factor, p = 30%.

It can be noted that the excess runoff harvesting potential in only sorghum cropping system is relatively higher than the maize or sunflower cases. However, in the mixed farming case, relatively better harvesting potential can be harnessed reaching as far as an average of 22 $m^3 \cdot ha^{-1} \cdot d^{-1}$ say in more than 15% of the days of the combined cropping period in days.

5.5. Soil Moisture Enhancement

Relevant mitigation practices might entail enhancing the available soil moisture availability to full utilization by crops. The potential to be explored is the harvesting of excess runoff during days where precipitation excesses are prevalent above the infiltration capacity and soil moisture holding capacity of soils.

The potential for enhancement of soil moisture in the rainfed agricultural fields of the study site is investigated. Results are summarized in Table 4. The corresponding soil moisture content for the post climate change scenarios, as percentile duration curves are presented for the baseline period as well as the three climate changes scenarios is presented in Figure 7.

6. DISCUSSION

The yield response to changes in precipitation variability can depend on the qualitative nature of the variability and it is used to understand the impacts of climate changes. The results also suggest that the current atmospheric general circulation models (GCMs) yield inconsistent results concerning changes in mean precipitation at particular locations, and the sign and magnitude of any change in the interannual variability is even less clear. However, the present results suggest that the qualitative nature of these changes may be a significant determinant of important agricultural impacts and which can help to devise mitigation mechanisms and develop adaptation plans.

The study also suggested that there are opportunities to improve agriculture water productivity and implement integrated soil-water management strategies in rain-fed farming systems. This is so important especially for the development of mitigation measures and adaptation measures in similar farming systems. For instance where livelihood improvement, food security and productivity increment is sought in rainfed agriculture-practicing areas of the region in general and in the study area considered, the Pandamatenga plains, in particular. These will also have a direct linkage in the vast lower rainfall and rainfed plains extending to southern Zambia, northern Namibia and northwestern Zimbabwe.

Table 3. Direct runoff in mm as percentage of days exceeded that can be harvested for the baseline climatology.

Indicator: % No of days exceeding	1971-2000				1971-2000			
	Baseline [mm]				Baseline [$m^3 \cdot ha^{-1} \cdot d^{-1}$]			
	Maize	Sorghum	Sunflower	Average	Maize	Sorghum	Sunflower	Average
10%	2.79	4.39	2.95	3.56	28	44	29	36
15%	1.70	2.32	2.03	2.25	17	23	20	22
20%	0.69	1.31	1.04	1.29	7	13	10	13

Table 4. Soil moisture in mm as percentage of days exceeded per cropping season for the baseline climatology.

Indicator: % No of days exceeding	1971-2000 (Baseline period)		
	Maize	Sunflower	Sorghum
70%	227	42	307
80%	219	38	298
90%	211	34	289

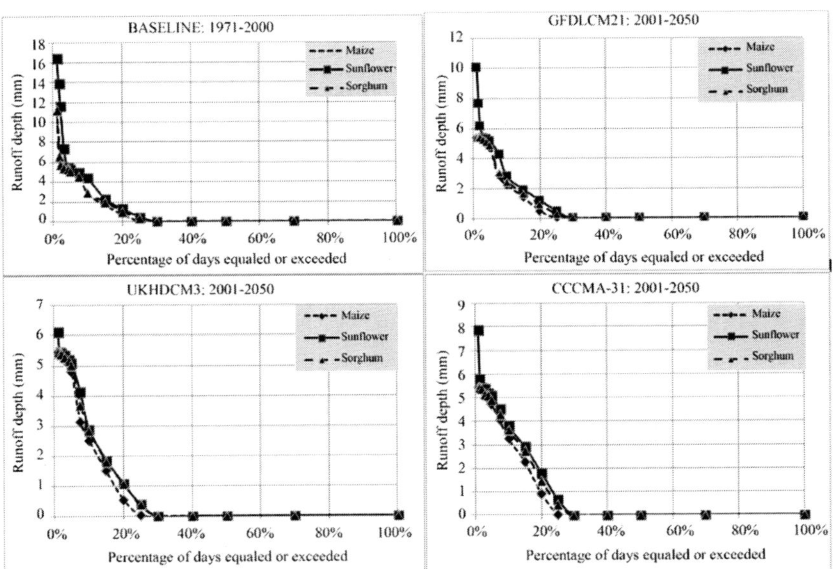

Figure 6. Rainwater harvesting potential and runoff depth under rainfed conditions.

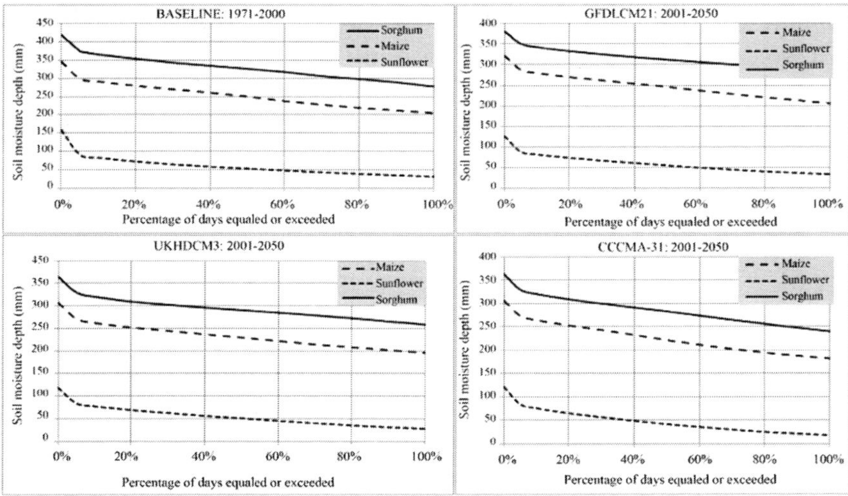

Figure 7. Soil moisture availability under rainfed conditions.

Integrated soil-water management strategies could include: 1) strengthening of conservation tillage to improve soil-water productivity and land fertility; 2) improving near real-time weather forecasting and advisory services to support farmers to adjust cropping pattern and planting dates of cultivars; 3) encouragement and incentive measures for effectively use available rainwater and harvest excess runoff; and 4) strengthening contribution to awareness and public policy processes in an effort demonstrate the potential benefits of developing adaptation strategies in terms of the socio-economic, economic diversification benefits and ultimately improvement of food security of a nation or the region at large.

In a recent survey conducted in the study area including farming communities in three districts in southern Zambia (Chogwe, Chisamda and Chipembi Districts), it was found that a mix of various Conservation Farming and Climate Adaptation Practices [55] . The farmers have developed local and indigenous knowledge systems and they also appreciate how improvements in conservation farming, water harvesting, water access and climate resiliency could help them protect their deteriorating agricultural and livestock yields in their community. These practices included: 1) Dry-season land preparation using minimum tillage; 2) Crop residue retention; 3) Seeding and input application in fixed planting stations; 4) Nitrogen-fixing crop rotation; 5) Infield water conservation; 6) Crop-livestock system for soil fertility; 7) Crop-livestock system for income generation; and 8) Mixed farming of maize with soybeans, groundnuts, etc. These practices are there for possible adoption by other communities in similar agro-ecological conditions, if they are given the means and support them in agricultural productivity and household food security enhancement.

7. CONCLUSIONS

As the regional impacts of climate change are wider than the local situation of this study area, this then highlights the need for outscaling of the mitigation measures. This outscaling becomes imperative to joint regional efforts of climate change mitigation and adaptation efforts especially in southern Africa [55] . These mitigation and adaptation mechanisms highlighted above can be considered for outscaling with the domain of Climate Smart Agriculture [56] , in which the modeling approach and tool can be harnessed as a decision support tool. This will help understand and assess the various rainfed farming systems and interventions employed to improve productivity and enhance resilience of smallholder farmers against impacts of climate change and variability. In turn, the interventions can be used to assess the contribution of and effectiveness of achieving maximum net benefits of mitigation, adaptation and emission reduction.

A typical modelling framework recommended in a study that attempts to understand climate change, agriculture and adaptation in a farming ecosystem is illustrated in Figure 8. For this, more computer simulations and research needs to be undertaken to address and understand the linkages of climate change, agriculture and adaptation including the complexity of feedback mechanisms that need to be incorporated across various spatial and temporal scales.

Sustainable rural development research has taken different approaches to the integration of management technologies in the search for a more holistic agricultural system, examples being Integrated Natural Resources Management (INRM), Integrated Water Resources Management (IWRM), Integrated Soil and Nutrient (Soil Fertility) Management (ISFM), Integrated Crop and Livestock Management (ICLM), Integrated Pest Management (IPM). These concepts are very common practices for farmers in southern Africa, who traditionally have implemented various mixed farming systems and farm practices appropriate to the local ecology. In order to enhance agricultural sustainability, a number of good localized examples of INRM exist [57] , such as the farming systems research approach instituted by FAO [58] . This approach basically aims at simultaneously improving livelihoods, agro-ecosystem resilience, agricultural productivity and the provision of environmental services by augmenting social, physical, human, natural and financial capital [59] .

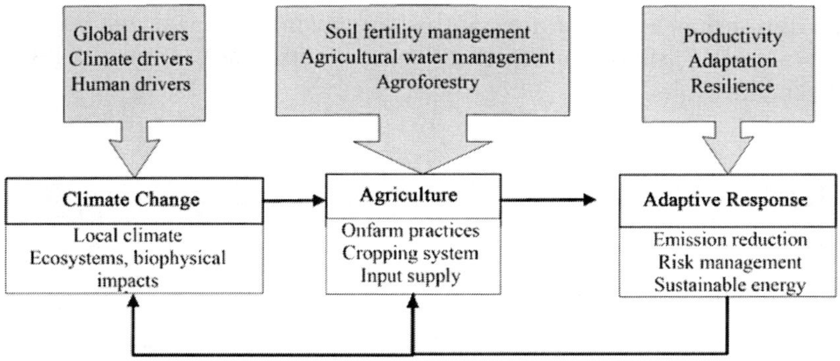

Figure 8. A typical modelling framework to understand climate change, agriculture and adaptation in a farming ecosystem.

It is also essential to note that a range of fundamental natural resources, including land, water, air, biological diversity including forests, fish, etc., provide the indispensable base for agricultural production system and sustenance of agricultural ecosystems. Due to population growth and expansion of agricultural activities, the physical and functional availability of natural resources is diminishing [59] . The latter can be attributed to loss of biodiversity, deforestation, loss of soil health, and water shortage. Given the multifunctional nature of agriculture, it is critical to consider linkages across ecosystems in which agricultural systems are embedded, as these have important implications for the resilience or vulnerability of these systems. These linkages between natural resource use and the social and physical environment across space and time are important issues for scaling out and the dissemination of agricultural knowledge, science and technology, with significant implications for sustainable development and the mitigation of adverse impacts [59] .

ACKNOWLEDGEMENTS

This research was supported by CCARDESA's CCARPASA Project, jointly implemented by University of Botswana and Continental Consultants based in Gaborone, Botswana. The authors also appreciate the support of the University of Botswana for funding the daily weather generation model development under Research Project Grant No: R025, "Development of Daily Precipitation Model for Botswana". The authors also acknowledge the Department of Metrological Services for providing daily rainfall time series data used in the weather generation modeling, and the Ministry of Agriculture for proving soil and related information.

CITE THIS PAPER

Berhanu F. Alemaw, Timothy Simalenga (2015) Climate Change Impacts and Adaptation in Rainfed Farming Systems: A Modeling Framework for Scaling-Out Climate Smart Agriculture in Sub-Saharan Africa. *American Journal of Climate Change*,**04**,313-329.

REFERENCES

1. SADC (2011) Climate Change Adaptation in SADC: A Strategy for the Water Sector. SADC Secretariat, Gaborone, Botswana, 40 p.
2. Meehl, G.A., Stocker, T.F., Collins, W.D., Friedlingstein, P., Gaye, A.T., Gregory, J.M., Kitoh, A., Knutti, R., Murphy, J.M., Noda, A., Raper, S.C.B., Watterson, I.G., Weaver, A.J. and Zhao, Z.-C. (2007) Global Climate Projections. In: Solomon, S., Qin, D., Manning, M., Chen, Z., Marquis, M., Averyt, K.B., Tignor, M. and Miller, H.L., Eds., Climate Change 2007: The Physical Science Basis, Contribution of: Working Group I to the Fourth Assessment Report of the Intergovernmental Panel on Climate Change, Cambridge University Press, Cambridge, 747-846.
3. Davidson, O., Halsne, S.K., Huq, S., Kok, M., Metz, B., Sokona, Y. and Verhagen, J. (2003) The Development and Climate Nexus: The Case of Sub-Saharan Africa. Climate Policy, 3, S97-S113.
4. Reich, P.F., Numben, S.T., Almaraz, R. and Eswaran, H. (2001) Land Resources Stress and Desertification in Africa. Agro-Science, 2, 1-10.
5. Richard, Y., Fauchereau, N., Poccard, I., Rouault, M. and Trzaska, S. (2001) 20th Century Droughts in Southern Africa: Spatial and Temporal Variability, Teleconnections with Oceanic and Atmospheric Conditions. International Journal of Climatology, 21, 873-885.
6. New, M., Hewitson, B., Stephenson, D.B., Tsiga, A., Kruger, A., Manhique, A., Gomez, B., Coelho, C.A.S., Masisi, D.N. and Kululanga, E. (2006) Evidence of Trends in Daily Climate Extremes over Southern and West Africa. Journal of Geophysical Research, 111, D14102.
7. Milgroom, J. and Giller, E. (2013) Courting the Rain: Rethinking Seasonality and Adaptation to Recurrent Drought in Semi-Arid Southern Africa. Agricultural Systems, 118, 91-104.
8. Ngongondo, C.S., Li, L., Gong, L.B., Xu, C.-Y. and Alemaw, B.F. (2013) Flood Frequency under Changing Climate in the Upper Kafue River Basin, Southern Africa: A Large Scale Hydrological Model Application. Journal of Stochastic Environmental Research and Risk Assessment, 27, 1883-1898.
9. Niang, I., Ruppel, O.C., Abdrabo, M.A., Essel, A., Lennard, C., Padgham, J. and Urquhart, P. (2014) Africa. In: Barros, V.R., Field, C.B., Dokken, D.J., Mastrandrea, M.D., Mach, K.J., Bilir, T.E., Chatterjee, M., Ebi, K.L., Estrada, Y.O., Genova, R.C., Girma, B., Kissel, E.S., Levy, A.N., MacCracken, S., Mastrandrea, P.R. and White, L.L., Eds., Climate Change 2014: Impacts, Adaptation, and Vulnerability. Part B: Regional Aspects. Contribution of Working Group II to the Fifth Assessment Report of the Intergovernmental Panel on Climate Change, Cambridge University Press, Cambridge, 1199-1265.
10. Thomas, D.S.G., Twyman, C., Osbahr, H. and Hewitson, B. (2007) Adaptation to Climate Change and Variability: Farmer Responses to Intraseasonal Precipitation Trends in South Africa. Climatic Change, 83, 301-322.

11. Stringer, L.C., Dyer, J.C., Reed, M.S., Dougill, A.J., Twyman, C. and Mkwambasi, D. (2009) Adaptations to Climate Change, Drought and Desertification: Local Insights to Enhance Policy in Southern Africa. Environmental Science and Policy, 12, 748-765.
12. Alemaw, B.F., Onema, J.M.K. and Love, D. (2013) Regional Drought Severity Assessment at a Basin Scale in the Limpopo Drainage System. Journal of Water Resource and Protection, 5, 1110-1116.
13. Ainsworth, E.A. and Long, S.P. (2005) What Have We Learned from 15 Years of Free-Air CO2 Enrichment (FACE)? A Meta-Analytic Review of the Responses of Photosynthesis, Canopy Properties and Plant Production to Rising CO2. New Phytologist, 165, 351-371.
14. Muller, C., Cramer, W., Hare, W.L. and Lotze-Campen, H. (2011) Climate Change Risks for African Agriculture. Proceedings of the National Academy of Sciences of the United States of America, 108, 4313-4315.
15. Dixon, R.K., Smith, J. and Guill, S. (2003) Life on the Edge: Vulnerability and Adaptation of African Ecosystems to Global Climate Change. Mitigation and Adaptation Strategies for Global Change, 8, 93-113.
16. Lobell, D.B., Burke, M.B., Tebaldi, C., Mastrandrea, M.D., Falcon, W.P. and Naylor, R.L. (2008) Prioritizing Climate Change Adaptation Needs for Food Security in 2030. Science, 319, 607-610.
17. Fischer, G., Shah, M., Tubiello, F.N. and Van Velhuizen, H. (2005) Socio-Economic and Climate Change Impacts on Agriculture: An Integrated Assessment, 1990-2080. Philosophical Transactions of the Royal Society B, 360, 2067-2083.
18. Challinor, A., Wheeler, T., Garforth, C., Crauford, P. and Kassam, A. (2007) Assessing the Vulnerability of Food Crop Systems in Africa to Climate Change. Climatic Change, 83, 381-399.
19. World Bank (2009) Agriculture and Rural Sector South Asia: Shared Views on Development and Climate Change. World Bank, Washington DC, 97-107.
20. Connolly-Boutin, L. and Smit, B. (2015) Climate Change, Food Security, and Livelihoods in Sub-Saharan Africa. Regional Environmental Change.
21. Makondo, C.C., Chola, K. and Moonga, B. (2014 Climate Change Adaptation and Vulnerability: A Case of Rain Dependent Small-Holder Farmers in Selected Districts in Zambia. American Journal of Climate Change, 3, 388-403.
22. Nakicenovic, N. and Swart, R. (2000) Special Report on Emissions Scenarios: A Special Report of Working Group III of the Intergovernmental Panel on Climate Change. Cambridge University Press, Cambridge, 599 p.
23. Jones, J.W. (1986) Decision Support System for Agrotechnology Transfer. Agrotechnology Transfer, 2, 1-5.
24. Williams, J.R. (1995) The EPIC Model. In: Singh, V.P., Ed., Computer Models of Watershed Hydrology, Water Resources Publications, Highlands Ranch, 909-1000.
25. Jones, C.A. and Kiniry, J.R. (1986) CERES-Maize: A Simulation Model of Maize Growth and Development. Texas A&M University Press, College Station, 194 p.
26. McCown, R.L., Hammer, G.L., Hargreaves, J.N.G., Holzworth, D. and Huth, N.I. (1995) APSIM—An Agricultural Production System Simulation Model for Operational Research. Mathematics and Computers in Simulation, 39, 225-231.
27. FAO (1992) Cropwat—A Computer Program for Irrigation Planning and Management. FAO Irrigation and Drainage Publications, No. 46, Rome.
28. Alemaw, B.F., Chaoka, T.R. and Totolo, O. (2006) Investigation of Sustainability of Rain-Fed Agriculture through Soil Moisture Modeling in the Pandamatenga Plains of Botswana. Physics and Chemistry of the Earth, 31, 960-996.
29. Fant, C. (2009) CliCrop: A One-Dimensional Model to Calculate Water Stress on Crops. University of Colorado at Boulder, UMI Dissertations Publishing, Ann Arbor.
30. SCS (1985) National Engineering Handbook, Section 4: Hydrology, Soil Conservation Service. USDA, Washington DC.

31. Crespo, O., Hachigonta, S. and Tadross, M. (2011) Sensitivity of southern African Maize Yields to the Definition of Sowing Dekad in a Changing Climate. Climate Change, 106, 267-283.
32. De Wit, C.T. (1958) Transpiration and Crop Yields. Agricultural Research Reports 64.6 Pudoc, Wageningen, 88 p.
33. Hanks, R.J. (1983) Yield and Water-Use Relationships: An Overview. In: Taylor, H.M., Jordan, W.R. and Sinclair, T.R., Eds., Limitations to Efficient Water Use in Crop Production, American Society of Agronomy, Madison, 393-411.
34. Nonhebel, S. (1994) The Effects of Use of Average Instead of Daily Weather Data in Crop Growth Simulation Models. Agricultural Systems, 44, 377-396.
35. Simane, B., van Keulen, H. and Stol, W. (1994) Application of a Crop Growth Model (SUCROS-87) to Assess the Effect of Moisture Stress on Yield Potential of Durum Wheat in Ethiopia. Agricultural Systems, 44, 337-352.
36. Doorenbos, J. and Kassam, A.H. (1979) Yield Response to Water. FAO Irrigation and Drainage Paper No. 33, Rome.
37. Ministry of Agriculture of Botswana (1990) Pandamatenga Development Study. Main Report—Annex B-Hydrology and Main Drainage. Arup-Atkins. Ministry of Agriculture, Gaborone.
38. Iglesias, A., Quiroga, S. and Diz, A. (2011) Looking into the Future of Agriculture in a Changing Climate. European Review of Agricultural Economics, 38, 427-447.
39. Ainsworth, E.A. (2008) Rice Production in a Changing Climate: A Meta-Analysis of Responses to Elevated Carbon Dioxide and Elevated Ozone Concentration. Global Change Biology, 14, 1642-1650.
40. Hillel, D. and Rosenzweig, C. (2010) Handbook of Climate Change and Agroecosystems: Impacts, Adaptation, and Mitigation. Imperial College Press, London, 440 p.
41. Wigley, T.M.L. (2008) MAGICC/SCENGEN 5.3 User Manual. NCAR, Boulder.
42. Alemaw, B.F. (2012) Resilience, Reliability and Risk Analyses of Maize, Sorghum and Sunflower in Rain-Fed Systems Using a Soil Moisture Modeling Approach. Agricultural Sciences, 3, 114-123.
43. Klaij, M.C. and Vachaud, G. (1992) Seasonal Water Balance of a Sandy Soil in Niger Cropped with Pearl Millet, Based on Profile Moisture Measurements. Agricultural Water Management, 21, 313-330.
44. Hashimoto, T., Stedinger, J.R. and Loucks, D.P. (1982) Reliability, Resiliency, and Vulnerability Criteria for Water Resource System Performance Evaluation. Water Resources Research, 18, 14-20.
45. Maier, H.R., Lence, B.J. Tolson, B.A. and Foschi, R.O. (2001) First-Order Reliability Method for Estimating Reliability, Vulnerability, and Resilience. Water Resources Research, 37, 779-790.
46. Fowler, H.J., Kilsby, C.G. and O'Connell, P.E. (2003) Modeling the Impacts of Climatic Change and Variability on the Reliability, Resilience, and Vulnerability of a Water Resource System. Water Resources Research, 39, 1222.
47. Apipattanavis, S., Podestá, G., Rajagopalan, B. and Katz, R.W. (2007) A Semiparametric Multivariate and Multisite Weather Generator. Water Resources Research, 49, 7205-7220.
48. Grantz, K., Rajagopalan, B., Clark, M. and Zagona, E. (2005) A Technique for Incorporating Large-Scale Climate Information in Basin-Scale Ensemble Streamflow Forecasts. Water Resources Research, 41, 1-14.
49. Gebretsadik, Y., Strzepek, K. and Schlosser, C.A. (2014) A Hybrid Approach to Incorporating Climate Change and Variability into Climate Scenario for Impact Assessments. World Institute for Development Economics Research, WIDER Working Paper 2014/112, 1-18.
50. Yates, D., Gangopadhyay, S., Rajagopalan, B. and Strzepek, K. (2003) A Technique for Generating Regional Climate Scenarios Using a Nearest-Neighbor Algorithm. Water Resources Research, 39, 1-15.

51. Riha, S.J., Wilks, D.S. and Simoens, P. (1996) Impact of Temperature and Precipitation Variability on Crop Model Predictions. Climatic Change, 32, 293-311.
52. Wilks, D.S. (1992) Adapting Stochastic Weather Generation Algorithms for Climate Change Studies. Climatic Change, 22, 67-84.
53. Alemaw, B.F. and Chaoka, T.R. (2008) A Parametric Daily Precipitation Model Application in Botswana. Botswana Journal of Technology, 17, 41-47.
54. Alemaw, B.F. and Chaoka, T.R. (2003) A Continental Scale Water Balance Model: A GIS-Approach for Southern Africa. Physics and Chemistry of the Earth, 28, 957-966.
55. CCARDESA (2014) Enhancing Evidence-Based Climate Change Adaptation Research and Policy for Agriculture in Southern Africa, CCARPASA Project, Final Report. CCARDESA Secretariat, Gaborone, 70.
56. FAO (2010) Climate-Smart Agriculture—Policies, Practices and Financing for Food Security, Adaptation and Mitigation. UN Food and Agriculture Organization (FAO), Rome, 41.
57. Palm, C.A., Vosti, S.A., Sanchez, P.A. and Ericksen, P. (2005) Slash-and-Burn Agriculture: The Search for Alternatives. Columbia University Press, New York.
58. Norman, D.W., Worman, F.D., Siebert, J.D. and Modiakgotla, E. (1995) The Farming Systems Approach to Development and Appropriate Technology Generation. Food and Agriculture Organization of the United Nations, Rome. http:// www.fao.org/docrep/v5330e/v5330e00.HTM.
59. McIntyre, B.D., Herren, H.R., Wakhungu, J. and Watson, R.T. (2009) International Assessment of Agricultural Knowledge, Science and Technology for Development (IAASTD): Global Report. Island Press, Washington DC, 590.

CHAPTER 9

Climate Change Impacts on Agriculture in 2050 Under a Range of Plausible Socioeconomic and Emissions Scenarios

Keith Wiebe[1], Hermann Lotze-Campen[2,3], Ronald Sands[4], Andrzej Tabeau[5], Dominique van der Mensbrugghe[6], Anne Biewald[2], Benjamin Bodirsky[2], Shahnila Islam[1], Aikaterini Kavallari[7], Daniel Mason-D'Croz[1]

[1] *International Food Policy Research Institute, 2033 K St NW, Washington, DC 20006, USA*

[2] *Potsdam Institute for Climate Impact Research, Telegraphenberg A 31, 14473 Potsdam, Germany*

[3] *Humboldt University Berlin, Unter den Linden 6, 10099 Berlin, Germany*

4 US Department of Agriculture, Economic Research Service, 1400 Independence Ave. SW, Mail Stop 1800, Washington, DC 20250- 0002, USA

[5] *LEI Wageningen UR, Droevendaalsesteeg 4, 6708 PB Wageningen, The Netherlands*

[6] *Global Trade Analysis Project, Purdue University, 403 W State St, West Lafayette, IN 47907, USA*

[7] *Food and Agriculture Organization of the United Nations, Via delle Terme di Caracalla, 00153 Rome, Italy*

[8] *Institute of Development Studies, University of Sussex, Brighton, East Sussex BN1 9RE, UK*

ABSTRACT

Previous studies have combined climate, crop and economic models to examine the impact of climate change on agricultural production and food security, but results have varied widely due to differences in models, scenarios and input data. Recent work has examined (and narrowed) these differences through systematic model intercomparison using a high-emissions pathway to highlight the differences. This paper extends that analysis to explore a range of plausible socioeconomic scenarios and emission pathways. Results from multiple climate and economic models are combined to examine the global and regional impacts of climate change on agricultural yields, area, production, consumption, prices and trade for coarse grains, rice, wheat, oilseeds and sugar crops to 2050. We find that climate impacts on global average yields, area, production and consumption are similar across shared socioeconomic pathways (SSP 1, 2 and 3, as we implement them based on population, income and productivity drivers), except when changes in trade policies are included. Impacts on trade and prices are higher for SSP 3 than SSP 2, and higher for SSP 2 than for SSP 1. Climate impacts for all variables are similar across low to moderate emissions pathways (RCP 4.5 and RCP 6.0), but increase for a higher emissions pathway (RCP 8.5). It is important to note that these global averages may hide regional variations. Projected reductions in agricultural yields due to climate change by 2050 are larger for some crops than those estimated for the past half century, but smaller than projected increases to 2050 due to rising demand and intrinsic productivity growth. Results illustrate the sensitivity of climate change impacts to differences in socioeconomic and emissions pathways. Yield impacts increase at high emissions levels and vary with changes in population, income and technology, but are reduced in all cases by endogenous changes in prices and other variables.

1. INTRODUCTION

Studies to date suggest that climate change has reduced growth in crop yields by 1–2 percent per decade over the past century, and adverse impacts are projected to increase in the future (Gourdji *et al* 2013, Intergovernmental Panel on Climate Change (IPCC) 2014). But understanding the magnitude of these impacts is complicated by the interaction of numerous biophysical and socioeconomic factors. Previous studies have combined climate, crop and economic models to examine the impact of climate change on agricultural production and food security, but results have varied widely due to differences in models, scenarios and input data (Rosenzweig and Parry 1994, Parry *et al* 2004, Nelson *et al* 2010).

Recent work as part of the agricultural model intercomparison and improvement project (AgMIP) has examined (and narrowed) these differences through systematic model intercomparison (Nelson *et al* 2014a, 2014b, von Lampe *et al* 2014). The explicit goal of that intercomparison of global economic models was 'to understand the differences in model projections and model behavior and to identify their sources; not to choose scenarios for their plausibility' (von Lampe *et al* 2014). Accordingly that exercise focused on climate impacts for a single high-emission representative concentration pathway (RCP 8.5), and a single 'middle-of-the-road' shared socioeconomic pathway (SSP 2). Running these scenarios with multiple climate, crop and economic models, they found global average yield reductions of 11 percent and price increases of 20 percent due to climate change relative to baseline values for major crops in 2050 (Nelson *et al* 2014a). However, these studies did not examine the impact of more plausible climate change scenarios, the sensitivity of the impact of climate change to different socioeconomic environments, or whether agricultural and trade policies could mitigate the impact on the agricultural sector and food prices by improving access to international supplies (Schmidhuber and Tubiello 2007).

This paper extends the literature by focusing on these three areas. As results are expected to differ widely between models, we use multiple models to obtain more robust results. Results from three general circulation models (GCMs) and five global economic models are used to examine the global and regional impacts of climate change on agricultural yields, area, production, prices and trade for coarse grains, rice, wheat, oilseeds and sugar (which together account for two thirds of global calorie availability and three quarters of harvested cropland area) to 2050. While climate change impacts are expected to increase more sharply after 2050 (Intergovernmental Panel on Climate Change (IPCC) 2014), an intermediate timeframe is critical for agricultural research and development planning that is currently under way.

2. METHODS

We begin by extending the range of SSPs considered. The full set of SSP storylines has been described in Kriegler *et al* (2012) and O'Neill *et al* (2014, 2015). Within the SSP process, standardized input data on country-specific changes in population and income have been developed

(International Institute for Applied Systems Analysis (IIASA) 2013a)9 . The SSPs are not intended to reflect specific policy or management choices, but rather different views about how the future may develop, given uncertainty about long-term changes in population, socioeconomic development, behavior and the role of global policy coordination. To date, the SSP framework has been mainly used to assess future scenarios relating to greenhouse gas emissions, mitigation options and policies, and changes in energy use and land use. In this paper, we use the SSP input data on population and income drivers to examine climate impacts on agricultural production and markets under three socioeconomic pathways (our specific implementation of SSP 1, SSP 2, and SSP 3). SSP 2 is a pathway with modest overall growth in population and incomes, and a slow pace of overall trade liberalization. SSP 1 features lower population growth, higher growth in per capita incomes, faster globalization and more integration of international markets, while SSP 3 describes a more fragmented world with less international trade, higher population growth, and lower growth in per capita incomes.

Building on these general scenario narratives, we add specific dimensions relevant to the agricultural sector, namely changes in agricultural productivity and agricultural trade policy regimes. Crop- and region-specific changes in agricultural productivity ('intrinsic productivity growth rates', or IPRs) had previously been taken from the IMPACT model's 'business as usual' scenario for SSP 2 (Nelson *et al* 2014a, 2014b, von Lampe *et al*2014). These IPRs summarize the improvements that can be achieved in agricultural productivity from a variety of advances in management practices, crop improvement and agricultural extension. For this paper, two additional sets of IPRs were developed for SSP 1 and SSP 3, using country-specific differences in economic growth between SSPs (relative to SSP 2) to adjust the IPRs upward or downward. Growth in income (as measured by gross domestic product, or GDP) was chosen as the primary driver to adjust the IPRs because public investment continues to be the primary driver of agricultural research and development, accounting for almost 80 percent globally, and recent trends in public investment in agricultural R&D have followed GDP growth, especially in developing countries (Beintema *et al* 2012). Accordingly, agricultural productivity grows fastest in SSP 1 and slowest in SSP 3. (See supplementary material for more details on IPR adjustments, available at stacks.iop.org/ERL/10/085010/mmedia.) Assumptions for each of the three SSPs as implemented in this analysis (and harmonized across the global economic models) are summarized in figure 1.

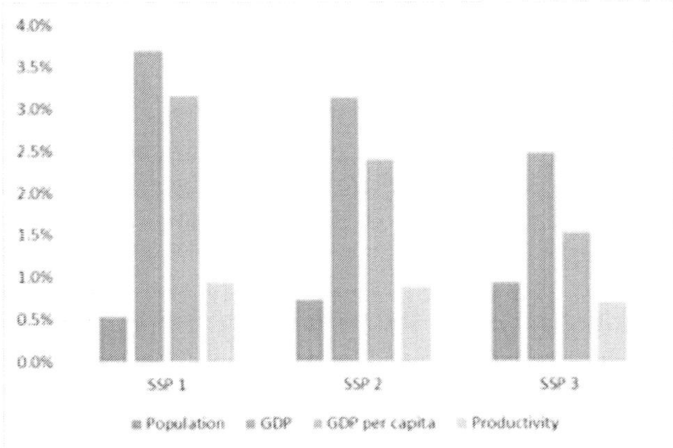

Source: Population from IIASA and GDP from OECD (IIASA 2013a), and the authors for agricultural productivity.

FIGURE 1. Population, GDP and agricultural productivity assumptions by SSP (annual percentage change, 2010–2050).

In principle, each SSP is consistent with multiple greenhouse gas emission pathways, depending on specific levels of emission mitigation efforts. For simplicity and in order to limit the number of scenarios, we combine each SSP with climate impacts for a unique RCP (Moss *et al* 2010, van Vuuren *et al* 2011, Kriegler *et al* 2012): SSP 1 with climate impacts for RCP 4.5, SSP 2 with RCP 6.0, and SSP 3 with RCP 8.5. While this approach may entail some inconsistencies (e.g., we do not include the effect of greenhouse gas mitigation policies on land use), our scenario selection assures an analysis of the range of plausible climate outcomes by the middle of the century. To illustrate the effects of the different socioeconomic and climate impact pathways in isolation, additional SSP × RCP combinations were analyzed using one global economic model (FARM, see below).

For each RCP, we use climate input data from three GCMs, namely HadGEM2-ES, IPSL-CM5A-LR, and MIROC-ESM-CHEM (Jones *et al* 2011, Watanabe *et al* 2011, Dufresne *et al* 2013), in order to account for differences in spatial distribution of projected climate change patterns, especially rainfall patterns, which are particularly important for agriculture (Intergovernmental Panel on Climate Change (IPCC) 2013). These models represent three of the five ISI-MIP GCMs that were selected by first availability (Hempel *et al* 2013). Of the ISI-MIP forcing data set (five GCMs), these three represent the ones with the strongest temperature response and cover the full range of precipitation changes, where HadGEM2-ES is the driest and IPSL-CM5A-LR the wettest (Warszawski *et al* 2014)10 .

These climate data are used as inputs in a global gridded crop model, LPJmL (Bondeau *et al* 2007, Müller and Robertson 2014), to generate quantitative estimates of the additional 'exogenous' impacts of climate change on crop yields for coarse grains, rice, wheat, oilseeds and sugar crops, i.e. before accounting for adjustments in area, prices or other variables. These exogenous climate impacts are represented as multiplicative shocks, and are derived using a delta approach as in previous AgMIP work (Müller and Robertson 2014, Nelson *et al* 2014a) to estimate relative yield changes per crop and region, allowing for more consistent application of climate change across models with different base yields and trends (Robinson *et al* 2014). Figure 2 shows these exogenous yield impacts, i.e. the difference between IPRs with and without climate change, for the five commodities under SSP 2 and climate impacts from RCP 6.0. (Similar results are found for the other RCPs; see figure S1 in the supplementary material.) The three dots for each commodity and region represent results from the three GCMs, and are in general relatively closely clustered. Exogenous yields for four of the five commodities are adversely affected by climate change, with the strongest impacts in most regions being on oilseeds. By contrast, climate change increases sugar yields in most regions, because for sugarcane the whole plant (including the stalk) is utilized, and total plant biomass of C_4 plants is only negatively affected at daily mean temperatures above 35 °C (see Müller and Robertson 2014 for more details). The performance of LPJmL has been compared to other global crop models in previous studies and typically projects moderate changes compared to the AgMIP/ISI-MIP ensemble (see Rosenzweig *et al* 2014, figure 4) if CO_2 fertilization effects are excluded, as we do here (Müller and Robertson 2014, Rosenzweig *et al* 2014). In line with previous work, we do not consider fertilization effects on yields with higher CO_2 concentrations. The CO_2 effect is still very uncertain owing to many complex interaction mechanisms, and is widely discussed in the research community (Long *et al* 2006, Tubiello *et al* 2007, Wang *et al*2012, Boote *et al* 2013).

Another key element in the SSP narratives is the pace of globalization and openness to international trade. Since this is an important aspect of agricultural markets and is potentially relevant for food security questions (Smith 1998, Schmidhuber and Tubiello 2007, Burnett and Murphy 2014), we consider two additional synthetic scenario variants on trade policy changes. In principle, changes in trade policy should already be part of each underlying SSP storyline, i.e. part of each SSP baseline run without climate impacts. However, in this study we separate agricultural trade policy changes from our specific SSP baselines in order to disentangle the specific effects on economic model outputs. Hence the specific SSP 1, SSP 2 and SSP 3 baselines in this paper differ only with respect to changes in population, income and agricultural productivity (IPRs). Climate impacts in 2050 are then analyzed relative to the specific SSP baseline. To examine the impact of explicit changes in agricultural trade policy, we run two additional scenarios: SSP 1 with a liberalized trade variant and SSP 3 with a more restricted trade variant. The actual size of the trade 'shocks' reflects a specific interpretation of the SSP storylines as described in O'Neill *et al* (2015), but other plausible interpretations could be envisioned to assess the sensitivity of simulation outcomes relative to the trade shocks11 . In addition, one could envision as well a change in trade preferences, for example a 'buy local' shift in preferences, that could be combined with (and possibly counteract) a reduction in trade barriers, but this is beyond the scope of this paper, which is to provide an illustrative example.

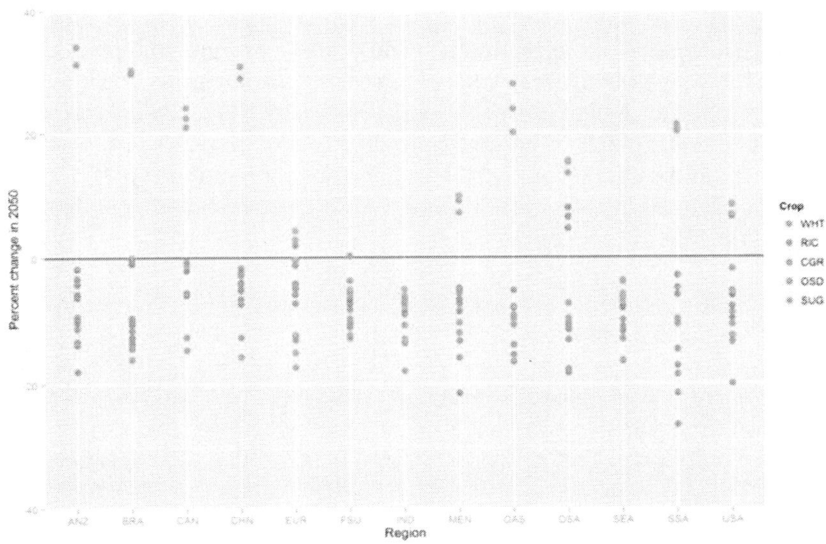

Note: Based on the three selected GCMs as represented by the LPJmL crop model (n = 15). Each dot depicts the result for one crop and one GCM.

Crops: WHT = wheat, RIC = rice, CGR = coarse grains, OSD = oilseeds, SUG = sugar.

Regions: ANZ = Australia/New Zealand, BRA = Brazil, CAN = Canada, CHN = China, EUR = Europe, FSU = Former Soviet Union, IND = India, MEN = Middle East/North Africa, OAS = Other Asia, OSA = Other Latin America, SEA = Southeast Asia, SSA = Sub-Saharan Africa, USA = United States of America.

Source: The authors.

FIGURE 2. Exogenous impacts of climate change on crop yields, by region, under SSP 2 and climate impacts for RCP 6.0 (% change relative to SSP 2 baseline values in 2050 without climate change).

Five global economic models with a focus on agriculture are then used to analyze climate impacts in combination with the three SSPs and their associated changes in IPRs. Three are computable general equilibrium (CGE) models (ENVISAGE, FARM, and MAGNET), which examine economy-wide effects, and two are partial equilibrium (PE) models (IMPACT and

MAgPIE), which include greater detail on the agriculture sector (Meijl *et al* 2006, Lotze-Campen *et al* 2008, Popp *et al* 2010, Rosegrant and the IMPACT Development Team 2012, Schmitz *et al* 2012, van der Mensbrugghe 2013, Sands *et al* 2014, Woltjer *et al* 2014; see also supplementary material). This model selection allows us to cover a range of different implementations of food demand, land use, trade, and productivity increase, which are all important factors for determining market changes based on climate impacts on crop yields. The performance of these models has been assessed and compared in previous studies (Nelson *et al* 2014a, 2014b, Schmitz *et al* 2014, Valin *et al* 2014, von Lampe *et al* 2014). These results have also been used in the latest IPCC 5th assessment report (Intergovernmental Panel on Climate Change (IPCC) 2014). The economic models are used to generate three different baseline projections of agricultural yields, area, production, prices and trade for coarse grains, rice, wheat, oilseeds and sugar to 2050 in the absence of climate change. These three baselines are then used as reference scenarios against which to compare the impacts of changes in climate as well as trade policy (table 1). As the socioeconomic drivers, climate patterns and trade policies are very different between the scenarios, we expect significant differences in results.

Table 1. Scenario definition.

Scenario	SSP	Radiative forcing	GCM	Trade policy	Economic model
1.0	SSP 1	No change	None	No change	ENVISAGE, FARM, IMPACT, MAGNET, MAgPIE
1.1		RCP 4.5	HadGEM		
1.2			IPSL		
1.3			MIROC		
1.4			HadGEM	Liberalized	ENVISAGE, FARM, MAGNET, MAgPIE
2.0	SSP 2	No change	None	No change	ENVISAGE, FARM, IMPACT, MAGNET, MAgPIE
2.1		RCP 6.0	HadGEM		
2.2			IPSL		
2.3			MIROC		
3.0	SSP 3	No change	none	No change	ENVISAGE, FARM, IMPACT, MAGNET, MAgPIE
3.1		RCP 8.5	HadGEM		
3.2			IPSL		
3.3			MIROC		
3.4			HadGEM	Restricted	ENVISAGE, FARM, MAGNET, MAgPIE

Source: The authors.

3. RESULTS

3.1. SSP baselines to 2050

We are interested in the impacts of climate change and trade policy in 2050, so we need to first account for the changes in yield and other variables that will take place between now and 2050 due to other factors such as changes in population, income and technology, as described above (see figure 1) and in the supplementary material. Growth in these factors will be the major drivers of long-term changes in food demand, prices and supply over this time period, especially in the developing countries. Comparison of baseline scenarios 1.0, 2.0 and 3.0 (harmonized across the economic models on country-specific growth rates in population and income, and on crop- and country-specific growth rates in productivity, combined with current climate conditions), allows us to consider the effects of these drivers in the absence of climate change to 2050. All five economic models use the same GDP and population projections as exogenous model inputs. With the exception of MAgPIE, IPRs are implemented in similar ways in the economic models. In the CGE models (FARM, MAGNET, ENVISAGE), the IPRs are implemented as land-augmenting shocks to production functions, while in the PE model IMPACT they are implemented directly as land productivity shocks. In the MAgPIE model, technology-based yield changes are determined endogenously, so here only the final yields are comparable to the other models (see also Nelson *et al*2014a and 2014b, Robinson *et al* 2014 and supplementary material for details on IPRs for different SSPs).

As expected, baseline results vary across the three SSPs as we implement them in this paper (figure 3; see also the last row in figure 5). Consistent with higher economic growth driven by higher technological change and inelastic demand for agricultural products, SSP 1 shows larger yield increases and smaller price increases than the other two SSPs. The opposite is true for SSP 3, where yields increase less and prices increase more than in the SSP 1 scenario. Final yields (YTOT) are higher than exogenous yields (YEXO) as endogenous management decisions lead to intensification in times of economic growth (Meijl *et al* 2006, Eickhout *et al* 2009). Land is a relatively scarce production factor and land prices increase at a higher rate than labor and capital prices (it is a fixed factor and substitution is limited due to low land-price elasticities), which induces land substitution by using more labor and capital. The use of more labor and capital per unit of land leads to an additional endogenous yield increase (intensification).

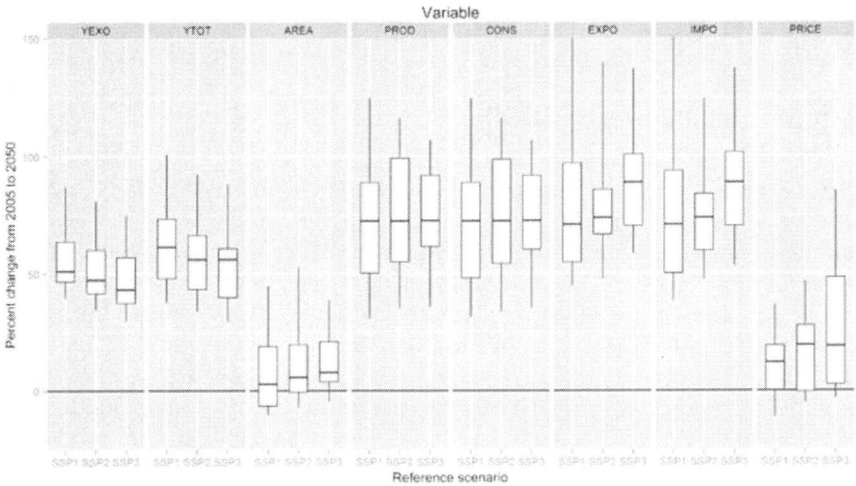

Note: The plots show pooled results for the five commodities from the five economic models, aggregated across thirteen regions (n = 25). All pooled data are combined into the sample for each boxplot, and cannot be distinguished individually. The boxes and whiskers in all figures depict 5th, 25th, 50th, 75th and 95th percentile values.

Variables: YEXO = exogenous yield shocks, YTOT = realized yields after management adaptation, AREA = agricultural area in production, PROD = total production, CONS = total consumption, EXPO = exports, IMPO = imports, PRICE = price.

Reference scenarios: SSP1 = SSP 1 with no climate change, SSP2 = SSP 2 with no climate change, SSP3 = SSP 3 with no climate change.

Source: The authors.

FIGURE 3. Baseline increases in global yields, area, production, consumption, exports, imports and prices of coarse grains, rice, wheat, oilseeds and sugar in 2050 (% change relative to 2005 values).

In the absence of climate change, global production and consumption of coarse grains, rice, wheat, oilseeds and sugar are projected to increase by an average (median) of 72–73 percent over 2005 levels by 205012 . In contrast to the other variables, growth in production and consumption is essentially the same across the three reference scenarios, suggesting that the effects of differences in population and income growth offset one another across the three SSPs, and that inelastic demand for food and other agricultural commodities forces adjustments primarily in yields and prices, and to a lesser extent in area13 . Finally, although we do not model explicit differences in trade policy in these three scenarios, we see that trade increases significantly across the three SSPs. A number of factors will influence trade patterns, even while global production and consumption levels remain largely invariant to the SSPs. For example, population increase in developing countries under SSP 3 combined with population decrease in developed countries would induce agricultural surpluses in the latter and deficits in the former, all else equal. Land-use changes—both expansion of crop land and re-allocation of cropping patterns—differ across regions and will drive differential supply response to climate-induced changes in yields. Land-scarce countries with a low land supply response will lose relative competitiveness and increase net imports as increased demand leads to higher land prices (and therefore agricultural prices) instead of using more area, with a lower land price increase in the case of land abundant countries (see Meijl 2006, Eickhout *et al* 2009).

The variation evident in figure 3 reflects differences in results across the five crops and the five economic models. Further disaggregation (see supplementary material, figure S2) reveals a number of differences in baseline results across the economic models, highlighting the importance of including multiple models in the analysis.

These three baseline projections (scenarios 1.0, 2.0 and 3.0) give us a picture of what could be expected in 2050 without changes in climate or trade policy. All of the subsequent results in this paper are presented as deviations from these baselines.

3.2. Impacts of climate change

These baseline results are affected to different degrees by changes in climate. We look first at the central case of SSP 2 combined with climate impacts for RCP 6.0 radiative forcing (figure 4). In this scenario, climate change is initially projected to reduce 'exogenous' yields by an average of 6.9 percent (median across crops, regions and models relative to the baseline value in 2050)—before adjustments in area, prices and other variables are considered. The initial exogenous yield shocks trigger endogenous increases in area and prices. Higher prices in turn lead to increases in other production factors applied to these crops—and thus endogenous yield growth—as well as lower demand for agricultural commodities, reducing the final impact of climate change on yields to a median decline of 3.8 percent. Area under the five crops increases by 2.9 percent, while production declines by 0.1 percent and prices increase by 4.2 percent. Median consumption is unchanged, but the variation around median consumption is much smaller than for the other variables. As we saw in the baseline scenarios (figure 3), inelastic demand for food and other agricultural commodities forces adjustments to exogenous shocks primarily through changes in yields, prices and area, while total production and consumption do not change significantly.

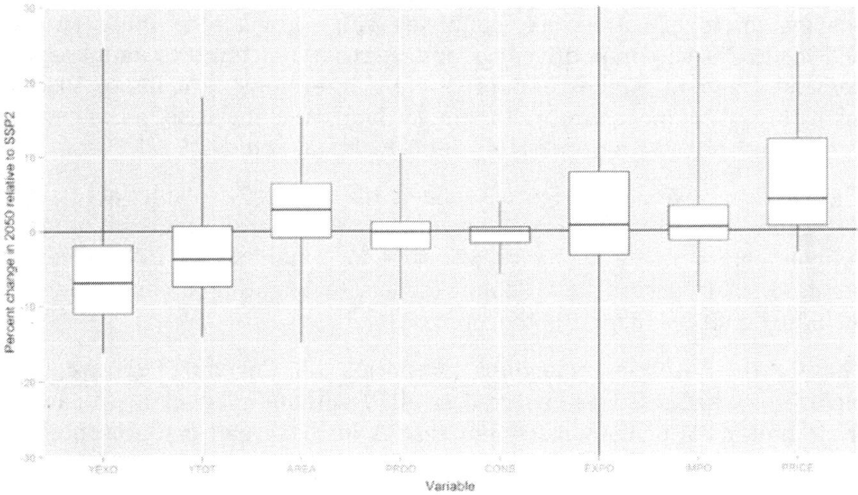

Note: The plots show pooled results for five commodities in thirteen regions from three GCMs and five economic models (n = 975). All pooled data are combined into the sample for each boxplot, and cannot be distinguished individually.
Variables: YEXO = exogenous yield shocks, YTOT = realized yields after management adaptation, AREA = agricultural area in production, PROD = total production, CONS = total consumption, EXPO = exports, IMPO = imports, PRICE = price.
Source: The authors.

FIGURE 4. Impacts of climate change on yields, area, production, consumption, exports, imports and prices of coarse grains, rice, wheat, oilseeds and sugar under SSP 2 and GCM results for RCP 6.0 (% deviation from SSP 2 baseline values in 2050 without climate change).

Similar patterns are evident for other SSP x RCP/GCM combinations. Figure 5 shows global climate impacts under SSP 1 and 3, each with climate impacts for a unique RCP. In the more 'sustainable' world of SSP 1, climate change driven by RCP 4.5 has impacts that are similar but slightly smaller than those under SSP 2 with climate impacts for RCP 6.0. The differences are small because GCM results for RCP 4.5 and RCP 6.0 are similar up to 2050 (with median YEXO declines of 8.2 percent and 8.5 percent for RCP 4.5 and RCP 6.0, respectively), and differences in population and income growth between SSP 1 and SSP 2 work to offset each other in terms of their impact on food demand (see also figure 1). In SSP 3, by contrast, climate change driven by RCP 8.5 is projected to have greater impacts, with a median YEXO decline of 12.8 percent. Final adapted yields (YTOT) are projected to decline by a median of 7.2 percent, while area increases by 3.8 percent, production and consumption decline by 0.9 percent, exports and imports increase by 4.0 and 5.3 percent, respectively, and prices increase on average by 15.5 percent. The price effects vary more widely across the different crops and models than was the case for the other two scenarios.

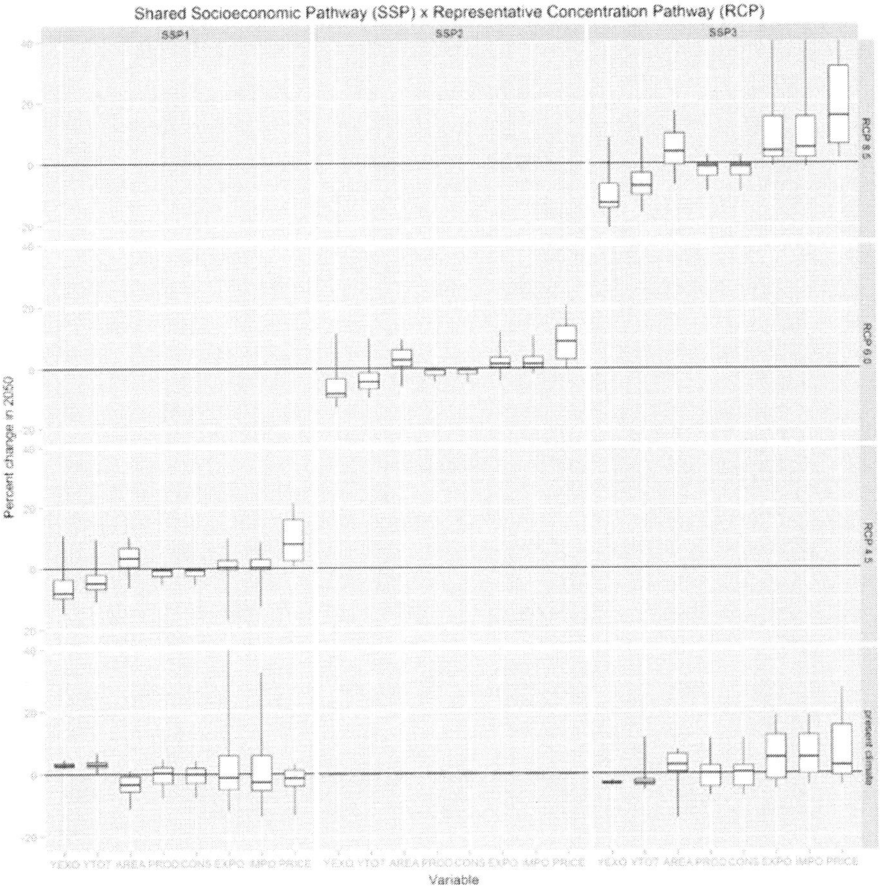

Note: The plots show pooled results for five commodities from three GCMs and five economic models, aggregated across thirteen regions (n = 75). All pooled data are combined into the sample for each boxplot, and cannot be distinguished individually. The first three rows depict % deviation from respective baseline values in 2050 without climate change (from Figure 3). The fourth row depicts % deviations from the SSP2 baseline values in 2050 without climate change.
Variables: YEXO = exogenous yield shocks, YTOT = realized yields after management adaptation, AREA = agricultural area in production, PROD = total production, CONS = total consumption, EXPO = exports, IMPO = imports, PRICE = price.
Source: The authors.

FIGURE 5. Impacts of climate change on global yields, area, production, consumption, exports, imports and prices of coarse grains, rice, wheat, oilseeds and sugar under different SSP × RCP/GCM combinations (% deviation from respective baseline values in 2050 without climate change).

The bottom row in figure 5 compares the SSP 1 and SSP 3 baselines directly with that of SSP 2. The median price changes and also the variation in price changes across crops, regions and models are lower across the last row than for the climate impacts moving up the columns. This suggests that the direct climate impacts in 2050 are stronger than the differences between the underlying socioeconomic trends, at least at the global level.

Aggregating across crops, regions and GCMs to isolate differences across the five economic models, we find that they generally agree on the direction of change in yields, area, production and prices in each scenario, although they differ in the magnitude of impacts, particularly in the scenario with the most rapid growth in population and emissions (SSP 3 with climate impacts for RCP 8.5). (See figure S3 in the supplementary material.) In all cases impacts on prices are larger than those on production because demand for food and other agricultural commodities is relatively inelastic. The variance across the economic models is also larger for prices than it is for the other variables, which is in line with results from previous studies, as the models differ substantially in their representation of demand and supply response, trade, and land supply (Nelson *et al* 2014a, 2014b, Schmitz *et al*2014, Valin *et al* 2014).

Because the matrix in figure 5 is incomplete, we are limited in our ability to distinguish the effects of differences across SSPs from the climate change differences across RCPs. In order to distinguish the effects of socioeconomic drivers from those of climate change, we ran a wider set of SSP × RCP/GCM scenarios for one of the economic models, namely FARM. Results are presented in figure 6. It shows relatively small differences when moving along the rows, i.e. changing SSPs for a given RCP-related climate, especially between SSP 1 and SSP 2, but with a larger increase in impacts for SSP 3, most notably for exports, imports and prices. Moving up the columns, there is relatively little difference between climate impacts for RCP 4.5 and RCP 6.0 for a given SSP, consistent with expectations given the similarity between these concentration pathways until 2050. By contrast, climate impacts on all variables are greater under RCP 8.5 for both SSP 2 and SSP 3.

Compared to the projected changes in the baseline scenarios due to other drivers such as population, income and technology (figures 1 and 3), average climate impacts on yields projected by 2050 are relatively modest, although these conceal differences across crops and regions, which we will consider below. It is also important to note that this analysis does not capture the potentially significant impacts of inter- and intra-annual variability in temperature and precipitation, which is currently beyond the capabilities of most crop models, but also limited by available input data from climate models. For these particular scenarios, differences across the three GCMs are also relatively small—probably because the impacts of climate change are less significant (at least at the global scale) in the period leading up to 2050 than they are projected to be later in the century.

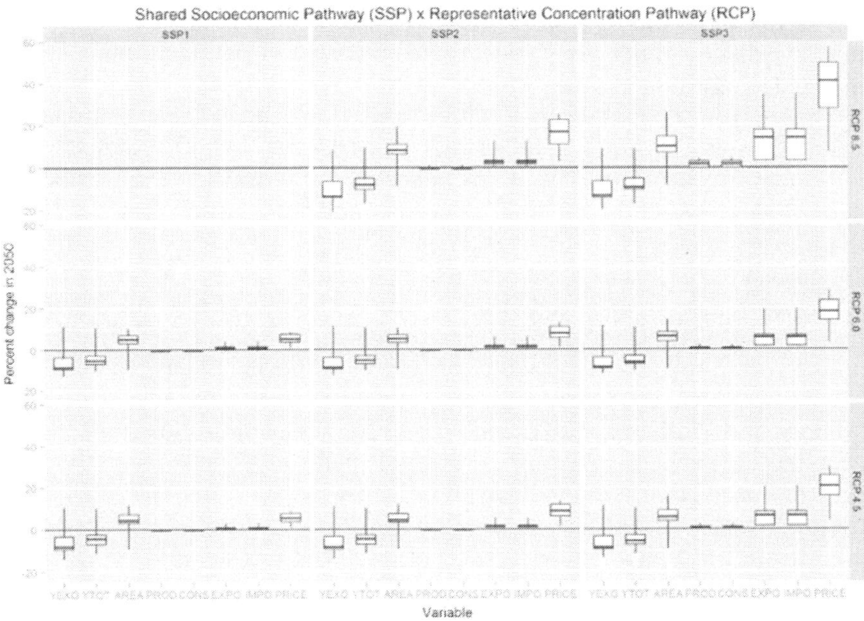

Note: The plots show pooled results for five commodities from three GCMs and the FARM economic model, aggregated across thirteen regions (n = 15). All pooled data are combined into the sample for each boxplot, and cannot be distinguished individually.

Variables: YEXO = exogenous yield shocks, YTOT = realized yields after management adaptation, AREA = agricultural area in production, PROD = total production, CONS = total consumption, EXPO = exports, IMPO = imports, PRICE = price.

Source: The authors.

FIGURE 6. Impacts of climate change on global yields, area, production, consumption, exports, imports and prices of coarse grains, rice, wheat, oilseeds and sugar under a range of SSP × RCP/GCM combinations using the FARM model (% deviation from respective SSP baseline values in 2050 without climate change).

3.3. Differences by crop and region

In broad terms crop productivity often increases in areas with strong limitations by low temperatures (e.g. mountainous and high latitude areas) and decreases elsewhere, but there are multiple deviations from that broad pattern depending on local conditions. We find that adverse impacts of climate change appear to be particularly strong for oilseeds and rice, and more moderate for coarse grains and wheat across the scenarios examined (figure 7), which is also consistent with the AgMIP/ISI-MIP global gridded crop model ensemble analysis for RCP 8.5 (Rosenzweig *et al* 2014), where all individual models except for one (GAEZ-IMAGE) project larger yield declines for soy and rice than for maize and wheat, when assuming no

CO_2 fertilization effects. Across regions, these patterns of crop-specific impacts also hold, but are subject to some variation, as for example soy is more strongly temperature limited in cooler regions and may see yield increases from climate change where wheat, which can deal with cooler temperatures, does not (Müller and Robertson 2014). Our results are similar to average historic impacts reported for recent decades for maize and wheat, and larger for oilseeds and rice (Lobell *et al* 2011, Intergovernmental Panel on Climate Change (IPCC) 2014), although comparison is difficult due to differences in methodology. Yield reductions are consistently greatest for oilseeds, while price increases are generally largest for rice, for which demand is more inelastic. By contrast, sugar crops are projected to have increased yields (as noted earlier), reduced area and comparatively modest price increases relative to the baseline (see also Müller and Robertson 2014 for a more detailed discussion). This contributes to mean yield impacts that are somewhat smaller than those in previous work that excluded sugar (Nelson *et al* 2014a). It also highlights the importance of further research on improved technologies to enhance adaptation, for example through increased tolerance of drought and heat.

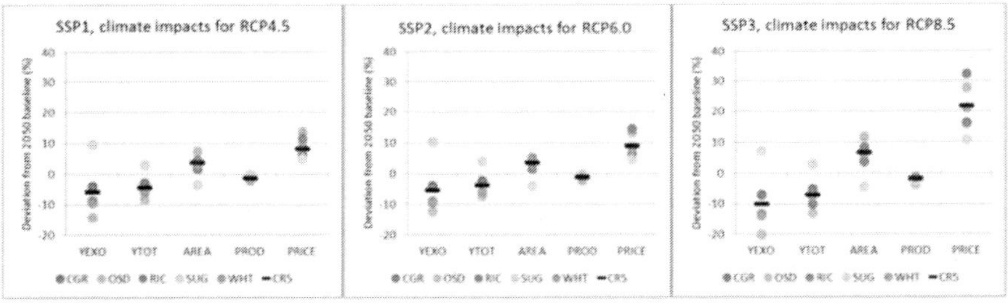

Note: The plots show pooled results for five commodities and five global economic models, aggregated across three GCMs and thirteen regions (n = 5).

Crops: CGR = coarse grains, OSD = oilseeds, RIC = rice, SUG = sugar, WHT = wheat, CR5 = 5-commodity aggregate.

Variables: YEXO = exogenous yield shocks, YTOT = realized yields after management adaptation, AREA = agricultural area in production, PROD = total production, PRICE = price.

Source: The authors.

FIGURE 7. Impacts of climate change on global yields, area, production and prices relative to baseline values in 2050 for each SSP, compared across the five crops.

Disaggregation by region shows that adverse yield impacts appear particularly strong for South Asia and Southeast Asia, and smaller in the other developing regions (see supplementary material, figure S4). Area increases are projected to be largest in sub-Saharan Africa and the Americas, consistent with other analyses (e.g. Alexandratos and Bruinsma 2012). Because of severe constraints on expansion of agricultural land, South Asia is projected to see larger price increases than the other regions under all three scenarios, with important implications for access to food.

3.4. Impacts of changes in trade policy

The results reported above assume no change in trade policies. Given the importance of trade to the global agricultural economy, we consider two additional scenario variants to explore sensitivity to trade policy assumptions: SSP 1 with more liberalized agricultural trade, and SSP 3 with more restricted trade, consistent with the underlying story lines of the respective SSPs. With more liberalized trade in SSP 1 we would expect smaller price increases on average compared to the no-change case reported earlier. This is confirmed in figure 8, the right panel of which shows a sharp increase in trade and a lower price effect, an average increase of 4.3 percent, compared to 8.0 percent under 'business-as-usual' trade policies as reported in the left panel. By contrast, with more restricted trade in SSP 3 (figure 9), trade declines and prices rise on average by 25.2 percent, compared to 15.5 percent with 'business-as-usual' trade policies. The spread across crops and models is also wider. This shows the critical role that trade policy can play in helping to alleviate the adverse impacts of climate change and reducing price increases on international markets. Note that the figures reflect the price and trade impacts of different trade policies of the relevant scenarios with the climate shocks. It still holds that the climate shocks lead to price increases that could be partially or fully offset or exacerbated by changes in trade policies. A more elaborate study of trade policy effects would require additional baseline scenarios with, for example, more restricted trade in SSP 3 without climate impacts. However, this is beyond the scope of this paper, and admittedly limits our ability to isolate trade effects from climate impacts at this point. Moreover, we note that the economic models have very different trade mechanisms and trade policy implementations, and further work is needed to improve harmonization in this area.

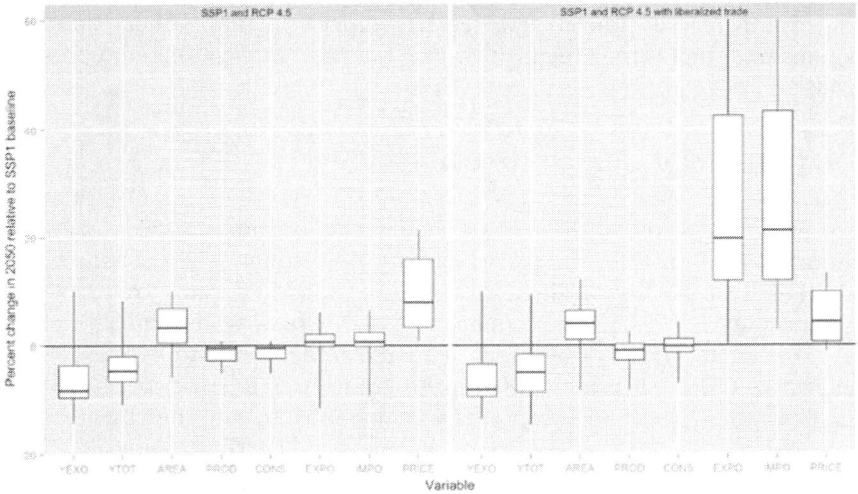

Note: The plots show pooled results for five commodities from one GCM (HadGEM2-ES) and four economic models (ENVISAGE, FARM, MAGNET and MAgPIE), aggregated across thirteen regions (n = 20). All pooled data are combined into the sample for each boxplot, and cannot be distinguished individually.
Variables: YEXO = exogenous yield shocks, YTOT = realized yields after management adaptation, AREA = agricultural area in production, PROD = total production, CONS = total consumption, EXPO = exports, IMPO = imports, PRICE = price.
Source: The authors.

FIGURE 8. Impacts of climate change with (right) or without (left) trade liberalization on yields, area, production, exports, imports and prices of five commodities under SSP 1 and climate impacts for RCP 4.5 (% deviation from SSP 1 baseline values in 2050 without climate change and with baseline trade policy).

4. DISCUSSION

We find that projected reductions in agricultural yields due to climate change by 2050 are larger for some crops than those estimated for the past half century (Intergovernmental Panel on Climate Change (IPCC) 2014), but smaller than projected increases to 2050 due to rising demand and intrinsic productivity growth (table 1). Estimated impacts follow a similar pattern but are somewhat smaller than those of earlier work focusing on SSP 2 and climate impacts for RCP 8.5 (Nelson *et al* 2014a), which projected an 11 percent decrease in yields and a 20 percent increase in prices for a similar set of crops. Impacts are smaller for SSP 1 and SSP 2 than they are for SSP 3, and smaller for climate impacts associated with RCP 4.5 and RCP 6.0 than they are for RCP 8.5. The differences may arise for several reasons. First, in

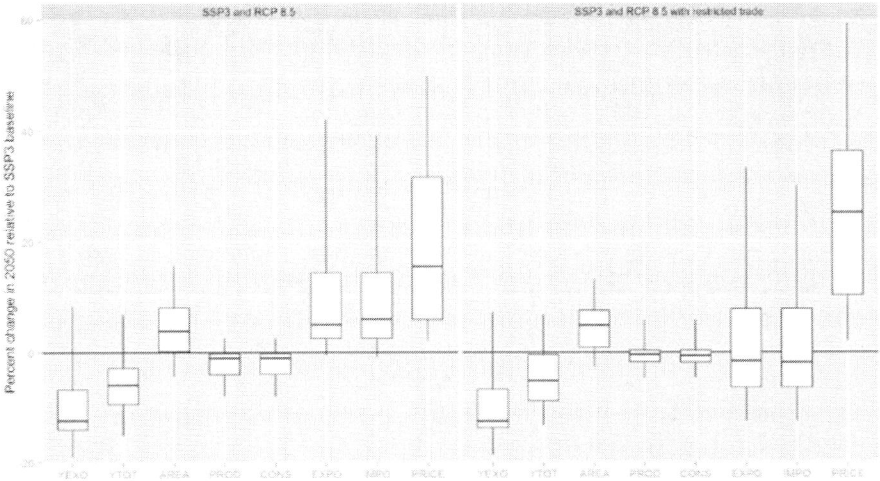

Note: The plots show pooled results for five commodities from one GCM (HadGEM2-ES) and four economic models (ENVISAGE, FARM, MAGNET and MAgPIE), aggregated across thirteen regions (n = 20). All pooled data are combined into the sample for each boxplot, and cannot be distinguished individually.

Variables: YEXO = exogenous yield shocks, YTOT = realized yields after management adaptation, AREA = agricultural area in production, PROD = total production, CONS = total consumption, EXPO = exports, IMPO = imports, PRICE = price.

Source: The authors.

FIGURE 9. Impacts of climate change with (right) or without (left) restricted trade on yields, area, production, consumption, exports, imports and prices of five commodities under SSP 3 and climate impacts for RCP 8.5 (% deviation from SSP 3 baseline values in 2050 without climate change and with baseline trade policy).

the previous work a second crop model (DSSAT) with more pronounced yield effects was used in addition to LPJmL, and a total of 10 economic models were applied in the analysis. LPJmL projections are representative for the middle of the crop model ensemble projections as analyzed for RCP 8.5 (Rosenzweig *et al* 2014) when CO_2 fertilization effects are excluded, as in the present analysis. Second, we apply climate impacts for different RCPs. Climate-induced yield changes for RCP 4.5 and RCP 6.0 by 2050 are relatively similar, but they are considerably smaller than for RCP 8.5 (see figures 5 and 6). Third, the economic models used in the present study have been further developed since the previous work, which led to changes (including greater flexibility) in their price responses (see supplementary material for details on specific models). Fourth, we use version 0.93 of the SSP drivers, which includes notable differences in population and income relative to version 0.5 that was used in the previous work (International Institute for Applied Systems Analysis (IIASA) 2013a, 2013b). For example under SSP 2, average per capita income is some 6.4 percent lower in 2050 under the latest projections compared with the earlier ones, and there is substantially more variation at the country level. And finally, we include sugar in addition to the four crops analyzed in earlier work (e.g. Nelson *et al*2014a). In contrast to the other crops, sugar yields are shown to increase on average as a result of climate change in the scenarios examined in this study.

Endogenous behavioral responses (e.g. in food demand, area and other production factors) are critical in differentiating final yield impacts from initial exogenous climate shocks. While the models generally agree on the direction of changes in yields, area, production and prices, differences across economic models in the magnitude of impacts are sometimes significant, due to different handling of demand, endogenous substitution effects, land supply, and international trade. We also show that changes in model outputs vary strongly across specific crops and regions. This holds especially for a SSP 3 world with increasingly diverse socioeconomic conditions and in which trade adjustments play a much greater role.

All scenarios show the potential for significant increases in prices for food and agricultural commodities, raising important concerns about food security, particularly for poor households. Differences in climate impacts between SSP 1 and SSP 2 are generally small, because exogenous yield shocks do not differ significantly between the two pathways, and because differences in population and income growth between the two pathways work to offset each other in terms of their impact on food demand. SSP 3 with RCP 8.5-related climate change generates impacts on yields and prices that are approximately twice as large as those under SSP 1 and SSP 2, as well as a wider spread across models, particularly in terms of price impacts.

The smaller differences we found between SSP 1 and SSP 2 may be due on the supply side to limited differences in IPRs (as we implemented them) across the scenarios. This question deserves further research. On the demand side, the smaller differences across these SSPs is also related to the fact that we focus primarily on staple crops, which are less responsive to income growth than are higher value commodities such as fruits, vegetables, and animal products. The relatively small differences between climate impacts for the RCP 4.5 and RCP 6.0 scenarios are related to the fact that large differences between these concentration pathways become evident only after 2050.

When trade is restricted, limiting options for adjustment to climate and other shocks, price increases are even greater and more widely spread across models. While harmonization of trade policies across models (let alone across countries) is a challenge, the results show that assumptions about trade policy matter significantly for diagnosing the projected adaptations and impacts. Response to assumptions about trade policy seems particularly sensitive in the case of SSP 3, in which trade adjustment plays a greater role than for the other SSPs, and for RCP 8.5, for which climate impacts are greater than for the other RCPs. The increased spread in price responses with more restricted trade also indicates the sensitivity of results with respect to the specific implementation of trade restrictions in the models. This reinforces the value of model comparison and the importance of more work on harmonizing the assumptions and implementation of trade policy assumptions. Moreover, while increased trade may help to alleviate pressures from combined social economic and/or climate impacts on agricultural production and prices, it may also entail other impacts and externalities potentially linked to them—both positive, for example increases in productivity embedded in increased inputs or investment (Huang *et al* 2011) and negative, for example increases greenhouse gas emissions due to deforestation (Schmitz *et al* 2014b). Some of these externalities are dealt with explicitly in the SSP storylines (see for example O'Neill *et al* 2015), but are beyond the scope of this paper. The current exercise involves only a partial implementation of the SSP storylines and limits the isolation of trade policy effects from climate impacts on agricultural markets. However, much more elaborate implementations of the SSP storylines are currently being developed and compared across integrated assessment models, including a focus on agriculture by the AgMIP modeling teams. These will highlight a broader set of themes and the interconnections across policies (for example the links between trade policies and environmental impacts such as greenhouse gas emissions and deforestation), and will allow a more comprehensive analysis of trade policy impacts in a multi-model setting.

It is important to note a number of other caveats with regard to this study. First, in line with previous work, we do not consider CO_2 fertilization effects on yields with higher CO_2 concentrations. The CO_2 effect is still very uncertain in the crop models owing to many complex interaction mechanisms, and it is widely discussed in the community. Omitting CO_2 fertilization effects would tend to suggest that our results err on the pessimistic side.

On the other hand, other omitted effects would tend to suggest that our results err on the optimistic side. For example, current crop models (together with limited inputs from GCMs) are not able to capture some potentially negative effects of climate change, such as the effects of extreme temperatures and precipitation, changes in pests and diseases, and ozone levels (Lobell and Gourdji 2012). These are likely to be significant (see e.g. Heft-Neal and Roland-Holst 2015), but incorporation in long-term projections awaits future modeling improvements.

Third, assumptions on adaptation in this study are limited to endogenous responses within the economic models, but improvements are needed in both economic and crop models in this regard. This also reflects the existing lack of conceptual and applied approaches for explicit assessments of adaptation options in this kind of aggregate economic models. While the economic models have some flexibility in factor substitution (e.g. using more labor, capital or intermediate inputs to compensate for less productive land), they do not cover large-scale endogenous investments for climate change adaptation (e.g. building new dams for increasing irrigation water availability or protective measures against sea-level rise). Moreover, endogenous R&D investments in climate-adapted crop varieties are not covered by most economic models.

Fourth, a number of other important dimensions are not yet incorporated in the scenarios we analyzed. Based on the RCP and SSP work in climate change mitigation assessments, it has been proposed to develop more detailed 'representative agricultural pathways' (RAPs) (e.g. Valdivia *et al* 2015). These need to be discussed and agreed upon in the wider agricultural modeling community (including biophysical as well as economic modelers), and should provide background information for a broad range of climate and global change impact studies at regional and global levels. In this study we take initial steps to better define scenario inputs with respect to agricultural productivity change and trade policy change for a subset of three SSPs. A comprehensive description and definition of global RAPs would require coverage of several key additional dimensions, potentially including agricultural land supply, environmental protection, and changes in dietary preferences, as well as a mapping of RAPs with all five SSPs and their particular storylines.

5. CONCLUSIONS

These findings illustrate the sensitivity of climate change impacts to differences in socioeconomic and emissions pathways, particularly for those characterized by slower economic growth (SSP 3) and higher emissions (RCP 8.5), and for which the impacts of climate change on yields, areas, prices and trade are relatively large. This in turn reinforces the importance of systematic comparison across more RCP/GCM × SSP combinations (including for SSP 4 and SSP 5, which were not included in the present study), as well as more detailed implementation of the various SSP storylines across several key dimensions.

Among those areas deserving further research are further development of RAPs, including dimensions such as agricultural land supply, changes in dietary preferences, and domestic farm policies as well as international trade (more systematically comparing baselines with climate impact scenarios); analysis of additional emissions pathways and SSP-specific mitigation strategies, especially bioenergy demand, biomass production and its effect on land use and crop prices; examination of regional results across models; and inclusion of additional climate, crop and economic models to further understand and potentially narrow differences in results. Topics deserving closer attention also include changes in inter- and intra-annual variability associated with climate change, improvement in our understanding of CO_2 fertilization effects as well as changes related to pests and diseases, and implications for food security at global, national and subnational scales. Work on many of these topics is currently under way through the Agricultural Model Intercomparison and Improvement Project (AgMIP) (Rosenzweig *et al* 2014).

ACKNOWLEDGMENTS

This study was made possible by a grant from the United States Department of Agriculture via the University Corporation for Atmospheric Research (UCAR Subaward No. Z14-13614). Support by and to the institutions where the authors are based, the Agricultural Model Intercomparison and Improvement Project (AgMIP), and comments from Ignacio Perez-Dominguez and three anonymous reviewers are also gratefully acknowledged. The views expressed are those of the authors and may not be attributed to the US Department of Agriculture, the Economic Research Service, or the supporting or other affiliated institutions.

FOOTNOTES

9 The core quantification of the SSP drivers has involved a number of different research groups. The demographic projections were prepared at the International Institute for Applied Systems Analysis (IIASA), see KC and Lutz (2012). Three independent sets of GDP projections were developed—all harmonized to the IIASA demographic projections. The different GDP projections were prepared by IIASA, the Organization for Economic Co-operation and Development (OECD) and the Potsdam Institute for Climate Impact Research (PIK). All the SSP scenarios described herein refer to the OECD-based GDP projections, version 0.93. A fifth team based at the National Center for Atmospheric Research (NCAR) prepared projections of urbanization rates. All the data are available for public download at https://secure.iiasa.ac.at/web-apps/ene/SspDb/dsd?Action

=htmlpage&page=about. IIASA (2013a) provides basic documentation of the main underlying features of the various projections.

10 In the 2050 timeframe analyzed here, global mean temperature change is not very different between GCM results for RCP 4.5 and RCP 6.0, and RCP 4.5 scenarios show slightly stronger warming and higher atmospheric CO_2 concentrations in 2050 than RCP 6.0 scenarios (Warszawski *et al* 2014).

11 SSP 1 is a more globalized world, so in the liberalized-trade variant of SSP 1 we remove tariffs and export subsidies on all trade in agricultural and food products. The trade measures are phased out over the period 2020–2035. Domestic support policies are not affected. For the restricted-trade variant of SSP 3 the world is divided into three trade blocks: East and South Asia; a broad European block that encompasses Western and Eastern Europe, Central Asia, the Middle East and Africa; and a Western Hemisphere block that includes North, Central and South America. Import tariffs between the blocks are doubled (with no change to export measures), but no changes are imposed on trade policies within blocks. (More details are provided in the supplementary material.)

12 This is broadly consistent with the AgMIP phase 1 results, where total demand for food crops increased between 55 and 97 percent depending on the model (see Valin *et al* 2014). It is slightly higher than the 60-percent increase reported in the most recent FAO projection (see Alexandratos and Bruinsma 2012). It should be noted that the results reported herein reflect version 0.93 of the SSP quantifications, not version 0.5 as in Phase 1, and also reflect individual model changes, for example to consumer behavior, including income and price elasticities (see supplementary material). Furthermore, the Alexandratos and Bruinsma (2012) projections are based on a dated World Bank scenario for GDP that projected lower income growth than even the relatively low-growth SSP 3 scenario included in the present analysis.

13 We do not assume differences in consumer preferences across SSPs here, although this is part of the SSP narratives (Kriegler *et al* 2012). One could imagine for example that the SSP 1 world might be one with less meat consumption as consumers' preferences towards more sustainable patterns of consumption lead to a decline in meat consumption in large parts of the world.

REFERENCES

1. *Alexandratos N and Bruinsma J 2012 World agriculture towards 2030/2050: the 2012 revision ESA Working Paper No. 12-03 (Rome: Food and Agriculture Organization of the United Nations)*
2. *Beintema N, Stads G-J, Fuglie K and Heisey P 2012 ASTI Global Assessment of Agricultural R&D Spending: Developing Countries Accelerate Investment (Washington, DC: International Food Policy Research Institute (IFPRI))*
3. *Bondeau A et al 2007 Modelling the role of agriculture for the 20th century global terrestrial carbon balance Glob. Change Biol.* ***13*** *679–706*
4. *Boote K J, Jones J W, White J W, Asseng S and Lizaso J I 2013 Putting mechanisms into crop production models Plant Cell Environ.* ***36*** *1658–72*
5. *Burnett K and Murphy S 2014 What place for international trade in food sovereignty? J. Peasant Stud.* ***41*** *1065–84*
6. *Dufresne J L et al 2013 Climate change projections using the IPSL-CM5 earth system model: from CMIP3 to CMIP5 Clim. Dyn.* ***40*** *2123–65*
7. *Eickhout B, van Meijl H, Tabeau A and Stehfest E 2009 The impact of environmental and climate constraints on global food supply Economic Analysis of Land Use in Global Climate Change Policy ed T Hertel, S Rose and R Tol (USA: Routledge)*
8. *Gourdji S M, Sibley A M and Lobell D B 2013 Global crop exposure to critical high temperatures in the reproductive period: historical trends and future projections Environ. Res. Lett.* ***8*** *024041*
9. *Heft-Neal S and Roland-Holst D 2015 Challenges and opportunities for climate adaptation in Thailand agriculture Building Resilience to Climate Change Through Climate Smart Agriculture ed D Zilberman, L Lipper, N McCarthy, S Asfaw and G Branca (Rome: United Nations Food and Agriculture Organization) (forthcoming)*
10. *Hempel S, Frieler K, Warszawski L, Schewe J and Piontek F 2013 A trend-preserving bias correction—the ISI-MIP approach Earth Syst. Dyn.* ***4*** *219–36*
11. *Huang H, von Lampe M and van Tongeren F 2011 Climate change and trade in agriculture Food Policy* ***36*** *S9–13*
12. *Intergovernmental Panel on Climate Change (IPCC) 2013 Climate Change 2013: The Physical Science Basis. Contribution of Working Group I to the Fifth Assessment Report of the Intergovernmental Panel on Climate Change ed T F Stocker (Cambridge: Cambridge University Press) et al*
13. *Intergovernmental Panel on Climate Change (IPCC) 2014 Climate Change 2014: Impacts, Adaptation, and Vulnerability. Part A: Global and Sectoral Aspects. Contribution of Working Group II to the Fifth Assessment Report of the Intergovernmental Panel on Climate Change ed C B Field et al (Cambridge: Cambridge University Press)*
14. *International Institute for Applied Systems Analysis (IIASA) 2013a Supplementary note for the SSP data sets, (prepared with NCAR, OECD and PIK) (*https://secure.iiasa.ac.at/web-apps/ene/SspDb/static/download /ssp_suplem entary %20text.pdf*)*
15. *International Institute for Applied Systems Analysis (IIASA) 2013b SSP Database (version 0.93)*
16. *Jones C D et al 2011 The HadGEM2-ES implementation of CMIP5 centennial simulations Geosci. Model. Dev.* ***4*** *543–70*
17. *KC S and Lutz W 2012 SSP Population Projections—Assumptions and Methods (Laxenburg, Austria: International Institute for Applied Systems Analysis (IIASA))*
18. *Kriegler E, O'Neill B C, Hallegatte S, Kram T, Lempert R J, Moss R H and Wilbanks T 2012 The need for and use of socio-economic scenarios for climate change analysis: a new approach based on shared socio-economic pathways Glob. Environ. Change* ***22*** *807–22*

19. *Lobell D B, Schlenker W and Costa-Roberts J 2011 Climate trends and global crop production since 1980 Science* ***333*** *616–20*
20. *Long S P, Ainsworth E A, Leakey A D B, Nösberger J and Ort D R 2006 Food for thought: lower-than-expected crop yield stimulation with rising CO2 concentrations Science* ***312*** *1918–21*
21. *Lotze-Campen H, Müller C, Bondeau A, Rost S, Popp A and Lucht W 2008 Global food demand, productivity growth and the scarcity of land and water resources: a spatially explicit mathematical programming approach Agric. Econ.* ***39*** *325–38*
22. *Lobell D B and Gourdji S M 2012 The influence of climate change on global crop productivity Plant Physiol.* ***160*** *1686–97*
23. *Meijl H van, van Rheenen T, Tabeau A and Eickhout B 2006 The impact of different policy environments on land use in Europe Agric. Ecosyst. Environ.* ***114*** *21–38*
24. *Moss R H et al 2010 The next generation of scenarios for climate change research and assessment Nature* ***463*** *747–56*
25. *Müller C and Robertson R D 2014 Projecting future crop productivity for global economic modeling Agric. Econ.* ***45*** *37–50*
26. *Nelson G C et al 2010 Food Security, Farming, and Climate Change to 2050: Scenarios, Results, Policy Options (Washington, DC: International Food Policy Research Institute)*
27. *Nelson G C et al 2014a Climate change effects on agriculture: economic responses to biophysical shocks* ***111*** *3274–9*
28. *Nelson G C et al 2014b Agriculture and climate change in global scenarios: why don't the models agree Agric. Econ.* ***45*** *85–101*
29. *O'Neill B C et al 2015 The roads ahead: narratives for shared socioeconomic pathways describing world futures in the 21st century Glob. Environ. Change at press ()*
30. *O'Neill B C, Kriegler E, Riahi K, Ebi K L, Hallegatte S, Carter T R, Mathur R and van Vuuren D P 2014 A new scenario framework for climate change research: the concept of shared socioeconomic pathways Clim. Change* ***122*** *387–400*
31. *Parry M L, Rosenzweig C, Iglesias A, Livermore M and Fischer G 2004 Effects of climate change on global food production under SRES emissions and socio-economic scenarios Glob. Environ. Change****14*** *53–67*
32. *Popp A, Lotze-Campen H and Bodirsky B 2010 Food consumption, diet shifts and associated non-CO2 greenhouse gases from agricultural production Glob. Environ. Change* ***20*** *451–62*
33. *Robinson S et al 2014 Comparing supply-side specifications in models of global agriculture and the food system Agric. Econ.* ***45*** *21–35*
34. *Rosegrant M W and the IMPACT Development Team 2012 International Model for Policy Analysis of Agricultural Commodities and Trade (IMPACT): model description (Washington, DC: International Food Policy Research Institute (IFPRI))*
35. *Rosenzweig C and Parry M L 1994 Potential impact of climate change on world food supply Nature****367*** *133–8*
36. *Rosenzweig C et al 2014 Assessing agricultural risks of climate change in the 21st century in a global gridded crop model intercomparison* ***111*** *3268–73*
37. *Sands R D, Jones C A and Marshall E 2014 Global Drivers of Agricultural Demand and Supply (Washington, DC: US Department of Agriculture, Economic Research Service) ERR-174 September 2014*
38. *Schmidhuber J and Tubiello F N 2007 Global food security under climate change* ***104*** *19703–8*
39. *Schmitz C et al 2014b Agricultural trade and tropical deforestation: interactions and related policy options Reg. Environ. Change ()*
40. *Schmitz C et al 2014 Land-use change trajectories up to 2050: insights from a global agro-economic model comparison Agric. Econ.* ***45*** *69–84*
41. *Schmitz C, Biewald A, Lotze-Campen H, Popp A, Dietrich J P, Bodirsky B, Krause M and Weindl I 2012 Trading more food—implications for land use, greenhouse gas emissions, and the food system Glob. Environ. Change* ***22*** *189–209*
42. *Smith L C 1998 Can FAO's measure of chronic undernourishment be strengthened? Food Policy* ***23****425–45*

43. *Tubiello F N, Amthor J S, Boote K J, Donatelli M, Easterling W, Fischer G, Gifford R M, Howden M, Reilly J and Rosenzweig C 2007 Crop response to elevated CO2 and world food supply—a comment on 'food for thought...' Science **312** 1918–21*
44. *Tubiello F N, Amthor J S, Boote K J, Donatelli M, Easterling W, Fischer G, Gifford R M, Howden M, Reilly J and Rosenzweig C 2006 Eur. J. Agron. **26** 215–23*
45. *Valdivia R O et al 2015 Representative agricultural pathways and scenarios for regional integrated assessment of climate change impact, vulnerability and adaptation Handbook of Climate Change and Agroecosystems: The Agricultural Model Intercomparison and Improvement Project Integrated Crop and Economic Assessments, Part 1 ed C Rosenzweig and D Hillel (London: Imperial College Press)*
46. *Valin H et al 2014 The future of food demand: understanding differences in global economic models Agric. Econ. **45** 51–67*
47. *van der Mensbrugghe D 2013 The Environmental Impact and Sustainability Applied General Equilibrium (ENVISAGE) Model Mimeo (Rome: FAO)*
48. *van Vuuren D P et al 2011 Representative concentration pathways: an overview Clim. Change **109** 5–31*
49. *von Lampe M et al 2014 Why do global long-term scenarios for agriculture differ? An overview of the AgMIP global economic model intercomparison Agric. Econ. **45** 3–20*
50. *Wang D, Heckathorn S A, Wang X Z and Philpott S M 2012 A meta-analysis of plant physiological and growth responses to temperature and elevated CO2 Oecologia **169** 1–1*
51. *Warszawski L, Frieler K, Huber V, Piontek F, Serdeczny O and Schewe J 2014 The Inter-Sectoral Impact Model Intercomparison Project (ISI–MIP): project framework Proc. Natl Acad. Sci. USA **111**3228–32*
52. *Watanabe S et al 2011 MIROC-ESM 2010: model description and basic results of CMIP5-20c3m experiments Geosci. Model Dev. **4** 845–7*
53. *Woltjer G, Kuiper M, Kavallari A, van Meijl H, Powell J, Rutten M, Shutes L and Tabeau A 2014 The MAGNET model: module description LEI Research Report (in print)*
54. Export references:

55. Citations

56. *Climate change and agriculture under CO 2 fertilization effects and farm level adaptation: Where do the models meet? Presley K. Wesseh and Boqiang Lin 2017 Applied Energy **195** 556*
57. *Impact of climate change on farms in smallholder farming systems: Yield impacts, economic implications and distributional effects Lemlem Teklegiorgis Habtemariam et al 2017 Agricultural Systems **152** 58*
58. *Cost of preventing workplace heat-related illness through worker breaks and the benefit of climate-change mitigation Jun'ya Takakura et al 2017 Environmental Research Letters **12** 064010*
59. *Focus on agriculture and forestry benefits of reducing climate change impacts Jeremy Martinich et al 2017 Environmental Research Letters **12** 060301*
60. *Kym Anderson 2016 259*
61. *Climate analogues suggest limited potential for intensification of production on current croplands under climate change T.A.M. Pugh et al 2016 Nature Communications **7** 12608*
62. *Projections of long-term food security with R&D driven technical change—A CGE analysis*
*Z. Smeets Kristkova et al 2016 NJAS - Wageningen Journal of Life Sciences **77** 39*
63. *Can sub-Saharan Africa feed itself? Martin K. van Ittersum et al 2016 Proceedings of the National Academy of Sciences **113** 14964*
64. *Fossil-fueled development (SSP5): An energy and resource intensive scenario for the 21st century Elmar Kriegler et al 2016 Global Environmental Change*
65. *Structural approaches to modeling the impact of climate change and adaptation technologies on crop yields and food security Shahnila Islam et al 2016 Global Food Security **10** 63*

66. *Impact of Climate Change on Cultivar Choice: Adaptation Strategies of Farmers and Advisors in German Cereal Production Janna Macholdt and Bernd Honermeier 2016 Agronomy* ***6*** *40*
67. *Mind the gap – the case for medium level emission scenarios Bas J. van Ruijven 2016 Climatic Change*
68. *Uncertainty in future agro-climate projections in the United States and benefits of greenhouse gas mitigation Erwan Monier et al 2016 Environmental Research Letters* ***11*** *055001*
69. *Climate variability, food production shocks, and violent conflict in Sub-Saharan Africa Halvard Buhaug et al 2015 Environmental Research Letters* ***10*** *125015*

CHAPTER 10

Effect of Sea Level Rise and Groundwater Withdrawal on Seawater Intrusion in the Gulf Coast Aquifer: Implications for Agriculture

Faye Anderson, Najla Al-Thani

Independent Scholars, Houston, TX, USA

ABSTRACT

The two main factors contributing to depletion of freshwater resources are climate change and anthropological variables. This study presents statistical analyses that are local in its specifics yet global in its relevance. The decline in Gulf Coast aquifer water quality and quantity has been alarming especially with the increased demand on fresh water in neighboring non-coastal communities. This study used seawater levels, groundwater use, and well data to investigate the association of these factors on the salinity of water indicated by chloride levels. Statistical analyses were conducted pointing to the high significance of both sea water level and groundwater withdrawals to chloride concentrations. However, groundwater withdrawal had higher significance which points to the need of water management systems in order to limit groundwater use. The findings also point to the great impact of increased groundwater salinity in the Gulf Coast aquifer on agriculture and socioeconomic status of coastal communities. The high costs of desalinization point to the increased signification of water rerouting and groundwater management systems. Further investigation and actions are in dire need to manage these vulnerabilities of the coastal communities.

Keywords: Sea Level Rise, Groundwater, Gulf Coast Aquifer, Coastal Vulnerability, Rerouting

1. INTRODUCTION

Groundwater makes up about one third of Earth's freshwater and is the primary source of clean water for more than 1.5 billion people worldwide and more than fifty percent of the United States (US) population. The main four issues with groundwater are overdraft, waterlogging, seawater intrusion, and pollution. The direct effects of climate change include groundwater recharge, discharge, storage, saltwater intrusion, biogeochemical reactions, chemical fate, and transport. Whereas climate change indirect effects are anthropological factors which exacerbate the direct effects [1] - [3] . Research has provided mixed results when comparing the impact of direct and indirect effects with spatial variations [4] .

The sharp interface between fresh groundwater and coastal salt water is affected by climate change due to the continual increase in sea level resulting from glacial melt and thermal expansion of water. By 2001, sea level has risen globally by 25 cm [3] . The increase in sea water level increases the height of water table which in turn results in subsurface flooding. This impacts septic systems, infrastructure, wetlands, and ecosystems [5] [6] . Another effect of sea level rise on groundwater is increased salinity. Fresh water pumping from coastal aquifers wells lowers the water table level below sea level which increases saltwater intrusion [4] [7] [8] .

The Gulf Coast aquifer is a major aquifer system extending from the Texas-Louisiana Border in the northeast to the Texas-Mexico border in the south. It is a heterogeneous complex system where the fluctuations of sea level throughout history resulted in sedimentary depositions in the coastal plains of the Gulf of Mexico Basin. Sediments of sand and clay continued to deposit with the change of sea level [9] . According to Texas Water Development Board (TWDB), freshwater saturated thickness in the aquifer averages about 1000 feet and the quality of its water varies with both depth and location being better in the north and declining to the south. Moreover, increased extraction has led to land subsidence in the most populated counties like Harris County and Fort Bend County [10] . The vulnerability of this coastal aquifer to increased extraction and climate change has impacted fresh water availability and caused increased flooding [7] . These lead to socioeconomic effects including displacement, evacuations, and public health issues. These impacts combined with other issues facing coastal communities in Texas and other places in the world, makes it more complicated to mitigate and manage the issues [11] - [18] . Disparity of impacts affect groups of lower income, non-English speakers, and vulnerable age groups (65 years or older, 5 years or younger) [14] . Seawater intrusion to the Gulf Coast aquifer is complicated and can take one or more of these three forms: lateral intrusion from the Gulf of Mexico; upward intrusion from deeper, more saline zones of a groundwater system; and downward intrusion from coastal waters [19] .

The objective of this study was to study the effect of sea level rise (direct climate change effect) and groundwater withdrawal (anthropologic indirect effect) on sea water intrusion in the Gulf Coast aquifer. The effects of these factors on the concentrations of chloride sampled from groundwater wells were investigated. Chloride is one of seawater intrusion indicators due to its chemical stability and similarity of its intruding rate to seawater's [20] . Chloride concentration in the Gulf Coast aquifer, among other constituents has been on the increase [21] .

Although chloride has a maximum contaminant level (MCL) of 250 mg/L [22] , a concentration of 100 mg/L or more has been assumed to indicate seawater intrusion [20] [23] .

2. METHODS

Monthly mean sea level data were downloaded from Sea Level Trends website developed by the National Oceanic and Atmospheric Administration (NOAA) [24] . Total chloride (Cl) in mg/L concentration along with other variables for the wells of Gulf Coast aquifer (Table 1) was extracted from the website of TWDB [25] . Estimated Use of Water in the United States County-Level Data for 2010 variables were downloaded from the United States Geological Survey (USGS) website [26] . Variables downloaded include total public supply groundwater withdrawals, domestic self-supplied fresh groundwater withdrawals, total industrial self-supplied groundwater withdrawals, fresh groundwater withdrawals for irrigation, fresh groundwater withdrawals for livestock, total groundwater withdrawals for aquaculture, and total groundwater withdrawals, in millions of gallons of water used per day (Mgal/d).

Statistical analyses were conducted at 0.05 significance level using GeoDa 1.6.7 [27] . ArcMap 10.3.1 was used for mapping [28] . Ordinary least squares with spatial dependence regression analyses were conducted. Spatial weights matrix was formed using wells longitude and latitude data [11] [29] - [31] . The spatial regression performed has the equation [Cl] = intercept + longitude + latitude +distance to the Gulf of Mexico + sea level + groundwater withdrawals + spatial residual.

3. RESULTS

Data for the twenty years from 1996 to 2015 were extracted and linked into one dataset. Summary statistics for chloride and pH are presented in Table 2. Annual average chloride has been higher than MCL except for year 2005 (247.28 mg/L). T-test and Fisher's F-test were conducted to test the hypothesis of changing means and variances respectively, between each two years (Figure 1). Tests' results led us to conclude that the means and variances for chloride have changed across the years. pH average remained near the neutral level of 7 (Figure 2). Figure 3 presents groundwater well locations along with their highest chloride concentrations during the twenty years from 1996 to 2015. The number of values was 177. Higher concentrations were concentrated in southern parts of the Gulf Coast Aquifer. Figure 4 presents the total groundwater withdrawals for 2010 in millions of gallons per day. Harris County (Houston, population of more than 5 million) was on top of the counties. Figure 5 presents a comparison of groundwater withdrawals percentage from the Gulf Coast aquifer per use. With more than sixty six percent groundwater withdrawal, irrigation is the major groundwater use in the communities of the Gulf Coast aquifer.

Kurtosis and skewness of chloride were 29.31 and 4.43 respectively. This indicates departure from normality and justified the application of logarithmic transformation in order to comply with regression assumptions. Pair-wise correlation coefficients are presented in Table 3. The pair-wise correlations between chloride and pH were low whereas the pair-wise association between the distance to the Gulf of Mexico and mean sea level was slightly higher. Correlation between chloride sampled from 1996 to 2014 and mean sea levels for the same years was 0.11 but kept increasing to 0.31 for the years 1900 and 1958. Table 4 presents the pair-wise associations between wells' chloride concentrations and their counties' groundwater withdrawal. All correlations except with withdrawal for livestock were negatively associated to chloride levels. Regression F-statistic came out significant and R-squared was 53%. Nevertheless, mean sea level was statistically significant with a p-value of 0.02 and a regression coefficient of 2.62, pH was not significant, and the distance to the Gulf of Mexico was significant

Table 1. List of variables downloaded and analyzed.

Variable	Source
Aquifer Name	TWDB
Latitude for the well location	TWDB
Longitude for the well location	TWDB
Sample Month	TWDB
Sample Day	TWDB
Sample Year	TWDB
County Name	TWDB
Chloride (mg/L)	TWDB
Well Depth (ft)	TWDB
Groundwater withdrawals	USGS
Sea level trends	NOAA

Table 2. Summary statistics for groundwater chloride and pH in the gulf coast aquifer.

	1996	1997	1998	1999	2000	2001	2002	2005	2006	2009	2010	2013	2015
Min Cl	46.00	8.00	320.00	148.00	68.30	8.44	309.00	28.00	50.00	3.16	228.00	4.56	217.00
Average Cl	293.19	356.22	320.00	865.48	449.72	470.20	1023.00	247.28	498.64	336.07	228.00	282.14	762.33
Std. Dev. Cl	301.13	825.67	203.00	926.43	535.12	996.77	858.22	209.83	679.83	429.81	305.00	348.41	907.56
Max Cl	853.30	6541.00	320.00	3920.00	1490.00	6840.00	2160.00	1110.00	2020.00	1490.00	228.00	1510.00	1810.00
Min pH	7.10	5.73	8.10	6.97	6.77	6.18	7.83	6.24	6.48	5.20	8.02	4.77	6.57
Average pH	7.78	7.38	8.10	7.44	7.20	7.23	7.87	7.33	7.21	7.30	8.02	7.32	7.24
Std. Dev. pH	0.36	0.50	0.70	0.46	0.26	0.51	0.06	0.54	0.46	0.75	0.01	0.76	0.59
Max pH	8.20	8.95	8.10	8.42	7.50	8.61	7.91	8.42	7.73	9.67	8.02	8.23	7.68

Table 3. Pair-wise associations between chloride concentration and the three variables.

	pH 1996-2014	Distance to the Gulf of Mexico 1996-2014	Mean Sea Level 1900-1958
Cl 1996-2014	0.04	0.15	0.31

Table 4. Pair-wise correlation coefficients between chloride concentration and the groundwater withdrawal variables.

Groundwater Withdrawal Variables	Cl (mg/L)
Public Supply, groundwater withdrawals, total, in Mgal/d	−0.089996493
Domestic, self-supplied groundwater withdrawals, fresh, in Mgal/d	−0.113263524
Industrial, self-supplied groundwater withdrawals, total, in Mgal/d	−0.129418839
Irrigation, groundwater withdrawals, fresh, in Mgal/d	−0.22733962
Livestock, groundwater withdrawals, fresh, in Mgal/d	0.047510713
Aquaculture, groundwater withdrawals, total, in Mgal/d	−0.100836603
Total groundwater withdrawals, total (fresh + saline), in Mgal/d	−0.219060339

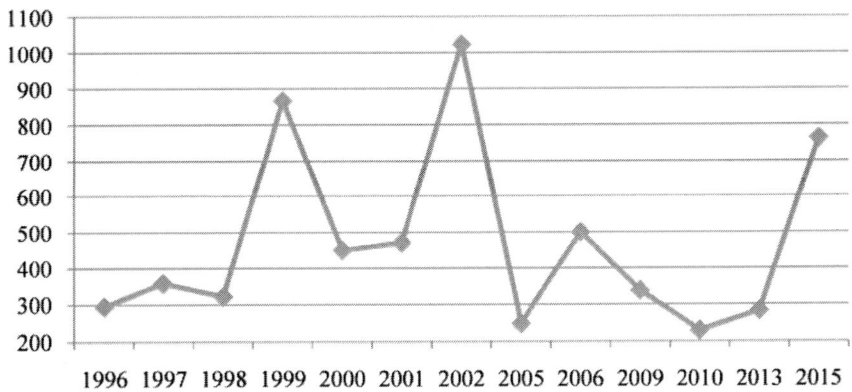

Figure 1. Annual average for chloride (mg/L) for groundwater sampled from the gulf coast aquifer.

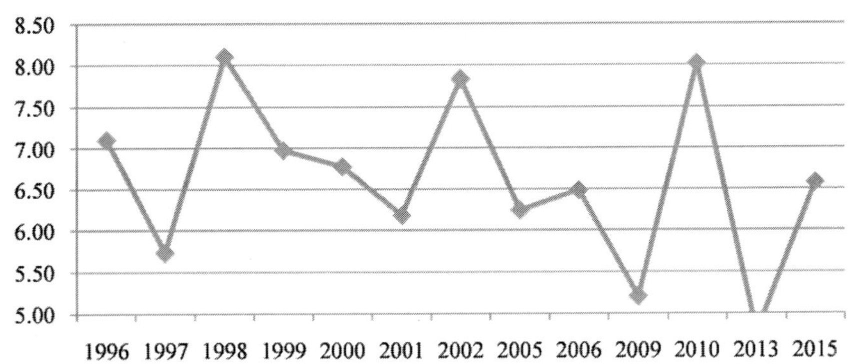

Figure 2. Annual average pH for groundwater sampled from the gulf coast aquifer.

with a regression coefficient of −1.02. The seven county groundwater withdrawal variables were highly significant to chloride levels. All except withdrawal for livestock and total withdrawals were negatively associated with the dependent variable (Table 5).

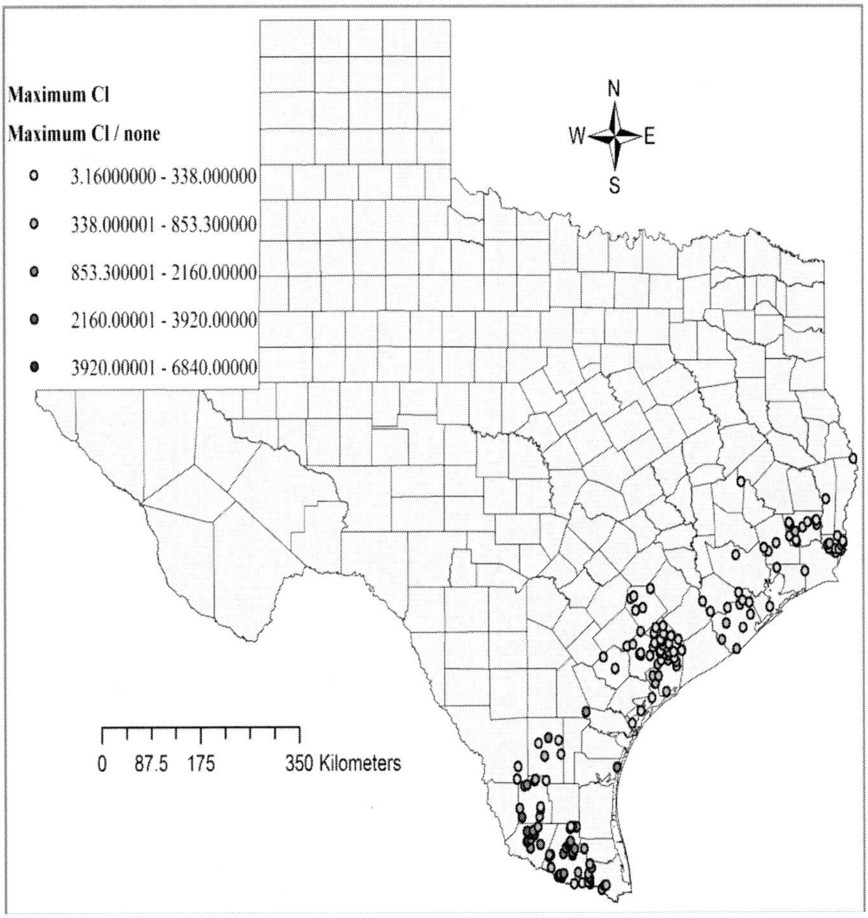

Figure 3. Maximum chloride concentrations, 1996-2014.

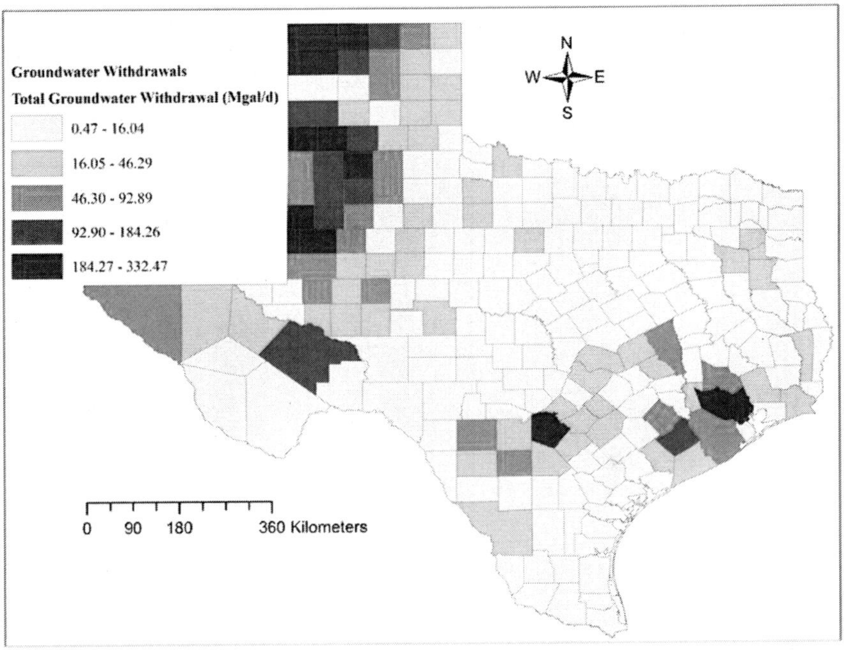

Figure 4. Total groundwater withdrawals for each county in 2010, in millions of gallons per day [26] .

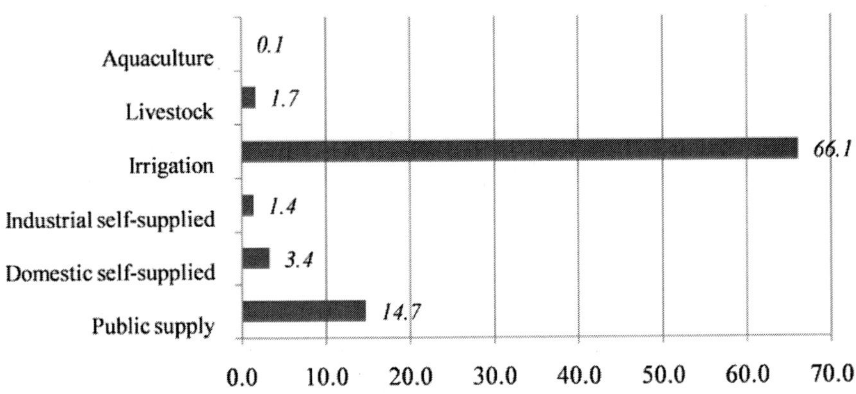

Figure 5. Groundwater withdrawal percentages per use in 2010 [26] .

Table 5. Ordinary least squares spatial regression coefficients.

Variable	Coefficient	Probability
Constant	6.96	0.00
pH	−0.05	0.40
Distance to the Gulf of Mexico	−1.02	0.00
Mean Sea Level	2.62	0.02
Public Supply, groundwater withdrawals, total, in Mgal/d	−0.12	0.00
Domestic, self-supplied groundwater withdrawals, fresh, in Mgal/d	−0.27	0.00
Industrial, self-supplied groundwater withdrawals, total, in Mgal/d	−0.24	0.00
Irrigation, groundwater withdrawals, fresh, in Mgal/d	−0.16	0.00
Livestock, groundwater withdrawals, fresh, in Mgal/d	1.43	0.00
Aquaculture, groundwater withdrawals, total, in Mgal/d	−1.45	0.00
Total groundwater withdrawals, total (fresh + saline), in Mgal/d	0.13	0.00

DISCUSSION

The high chloride concentration is a strong indicator of seawater intrusion [23] especially in wells within proximity of salt water bodies like the Gulf of Mexico. Annual average concentrations had high variations from 1996 to 2015. This is attributed to variations in the hydrologic cycle. According to the Intergovernmental Panel on Climate Change (IPCC), climate change increased precipitation variability in the US. Increased precipitation (with uncertain amounts) is caused by increased water vapor and evaporation associated with warmer temperatures. The change in precipitation varies from one region to another but has resulted in higher risks of flooding, decrease in summer water supply, and decrease of both surface and ground water quality [3] [18] . However, the United States Environmental Protection Agency (EPA) claims that the increase in precipitation in the US is estimated to be 6% whereas it is 2% worldwide [32] .

The spatial variation in chloride groundwater concentrations could be attributed to two factors: recharge and variations in population water use. Recharge is at its lowest in the southern parts of the Gulf Coast Aquifer [33] . The 95 year rolling window pair-wise association between chloride concentrations and sea water levels is supported by the characteristic timescale of 100 - 200 years of ocean thermal expansion due to global warming [34] .

Increased withdrawal of groundwater has deteriorated the quality of groundwater. This study contributes to the ongoing debate on the impacts of sea level rise versus withdrawal where the anthropological effects have more severe impacts [4] . This is supported by the higher statistical significance for groundwater withdrawal variables than that for sea level (Table 5) and the increased percentage of irrigation groundwater withdrawal (Figure 5).

Relevant mitigations include rerouting water, desalinization, seawater injection wells [8] , and demand management through legislation [35] [36] . Groundwater management systems are interpreted in decisions controlling the volume of water withdrawn annually from the aquifer, location of pumping and artificial recharge wells and their rates, and groundwater contamination control [37] . The high costs of desalinization make them a very expensive remedy for irrigation [38] . Hence, more efforts need to be directed towards water rerouting and management systems.

5. CONCLUSION

The primary goal of this study was to assess the effects of sea level and groundwater withdrawal on chloride levels in the Gulf Coast aquifer, as a proxy to sea water intrusion in the coastal aquifer. Data were extracted from three sources: NOAA, TWDB, and USGS. Statistical analyses demonstrated that temporal and spatial variability of chloride concentrations as well as its statistically significant associations with sea level and groundwater withdrawals. Validation demonstrated the robustness of the model. The findings support the calls for better water management especially in the southern communities where higher chloride concentrations were measured. Moreover, the viability of interventions like seawater barrier injection wells should be explored. Analyses point to the increased importance of water management and policies in order to protect freshwater sources especially for irrigation. Finally, in order to overcome the limitations of the study, further wider investigations are recommended for different coastal aquifers.

6. STRENGTHS AND LIMITATIONS

To our knowledge, this is the first study to estimate the statistical significance between groundwater salinity and both sea level and groundwater use for the Gulf Coast major aquifer. Despite the low R-squared value (53%) of the ordinary least square spatial regression, the analysis captured important predictors of groundwater chloride level.

Groundwater withdrawal data are at the county level [39] . Future research could benefit from groundwater extraction (pumpage) more accurate quantities at the well level. Also, the latest available data are for 2010 which is six years behind the timeframe of the study.

CITE THIS PAPER

Faye Anderson, Najla Al-Thani (2016) Effect of Sea Level Rise and Groundwater Withdrawal on Seawater Intrusion in the Gulf Coast Aquifer: Implications for Agriculture. *Journal of Geoscience and Environment Protection*,**04**,116-124.

REFERENCES

1. Winkler, A. (2013) Climate Change Effects on Groundwater Resources: A Global Synthesis of Findings and Recommendations. In: Treidel, M.-B.J. and Gurdak, J.J., Eds., Environmental Earth Sciences, Springer Science & Business Media B.V., 1937-1939.
2. Taylor, R.G., et al. (2013) Ground water and Climate Change. Nature Climate Change, 3, 322-329.
3. Romero-Lankao, P., et al. (2014) North America. In: Barros, V.R., Field, C.B., Dokken, D.J., Mastrandrea, M.D., Mach, K.J., Bilir, T.E., Chatterjee, M., Ebi, K.L. and Estrada, R.C.G.Y.O., Girma, B., Kissel, E.S., Levy, A.N., MacCracken, S., Mastrandrea, P.R. and White, L.L., Eds., Climate Change 2014: Impacts, Adaptation, and Vulnerability. Part B: Regional Aspects. Contribution of Working Group II to the Fifth Assessment Report of the Intergovernmental Panel on Climate Change, Cambridge University Press, Cambridge.
4. Ferguson, G. and Gleeson, T. (2012) Vulnerability of Coastal Aquifers to Groundwater Use and Climate Change. Nature Climate Change, 2, 342-345.
5. Anderson, F. (2015) The Development of Rural Sustainability Using Participatory Action Research: A Case Study from Guatemala. Journal of Human Resource and Sustainability Studies, 3, 28-33.
6. Anderson, F. and Al-Thani, N.N.J. (2015) Sustainability Atlas of Texas Ecoregions. Journal of Human Resource and Sustainability Studies, 3, 203-210.
7. Williams, K., et al. (2000) Sea-Level Rise and Coastal Forests on the Gulf of Mexico.
8. Johnson, T. (2007) Battling Seawater Intrusion in the Central & West Coast Basins. WRD Technical Bulletin.
9. Chowdhury, A.H. and Turco, M.J. (2006) Geology of the Gulf Coast Aquifer, Texas. In: Mace, R.E., et al., Eds., Aquifers of the Gulf Coast of Texas: Texas Water Development Board Report.
10. TWDB (2016) Gulf Coast Aquifer. http://www.twdb.texas.gov/groundwater /aquifer/majors/gulf-coast.asp
11. Anderson, F. (2016) Application of Multivariate Geostatistics in Environmental Epidemiology: Case Study from Houston, Texas. Journal of Geoscience and Environmental Protection, accepted.
12. Anderson, F. (2016) Assessing the Association between Heart Attack, High Blood Pressure, and Heart Disease Mortality Rates and Particulate Matter and Socioeconomic Status using Multivariate Geostatistical Model. Occupational

Diseases and Environmental Medicine, 4, 8-15. http://dx.doi.org/10.4236/odem.2016.41002
13. Anderson, F. (2016) Application of Multivariate Geostatistics in Environmental Epidemiology: Case Study from Houston, Texas. Journal of Geoscience and Environmental Protection, accepted.
14. Anderson, F. and Al-Thani, N. (2016) Case Study: Reviewing Methods of Assessing Community Adaptive Capacity for Jefferson County, Texas. American Journal of Climate Change, 5, 8-14.
15. Anderson, F. and Burau, K. (2015) Back to Basics: Is Civic Agriculture the Solution to Food Deserts in Texas? Open Journal of Social Sciences, 3, 82-89.
16. Anderson, F., Carson, A., Whitehead, L. and Burau, K. (2015) Age, Race and Gender Spatiotemporal Disparities of COPD Emergency Room Visits in Houston, Texas. Occupational Diseases and Environmental Medicine, 3, 1-9.
17. Anderson, F., Carson, A., Whitehead, L. and Burau, K. (2014) Spatiotemporal Analysis of the Effect of Ozone and Fine Particulate on CVD Emergency Room Visits in Harris County, Texas. Open Journal of Air Pollution, 3, 87-99.
18. Anderson, F. and Najla, N.J. (2016) Exploratory Analyses for Water Quality of the Gulf of Mexico Basin, Texas, 2005-2014. Journal of Geoscience and Environment Protection, 4, 58-63.
19. Barlow, P. and Reichard, E. (2010) Saltwater Intrusion in Coastal Regions of North America. Hydrogeology Journal, 18, 247-260.
20. USGS (2000) Is Seawater Intrusion Affecting Ground Water on Lopez Island, Washington?
21. Oden, J.H., Brown, D.W. and Oden, T.D. (2011) Groundwater Quality of the Gulf Coast Aquifer System. U.S. Geological Survey Data Series, USGS, Houston, TX.
22. EPA (2016) Secondary Drinking Water Standards: Guidance for Nuisance Chemicals. http://www.epa.gov/dwstandardsregulations/secondary-drinking-water-standards-guidance-nuisance-chemicals
23. Whiteman, K.J., et al. (1983) Occurrence, Quality, and Use of Ground Water in Orcas, San Juan, Lopez, and Shaw Islands, San Juan County, Washington, in Water-Resources Investigations Report.
24. NOAA (2016) Sea Level Trends. http://tidesandcurrents .noaa.gov/sltrends/sltrends.html
25. TWDB (2016) Groundwater Database Reports. http://www. twdb.texas.gov/groundwater/data/gwdbrpt.asp
26. USGS (2014) Estimated Use of Water in the United States County-Level Data for 2010.
27. Anselin, L., Syabri, I. and Kho, Y. (2006) GeoDa: An Introduction to Spatial Data Analysis. Geographical Analysis, 38, 5-22.
28. ESRI (2015) ArcMap 10.3.1. Redlands, CA.
29. Fotheringham, A.S., Charlton, M.E. and Brunsdon, C. (1998) Geographically Weighted Regression: A Natural Evolution of the Expansion Method for Spatial Data Analysis. Environment and Planning A, 30, 1905-1927.
30. Schabenberger, O. and Gotway, C.A. (2004) Statistical Methods for Spatial Data Analysis. Chapman and Hall/CRC, Boca Raton, 512 p.
31. Anderson, F. (2014) Multivariate Geostatistical Model for Groundwater Constituents in Texas. International Journal of Geosciences, 5, 1609-1617.
32. EPA (2015) Changing Rain and Snow Patterns. http://www3.epa.gov/climatechange/kids/impacts/signs/precip-patterns.html
33. TWDB (2011) Estimation of Groundwater Recharge to the Gulf Coast Aquifer in Texas, USA.
34. Jevrejeva, S., Moore, J.C. and Grinsted, A. (2010) How Will Sea Level Respond to Changes in Natural and Anthropogenic Forcings by 2100? Geophysical Research Letters, 37.
35. Taylor, R.G., et al. (2013) Ground Water and Climate Change. Nature Climate Change, 3, 322-329.

36. Winkler, A. (2013) Treidel H, Martin-Bordes JL, Gurdak JJ (eds) Climate Change Effects on Groundwater Resources: A Global Synthesis of Findings and Recommendations. Environmental Earth Sciences, 70, 1937-1939.
37. Alley, W.M., Healy, R.W., LaBaugh, J.W. and Reilly, T.E. (2002) Flow and Storage in Groundwater Systems. Science, 296, 1985-1990.
38. FAO (2004) Water Desalination for Agricultural Applications.
39. TWDB (2016) Historical Groundwater Pumpage. http://www.twdb.texas.gov/waterplanning/waterusesurvey/historical-pumpage.asp

CHAPTER 11

Impact of Climate Change on Agriculture Especially in Jessore and Sathkhira Districts According to Farmers Mitigation Strategies to Climate Change; Evidence from Farmer Level Data

Humayun Kabir MD*

Faculty of Urban and Regional Planning, International Islamic University Malaysia, Kulliyyah of Architecture and Environment Design (KAED), Jalan Gombak, 53100 Kuala Lumpur, Selangor, Malaysia

ABSTRACT

This study studies the adaptation of farmers to demeaning environmental situations likely to be caused or exacerbated under global climate change. It examines four central components: (1) The rate of self-reported acceptance of adaptive instruments (coping plans) consequently of changes in climate; (2) Ranking the potential coping plans based on their apparent importance to agricultural initiatives; (3) Documentation the socio-economic factors related with adoption of coping plans, and (4) Ranking potential limitations to adoption of coping plans based on farmers' reporting on the grade to which they face these restraints. As a preliminary matter, this paper also hearsays on the perceptions of farmers in the study about their involvements with climatic change. The study area is included of fifteen villages in the coastal region (Jessore and Sathkhira districts), a topographical region which climate change works has decorated as prone to accelerated degradation. Five hundred (500) farmers take part in the project's survey, from which the data was used to compute biased indexes for positions and to achieve logistic reversion. The rankings, model results, and expressive statistics, are stated here. Results showed that a common of the farmers self-identified as having involved in adaptive behavior. Out of 15 adaptation policies, irrigation ranked first among farm adaptive measures, while crop assurance has ranked as least utilized. The logit model clarified that out of eight factors surveyed, age, education, family size, farm size, family income, and involvement in collectives were meaningfully related to self-reported adaptation. Notwithstanding different support and technical interferences being available, lack of available water, lack of cultivable land, and random weather graded highest as the respondent group's restraints to coping with ecological poverty and change belongings. These results deliver policy makers and advance service providers with vital awareness, which can be used to better mark interventions which build endorse or facilitate the adoption of coping appliances with potential to build resiliency to altering climate and subsequent ecological effects.

Keywords: Farmers' adaptation; Climate change; Bangladesh; Jessore; Sathkhira

INTRODUCTION

This study asks how farmers adjust to changing climatic conditions, particularly drought. Agricultural lists first perceive the opposing effects of climate change at the farm manufacture level and then take adaptive actions to silent these adverse effects [1] as adaptive actions are able to reduce susceptibility [2]. Therefore, impact and adaptation are interlinked. It was evident that farmers incurred significant amounts of production losses due to climate change. This is focused on what mitigation strategies have been practiced by farmers in the study area to reduce their food production losses due to climate change.

The specific inquiry questions investigated in this part are: (a) What are farmers' perceptions of climate change in a very severe drought-prone area of Bangladesh? (b) What are the major mitigation strategies in the study area? (c) What are the determinants of farmers' mitigate choices? And (d) What are the barriers to effective mitigation to climate change? These four study questions will be addressed using micro data at the farm level. The organization of this part is as follows. Section 2 provides a brief overview of the literature. The theoretical framework is outlined in section 3. The methodology is presented in section 4. Section 5 reports and discusses the results while section 6 concludes the part.

CHANGE AND AGRICULTURE: A SHORT-TERM OVERVIEW OF THE LITERATURE

Climate change affects crop agriculture badly, particularly in countries in the lower latitudes of the world [1,3]. Alteration is seen as an essential policy option as compared to mitigation to limit the negative effects of climate change [4-6]. This is because mitigation has an insignificant impact on the current stock of greenhouse gases in the short run and, moreover, it requires collective and global actions [4].

There have been several studies examining the potential effects of climate change on agriculture globally [7-11]. The earlier studies assumed either no or little adaptation at an aggregate level. However, farmers' mitigation has been under researched especially at the farm level. Furthermore, an analysis of the determinants of adaptation strategies is limited in the climate change impact literature. Nevertheless, agroeconomic rationality implies that addressing climate change requires mitigate strategies and farmers usually make adjustments in their production processes to overcome any negatives experienced. Adaptation is very important if farmers are to counter the potential unfavourable impacts of climate change [4,10,12]. Adaptive measures are able to protect the livelihoods of poor farmers and ensure food security by reducing the potential negative impacts and reinforcing the advantages associated with climate change [1,2,13-15].

There is a growing number of study on farm level mitigation strategies and their determinants globally [5,1] However, mitigation in agriculture varies across countries. Different mitigation strategies are practiced by farmers depending on the climatic conditions, farm types and other conditions such as political, economic and institutional factors [2,16,17]. More precisely, mitigation choices are context specific and change from area to area and over time [18]. Therefore, country or area specific studies of climate change mitigation are required. In this context, research studies for Bangladesh are very limited [19-23].

Paul [19] documented some adjustment measures such as crop replacement, irrigation, gap filling and the inter-cropping of wheat and kaon (a local food crop). Ali [20] identified some adaptive measures such as the construction of embankments and cyclone shelters, and the introduction of new rice varieties suitable to higher salinity levels and temperatures. Rashid and Islam [23] identified drought, flood, soil salinity and cyclones as the major extreme climatic events which adversely affect agricultural operations and production. Changes in behavioural patterns, human practices and international actions are suggested as anticipatory adaptive measures. Based on focus group discussions and key informant interviews, Ahmed and Chowdhury [21] and FAO [22] identified the excavation of DTWs which facilitated irrigation, the excavation of ponds, switching to mango farming, the cultivation of short-duration and drought-tolerant crop varieties and home stead garden in gas major adaptation strategies for the Jessore and Sathkhira districts of southwest Bangladesh. However, none of these studies analyzed the determinants of **farmers' adaptation** strategies alongside the farmers' perception of **climate change** and the barriers to adaptation which are crucial for devising effective adaptation policies. Moreover, farm level adaptation strategies in the Jessore and Sathkhira districts have not been studied. Therefore, the objective of this chapter is to examine, using a detailed farm level dataset, farmers' perception of climate change, barriers to adaptation and factors affecting adaptation choices in rice production systems by using the case of farmers in greater Jessore, a severely drought-prone district of **Bangladesh**.

ASSUMED FRAMEWORK

Crop models or climate impact calculation techniques have been the most frequently used approaches to understanding the relationship between climate change and agriculture. Crop models are used to estimate the potential effects of future long-term climate change scenarios. Farmers' responses to climate variability and extreme climate events in these models are simply hypothetical, and either no adaptation or optimum adaptation is presumed [24]. Furthermore, climate scenarios under these models are inevitably not the scenarios to which farmers are most susceptible. A complementary approach, vulnerability theory, is used to explore the relationship between agricultural systems, the susceptibility to climate change and extreme events and farmers' adaptation explicitly [15]. The term 'vulnerability' generally represents 'the degree to which a system is susceptible to, or unable to cope with, adverse effects of climate change, including climate variability and extremes' [13]. The vulnerability of an agricultural system is explained as a function of exposure sensitivity, which indicates the susceptibility of a system to be affected by climate stimulus, and the adaptive capacity of the system [15]. According to the theory, vulnerability is positively related with exposure sensitivity while there is an adverse relationship between vulnerability and adaptive capacity which is the ability of a system, region or community to adapt to the impact of climate change [15]. More precisely, if exposure sensitivity increases, vulnerability also increases but increased adaptations are possible to alleviate vulnerability. Exposure sensitivity is not homogenous. It will vary from farm to farm as the characteristics of the farms that make them more or less vulnerable to particular climatic changes and extreme events are different. Vulnerabilities to climate change vary also because of sociodemographic, environmental, institutional and social characteristics that are either exogenous or endogenous to the community. Farmers'miigate capacity to climate change is influenced by socioeconomic, institutionalandsocialfactors.Socio-economiccharacterist icsincludeage,education, gender, household size, farm size, farming experience and wealth (i.e., household assets) [25,26]. Institutional factors consist of access to extension services, climate information and credit, and tenure status [1,26]. Finally, social capital includes farmerto- farmer extension services and the number of relatives living close by [16].These determinants may assist or restrict adaptation choices. In the case of agricultural systems, farmers are the first people confronting climate variability and change. It is thus essential to comprehend farmers' perceptions of and adaptations to climatic changes in order to diminish vulnerability and to enhance the overall resilience of the system [15].

METHODOLOGY

STUDY AREA

The extenuation part of this study took place in the same villages of **Jessore** and Sathkhira districts as earlier. District level analysis of climate data in Part 5 reveals that average annual rainfall across greater Jessore varies from 839 mm to 2241 mm. Moreover, the district average total rain fall for the 1964-2009 period is 1505 mm compared to 2408 mm for the whole country. Furthermore, the temperature in the district is as high as 44°C in May and as low as 6°C in January. In terms of extreme climate events, the district is severely drought- affected but is almost free from cyclones and floods [21,22].

DATA SOURCES

Micro data from a farm level survey conducted by the researcher is the main source of data for this analysis. The sample comprised the same 550 households who were selected randomly from the 15 selected villages. This part used data on socio-demographic features (age, gender, education and household size), farm characteristics (farm size and tenure status), institutional convenience (access to extension, weather information, credit, subsidy and irrigation facility) and farmers' perceptions about climate change, variation strategies and barriers to mitigation. Moreover, climate data from BMD (2010) is used to make a judgment of climate change in Jessore and **Sathkhira** with that of Bangladesh. The time period for this analysis is from 1964 to 2009 as climate data for the district is available for only this period.

THEORY OF CHANCE UTILITY AND A MICRO-ECONOMETRIC MODEL

Providing that various adaptive options are trained by farmers, the selection of the choice model can be either a multinomial probit (MNP) or a multinomial logit (MNL) model. This study uses the MNL model to analyze the determinants that affect farmers' choices of adaptation strategies. This is because this model gives more precise estimation results than the MNP model [27]. Moreover, the MNL model has been successfully and commonly used in some recent studies [5,12,16] while the MNP model is not usually used largely because of the practical difficulty involved in its estimation process.

Farmers' choice of mitigation strategies is a discrete and mutually exclusive choice. In the context of the current study, a farmer can select a strategy among eight alternatives:

(i) More irrigation, (ii) Short-duration rice, (iii) Supplementary irrigation, (iv)Changing planting date,(v)Agro-forestry, (vi)Use of different crop varieties, (vii) Non-rice crops and (viii) No adaptation. It is assumed that the selection of one of these strategies is independent of the other strategies. The choice of one strategy is categorized by various socio-demographic factors such as age, education, tenure status, access to climate information, extension services and subsidies.

The theoretical underpinning that a farmer chooses among different alternatives lies in the theory of random utility. In this theory, the utility of each alternative is exhibited as a linear function of observed characteristics (farmer and/or alternative specific) plus an additive error term. Furthermore, farmers are assumed to select the alternative that has the highest utility. More particularly, the utility a farmer from alternatives j and k is given by

$$Uij = +\varepsilon ij$$

$$= V + \varepsilon ik$$

respectively; where Vij and Vik imply the deterministic or systematic component of the utility, and εij and εik represent the stochastic component which represents the uncertainty.

This is the MNL model (Greene 2003). The MNL model significantly requires the assumption of independence of irrelevant alternatives (IIA) to hold in order to obtain unbiased and consistent parameter estimates. The IIA assumption necessitates that the probability of adopting a particular adaptation strategy by a given farm household requires independence from the probability of selecting another adaptation strategy.

The numerator is the utility (i.e., net benefit) from choice j and the denominator is the sum of utilities of all alternative choices. The probability of selecting a specific adaptation strategy is equal to the probability of that specific alternative being higher than or equal to the utilities of all other alternatives in the set of strategies. The parameters of this model can be estimated using maximum likelihood methods. However, the parameter estimates of the MNL model merely show the direction of the impact of the explanatory variables on the dependent variable. The real extent of changes or probabilities is not represented by the estimates. Moreover, parameter estimates are hard to interpret since they are derived from non-linear estimates (Greene 2003). Therefore, the MNL model parameters are transferred in to relative risk ratios (RRR). This RRR measures the effects on the relative odds of one outcome being selected relative to the base line outcome for a unit change in any of the explanatory variables.

STUDY OUTCOMES AND ARGUMENT

OVERVIEW OF CLIMATE CHANGE IN JESSORE AND SATHKHIRA DISTRICTS AND COMPARISON TO BANGLADESH: These Jessore and Sathkhira districts are in the southwestern climatic sub-zone (Zone E) which is characterized by very hot summers and relatively low rainfall [28]. Data on maximum temperature, minimum temperature and rainfall for the 1964-2009 period has been analyzed in this Chapter to assess the changes in these climate variables. A comparison of the Jessore and sathkhira districts with the whole country was also made. Data source was the Bangladesh Meteorological Department. **FIGURE 1** shows that there is an increasing trend in maximum temperature for both the Jessore and sathkhira districts and Bangladesh over the period. The growth in maximum temperatures in the greater jessoredistrictis higher than that for Bangladesh (**FIGURE 2**).

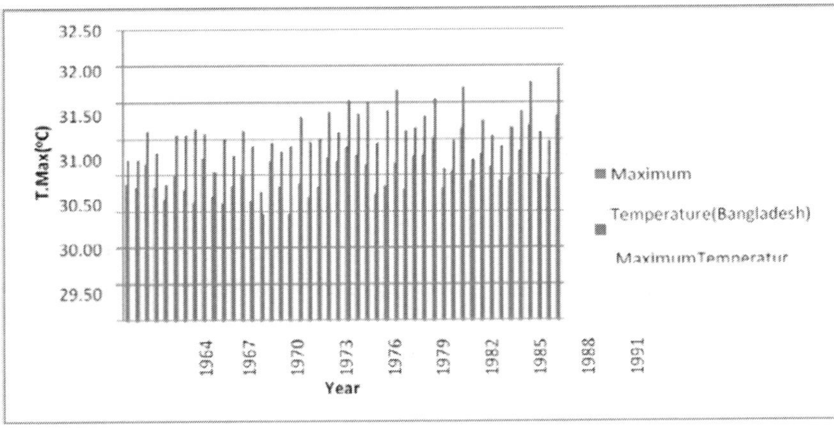

Source: Bangladesh Meteorological Department.

FIGURE 1: Trends in maximum temperature.

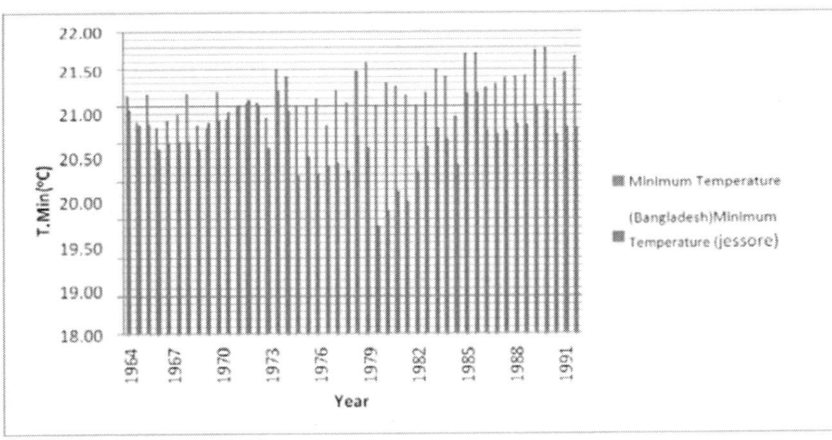

Source: Bangladesh Meteorological Department.

FIGURE 2: Trend in minimum temperature.

Although the minimum temperature in Bangladesh has increased over time, it has decreased in the Jessore and Sathkhira. The difference between maximum and minimum temperature is always higher for the Jessore and Sathkhiradistricts as compared to the whole of Bangladesh as depicted in **FIGURE 3**.

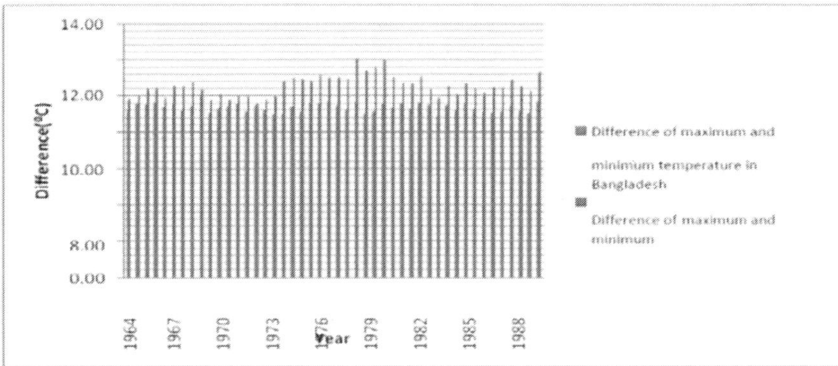

Source: Bangladesh Meteorological Department.

FIGURE 3: Difference of maximum and minimum temperature.

The total mean annual rainfall has increased for Bangladesh while it has decreased for greater Jessore. Annual total rainfall in greaterJessore is far below that of the whole country as illustrated in **FIGURE 4**.

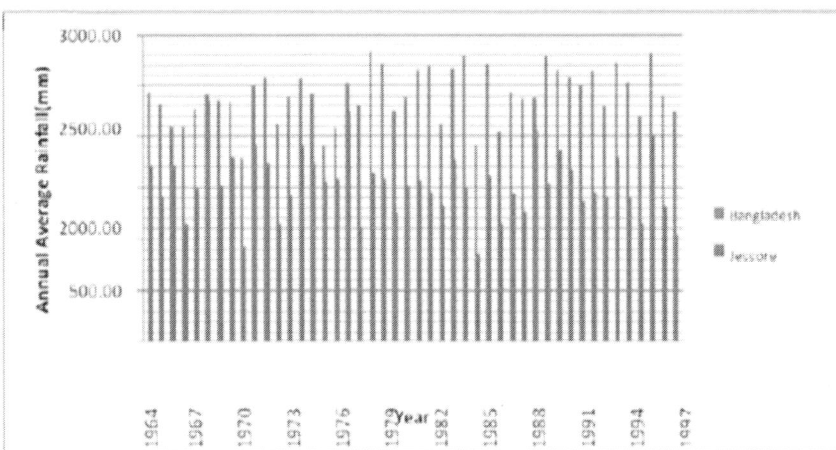

Source: Bangladesh Meteorological Department.

FIGURE 4: Trend in annual average total rainfall.

In addition to the changes in maximum temperature, minimum temperature and annual rainfall, the frequency and severity of droughts have increased in recent times in the Jessore and Sathkhira districts.

FARMER'S PERCEPTION OF CLIMATE CHANGE: Farmers should perceive first that there is climate change in order to take necessary adaptive strategies [1]. The surveyed farm household heads were asked about their perceptions of changes in various climate variables over the past 20 years. The major components were yearly temperature, rainfall, drought, and the availability of groundwater and surface water. Perceptions on climatic components were divided into four categories: increased, decreased, remaining same and don't know. Farmers' perceptions on each climatic parameter change are presented below.

TEMPERATURE CHANGES: The results in **FIGURE 5** signify that 97% of household heads have noticed rising temperatures while only an insignificant 0.55% noticed a decrease in temperature. Temperature remained unchanged for 1% of household heads while another 1%of household heads had no knowledge about it. Most of the farmers' perceptions are in accordance with the analysis of official data in the previous section.

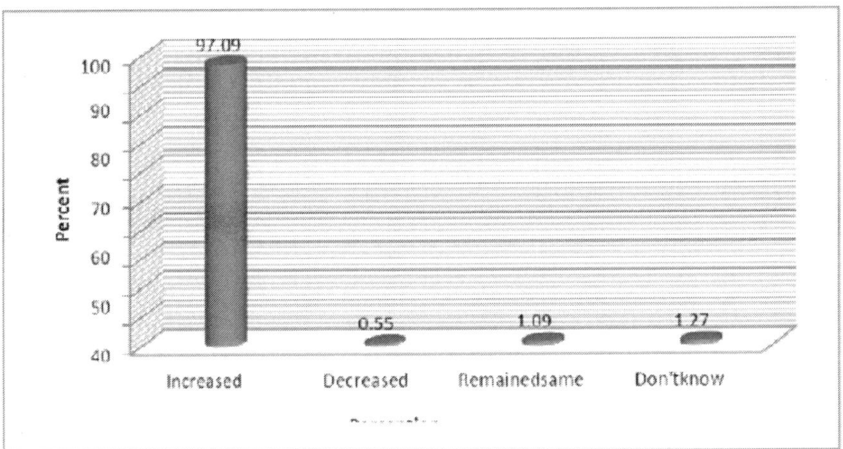

Source: Bangladesh Meteorological Department.

FIGURE 5: Farmers' perceptions of yearly temperature changes over the last 20years rainfall changes.

The results in **FIGURE 6** indicate that 99% of household heads observed a decline in total yearly rainfall. No household heads perceived an increase in rainfall while rainfall remained the same to 0.36% of households. Analysis of official rainfall data in Unit is consistent with the perception of the majority of household heads.

CHANGES IN DROUGHTS: The study area is a drought-prone area. Other extreme events such as cyclone and floods are almost nonexistent. Accordingly, farmers' perception of droughts is reported in **FIGURE 6**. Nearly 100% of households noticed that frequency of drought has increased over the last 20 years.

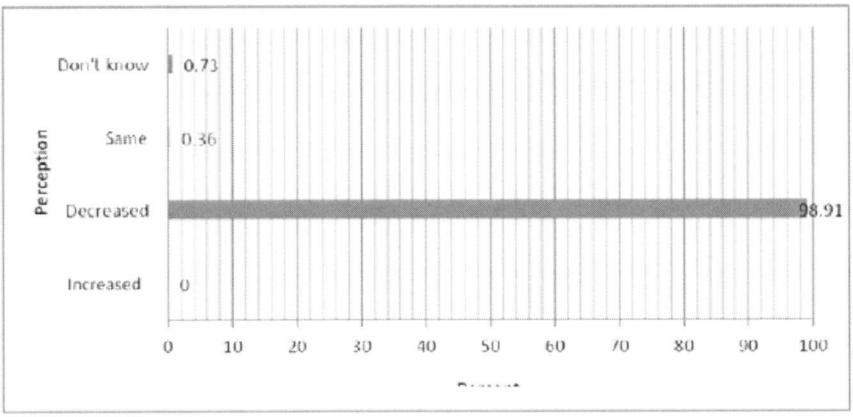

Source: Bangladesh Meteorological Department.

FIGURE 6: Farmers' perceptions of yearly rainfall changes over the last 20years changes in droughts.

CHANGES IN OTHER CLIMATIC PARAMETERS: Other important climate parameters include groundwater, surface water, heat waves and colder weather. Farmers were also asked about these over the past 20 years. Farmers' views on these parameters are shown in **TABLE 1**. Almost 100% of the household heads perceived that availability of both groundwater and surface water had decreased. The severity of heat waves had increased for nearly 100% of household heads while the perception on the severity of colder weather is diverse.

FARM-LEVEL ADAPTATION STRATEGIES: It is useful to discover adaptation strategies in order to obtain an understanding of an agricultural system's mitigate capacity [15]. Farmers in the study area were asked to reveal their major adaptive strategies in response to changing climate. These are summarized in **FIGURE 7**.

TABLE 1: Farmers perceptions of other climate parameters over last 20 years.

Other climate parameters (yearly)				
Farmers' perception	**Availability of ground water (%)**	**Availability of surface water (%)**	**Severity of heat wave (%)**	**Severity of colder weather (%)**
Increased	0	0	98.9	43.45
Decreased	99.64	98.9	0.36	40
Remained same	0.36	1.1	0.37	8
Don't know	0	0	0.37	8.55

Source: Bangladesh Meteorological Department

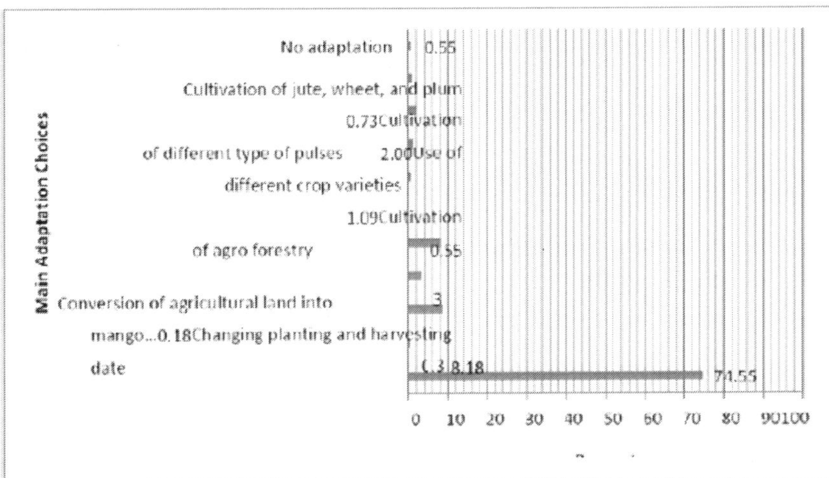

Source: Bangladesh Meteorological Department.

FIGURE 7: Farmers' main adaption strategies.

Farmers have adopted a variety of adaptation strategies including irrigation, direct seeded rice, greater emphasis on Aman rice with supplementary irrigation, short-duration rice varieties, changing planting and harvesting dates, the conversion of paddy land into mango or chards, agro-forestry, using different crop varieties, the cultivation of various pulses and the cultivation of jute and wheat. Irrigation is the most commonly used method (75%).Other main adaptive choices are changing the planting date and supplementary irrigation for Aman rice. In addition to the main mitigation strategies farmers were asked about their secondary adjustment measures which are presented in **FIGURE 8**.

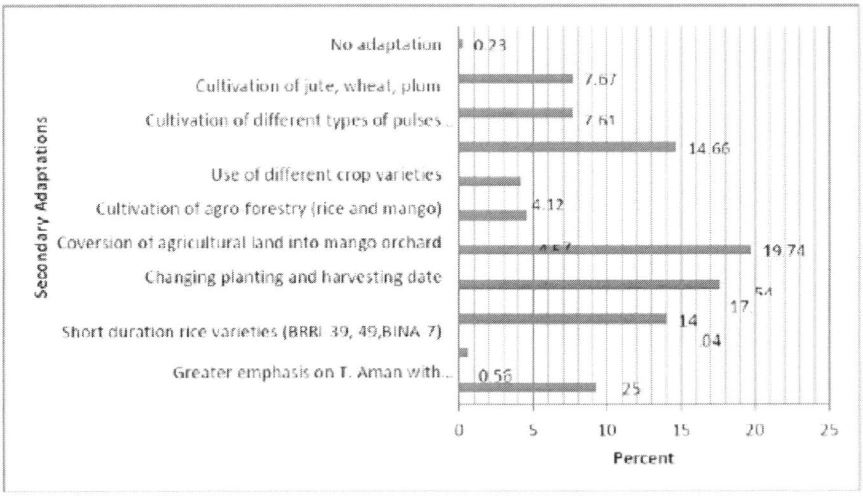

Source: Bangladesh Meteorological Department.

FIGURE 8: Farmers' secondary mitigation measures.

Farmers' secondary mitigation measures Changing planting date, cultivation of short-duration rice, using different crop varieties and supplementary irrigation for Aman rice are important secondary adaptation strategies. The main and secondary mitigation choices mentioned by farmers are very similar to those found in other studies for adjacent districts [21]. The adoption of these adaptation strategies implies that the farmers in the study area are risk–averse.

ADJUST TO THE ALLEVIATION

Famers take other adjustment measures after adaptation has taken place. This is because mitigation incurs costs. Mitigate strategies might resolve one problem but they sometimes create other problems which necessitate an 'mitigate to mitigations' [19]. Farmers in the study area do take other adjustment measures after mitigation (**FIGURE 9**). The results reveal that 28% of households took loans from rural usury lenders and relatives, 26% sold their livestock and nearly 17% used their previous savings in order to undertake mitigation measures. Other adapt to mitigation measures included the sale of other assets, mortgaging of land, borrowing institutional micro-credit, and family members migrating to urban areas in search of additional income sources.

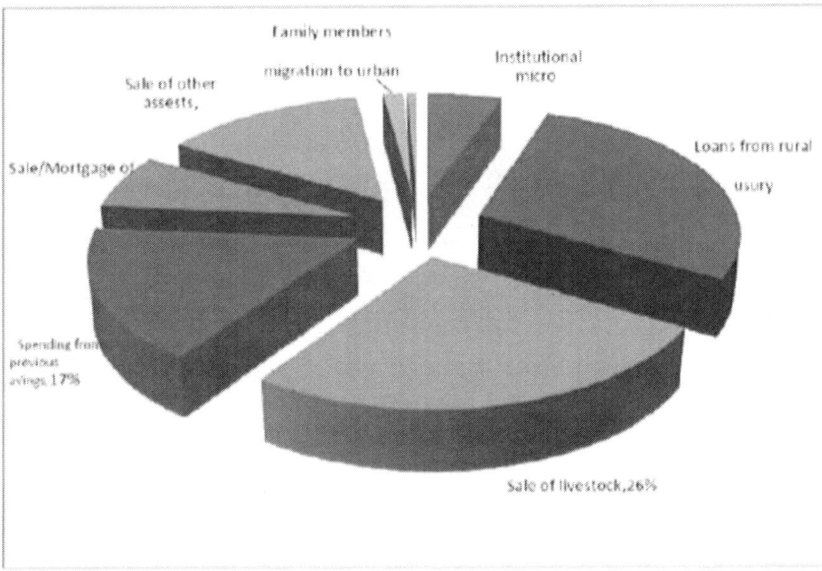

Source: Bangladesh Meteorological Department.

FIGURE 9: Measures for mitigate to mitigation.

OBSTACLES TO MITIGATION

Issues such as accessibility and usefulness of climate information, the institutional environment and the socio-economic situation of households affect farmers' capacity to adapt to climate change [29-31]. Farmers perceived barriers to the adoption of various mitigation strategies (**FIGURE 10**). Farmers outlined the most important barriers as a lack of weather information, a lack of knowledge on appropriate adaptation strategies and a lack of credit (moneyor saving). Other important barriers area lack of own land, a lack of irrigation water and labour shortages.

DETERMINANTS OF MITIGATIONS: EVIDENCE FROM THE MNL MODEL

MODEL VARIABLES: The mitigation MNL model with the 11 choices as exposed in **FIGURE 7** failed to produce realistic results in terms of arithmetical significance of the parameter estimates and on the edge effects. Following Gbetibouo, the model was reorganized by categorizing closely related strategies into the same group. The merging of direct-seeded rice with short- duration rice, the integration of conversion of agricultural land into shirm farm with Agro economy, and the cultivation of jute, wheat, plum and different types of pulses were grouped into non-rice crops. Therefore, the options finally included in theMNL model had eight categories: (i) More irrigation, (ii) Short-duration rice, (iii)Greater emphasison additional irrigation for Amanrice,(iv) Altering planting date,(v) Agro-forestry, (vi) Use of different crop varieties, (vii) Non-rice harvests and (viii)No variation (**FIGURE 11**). However, the last category is the reference category in this analysis. The dependent variable of the MNL model is thus the choice of mitigation having eight categories.

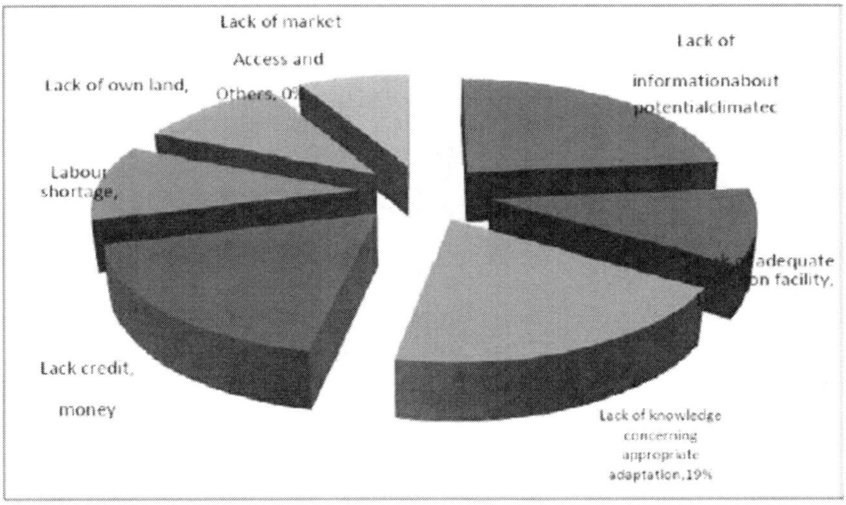

Source: Bangladesh Meteorological Department.

FIGURE 10: Barriers to adaptation.

The explanatory variables for this study have been selected on the basis of the available literature. They include household, farm and institutional characteristics of gender, age and education of household head, household size, household assets, farm income, farm size, tenure status, farming experience, livestock ownership, access to institutional extension services, farmer-to-farmer extension, information on climate change, access to credit, subsidy, electricity and distance to market (**TABLE 2**).

RESULTS FROM THE MNL MODEL AND DISCUSSIONS

The MNL model with eight categories of mitigation choices was run and tested for the IIA assumption by applying the Hausman test. The results of the Hausman test are set out in **TABLE 3**. AllP-values for omitted variables are 1.00 indicating that the model has passed the assumption. If the chi-square value is less than 0.00, the estimated model does not meet the asymptotic assumptions of the test. Negative test statistics are very common in quantitative work. Hausman and McFadden noted this possibility and concluded that a negative result was evidence that the assumption of IIA had not been violated.

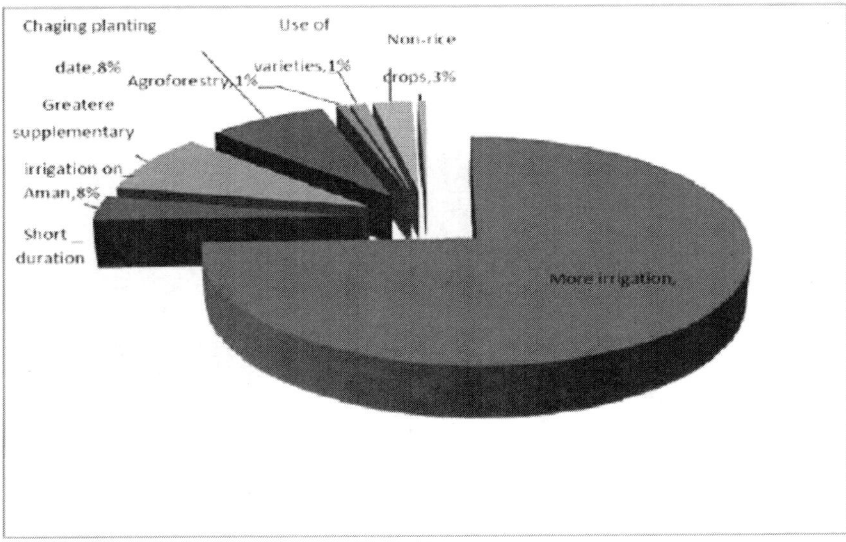

Source: Bangladesh Meteorological Department.

FIGURE 11: Farmers' main mitigation choices.

TABLE 2: Explanatory variables hypothesized to affect adaptation strategies.

Variables	Value	Expected sign	Citations
Gender of household head	1=male,0=female	+/-	Nhemacha [12], Deressa et al. [16] Gbetibouo [26]
Age of household head	Years	+/-	Nhemacha [12], Deressa et al. [16] Gbetibouo [26], Hisali et al. [17] Mendelsohn [5]
Education of household head	Years	+	Deressa et al. [16], Mendelsohn [5]
Household size	Number	+	Nhemacha [12], Deressa et al. [16] Gbetibouo [26], Hisali et al. [17] Mendelsohn [5], Bryan et al. [1]
Farm income	Tk.	+	Nhemacha [12], Deressa et al. [16]
Household assets	Tk.	+	Bryan et al. [1], Gbetibouo [26].
Farm/land size/land area	Decimal	+	Nhemacha [12], Deressa et al. [16] Gbetibouo [26], Bryan et al. [1]
Tenure status	1=own,0=otherwise	+	Gbetibouo [26], Bryan et al. [1]
Farming experience	Years	+	Gbetibouo [26]
Livestock ownership	1=Yes, 0=No	+	Deressa et al. [16]
Access to extension (institutional)	1=Yes, 0=No	+	Nhemacha [12], Deressa et al. [16], Gbetibouo [26], Hisali et al. [17], Bryan et al. [1]
Farmer-to-farmer extension	1=Yes, 0=No	+	Deressa et al. [16]
Information on climate change	1=Yes, 0=No	+	Deressa et al. [16], Gbetibouo [26], Gbetibouo [26]
Credit access	1=Yes, 0=No	+	Nhemacha [12], Deressa et al. [16], Gbetibouo [26], Hisali et al. [17], Bryan et al. [1]
Access to subsidies	1=Yes, 0=No	+	Kurukulasuriya [11]
Access to electricity	1=Yes, 0=No	+	Nhemacha [12]
Distance to market	Kilometres	-	Deressa et al. [16], Hisali et al. [17], Bryan et al. [1]

Source: Bangladesh Meteorological Department by Sarker

TABLE 3: Hausman test of IIA assumption for the MNL model.

Omitted	Chi-square	d.f.	P > chi-square	Evidence for H_0
More irrigation	0.052	6	1.000	Yes
Short-duration rice	- 82.386	18	1.000	Yes
Supplementary irrigation	- 49.955	13	1.000	Yes
Changing planting date	- 84.181	16	1.000	Yes
Agro-forestry	- 74.813	15	1.000	Yes
Use of different varieties	- 46.435	13	1.000	Yes
Non-rice crops	- 50.571	13	1.000	Yes
No adaptation	- 53.806	15	1.000	Yes

Ho: odds (outcome-J vs Outcome-K) are independent of other alternatives
Source: Bangladesh Meteorological Department by sarker

TABLE 4: Relative risk ratios of the MNL model for rice farmers' adaptation.

Explanatory variables/ Adaptive strategies	Irrigation		Short-duration rice		Supplementary irrigation		Changing planting date		Agro-forestry		Different crop varieties	Non-rice crop		
	RRR	P level	RRR	P level	RRR	P level	RRR	P level	RRR	P level	RRR	Plevel	RRR	P level
Gender of household head	6.69e-09*	0.096	5.05e-09*	0.095	8.026	-	1.09E-08	0.109	0.117	1	2.656	0.938	2.65e-09*	0.083
Age of household head	9.725	0.237	37.680*	0.097	16.578	0.157	18.084	0.182	10.947	0.457	53.192	0.139	14.205	0.262
Education of household head	2.577**	0.042	1.751	0.283	2.504*	0.061	4.017***	0.005	10.506***	0.009	3.414**	0.039	4.742***	0.01
Household size	0.3	0.244	0.148	0.103	0.255	0.213	0.573	0.616	0.524	0.652	0.128	0.128	0.182	0.188
Yearly farm income	12.531***	0.001	6.317**	0.026	8.559***	0.008	4.573	0.069	10.760**	0.031	5.304*	0.074	4.649*	0.075
Household asset	0.652	0.356	0.706	0.512	0.7	0.471	0.606	0.322	0.674	0.509	0.280**	0.028	0.747	0.599
Farm size	0.377	0.145	0.586	0.495	0.355	0.156	0.358	0.17	1.714	0.643	7.468**	0.048	0.532	0.458
Tenure status	4.332	0.133	7.169*	0.081	5.34	0.106	3.171	0.266	1.36E+08	0.998	12.909*	0.096	10.186*	0.073
Farming experience	0.129*	0.053	0.094**	0.041	0.116*	0.054	0.156	0.113	0.197	0.335	0.119	0.13	0.164	0.137
Livestock ownership	1.917	0.509	6.78	0.179	3.588	0.292	1.258	0.835	8.48E+07	0.998	0.963	0.976	1.375	0.79
Access to extension	1.676	0.55	1.102	0.923	1.383	0.73	0.274	0.23	7.425	0.112	1.352	0.789	1.516	0.694
Farmer-to-farmer extension	8.078	0.126	1.60e+10**	0.003	9.07E+08	0.046	9.169	0.146	17.459	0.167	12.211	0.214	1.42E+10	-
Information on climate change	1.096	0.915	0.881	0.901	0.533	0.508	0.642	0.653	0.937	0.959	1.333	0.814	1.611	0.673
Credit access	3.84	0.124	8.672**	0.031	2.204	0.414	1.9	0.531	1.715	0.679	2.689	0.394	3.88	0.198
Access to subsidies	0.172*	0.058	0.247	0.217	0.264	0.242	0.147	0.173	0.183	0.299	0.676	0.918	0.336	0.352
Access to electricity	2.36	0.267	1.254	0.809	2.63	0.262	7.648**	0.026	1.04E+08	0.998	7.432	0.154	2.48	0.398
Distance to market	0.389	0.181	0.472	0.35	0.717	0.656	0.973	0.971	1.694	0.598	0.265	0.156	0.251	0.108
Model Summary														
Base outcome	: No adaptation													
Number of observations	: 550													
LRchi-square	:244.32													
Prob>chi-square	: 0.000													
Loglikelihood	:-497.77													
Pseudo-R2	: 0.20													

***Significant at 1% probability level,**Significant at 5% probability level,*Significant at 10% probability level. RRR = Relative risk ratio
Source: Bangladesh Meteorological Department
Collected by Rashid

Therefore, the use of the MNL model for adaptation strategies is justified. Probabilities of chi-square values are positive which indicate that the use of MNL model for the dataset is valid.

As most of the explanatory variables are dummies, the RRR can be explained as the relative probability of choosing alternative j to no adaptation which is the base category (or comparison group). Following Yip et al. [32] and Hisali et al. [17], RRR is presented for each adaptation choice (choice j) given a particular characteristic (xi) in **TABLE 4** as well as factors that guide farm household choice of an mitigation choice in the face of climate change. The probability value of LR chi-square implies that all variables are jointly significant though some variables are not individually statistically significant. Following Bryan et al. [1], only the statistically significant variables affecting adaptation choices are discussed here.

GENDER OF HOUSEHOLD HEAD: The results show that male-led households increase the chances of more irrigation, the use of shortduration rice and non-rice crops as opposed to using no mitigation. This is probably because male-led households are more informed about new technology than female-led households [1,33].

AGE OF HOUSEHOLD HEAD: Age of the household head is a proxy for experience and affects mitigation strategies to climate change [1]. Our results reveal that age is significant for short- duration rice and the value of RRR indicates a unit increase in age of household head increases the possibility of the use of short-duration rice. This finding is consistent with Kebede et al. and Deressa et al. [1].

EDUCATION OF HOUSEHOLD HEAD: Higher levels of education are positively related to the adoption of improved technologies: farmers with more schooling are expected to adapt better to climatic changes and extreme climate events [1,34-36]. Years of education of household head is a significant determinant or all adaptation strategies excluding short-duration rice. The values of RRR indicate that the education of the household head increases the chances of adopting irrigation, supplementary irrigation, changing planting date, agro-forestry, different crop varieties and non-rice crop relative to the choice of no adaptation.

CONCLUSION

The study clarified that out of eight factors surveyed; age, education, family size, farm size, family income, and involvement in cooperatives were meaningfully related to self-reported adaptation. Notwithstanding different support and technical intrusions being available, lack of available water, lack of cultivable land, and random weather graded highest as the defendant group's restraints to coping with ecological poverty and change belongings. These results deliver policy makers and advance service providers with vital awareness, which can be used to better mark interventions which build endorse or facilitate the adoption of coping appliances with potential to build resiliency to changing climate and following biological possessions.

REFERENCES

1. Bryan E, Deressa TT, Gbetibouo GA, Ringler C (2009) Adaptation to climate change in Ethiopia and South Africa: options and constraints. Environmental scienceand policy 12: 413-26.
2. Reidsma P, Ewert F, Lansink AO, Leemans R (2010) Adaptation to climate change and climate variability in European agriculture: The importance of farm level responses. European Journal of Agronomy 32: 91-102.
3. Liu S, Mo X, Lin Z, Xu Y, Ji J et al. (2010) Crop yield responses to climate change in the Huang-Huai-Hai Plain of China. Agricultural water management 97: 1195-209.
4. Stern N (2006) What is the economics of climate change? World Economics 7: 1-10.
5. Kurukulasuriya P, Mendelsohn R (2008) A Ricardian analysis of the impact of climate change on African crop land. African Journal of Agricultural and Resource Economics 2: 1-23.
6. Reidsma P, Ewert F, Oude Lansin A, Leemans R (2009) Vulnerability and adaptation of European farmers: a multi-level analysis of yield and income responses to climate variability. Regional Environmental Change 9: 25-40.
7. Adams RM, Rosenzweig C, Peart RM, Ritchie JT, McCarl BA et al. (1990)Global climate change and US agriculture. Nature 345: 219-24.
8. Mendelsohn R, Nordhaus WD, Shaw D (1994)The impact of global warming on agriculture: a Ricardian analysis. The American Economic Review84: 753-71.
9. Adams RM, Fleming RA, Chang CC, McCarl BA, Rosenzweig C (1995)A reassessment of the economic effects of global climate change on US agriculture. Climatic Change30: 147-67.
10. Mariara JK, Karanja FK (2007) The economic impact of climate change on Kenyan crop agriculture: A Ricardian approach. Global and Planetary Change57: 319-30.
11. Kurukulasuriya P, Ajwad MI (2007)Application of the Ricardian technique to estimate the impact of climate change on smallholder farming in Sri Lanka. Climatic Change81: 39-59.
12. Nhemachena C, Hassan R (2007) Micro-Level Analysis of Farmers Adaption to climate change in Southern Africa. International Food Policy Research Institute, Washington, Washington DC.
13. IPCC (2001) Climate change 2001: impacts, adaptation, and vulnerability: contribution of Working Group II to the third assessment report of the Intergovernmental Panel on Climate Change. Cambridge University Press, Cambridge, UK.
14. Bradshaw B, Dolan H, Smit B (2004)Farm-level adaptation to climatic variability and change: crop diversification in the Canadian prairies. Climatic Change67: 119-41.
15. Reid S, Smit B, Caldwell W, Belliveau S (2007)Vulnerability and adaptation to climate risks in Ontario agriculture. Mitigation and Adaptation Strategies for Global Change12: 609-37.
16. Deressa TT, Hassan RM (2009)Economic impact of climate change on crop production in Ethiopia: Evidence from cross-section measures. Journal of African Economies18: 529-54.
17. HisaliE, Birungi P, Buyinza F (2011) Adaptation to climate change in Uganda: Evidence from micro level data. Global Environmental Change 21: 1245-61.
18. Smith B, Burton I, Klein RJT, Wandel J (2000)An anatomy of adaptation to climate change and variability. Climatic Change45: 223-51.
19. Paul BK (1998)Coping mechanisms practiced by drought victims (1994/5) in North Bengal, Bangladesh. Applied Geography18: 355-73.
20. Ali A (1999)Climate change impacts and adaptation assessment in Bangladesh. Climate Research12: 109-16.
21. Ahmed AK, Chowdhury EH (2006) Study on livelihood systems assessment, vulnerable groups profiling and livelihood adaptation to climate hazard and long

term climate change in drought-prone areas of NW Bangladesh. Rural Development Division, Food and Agriculture Organization, Rome.

22. FAO (2006) Livelihood adaptation to climate variability and change in drought-prone areas of Bangladesh. Food and Agriculture Organization, Rome.
23. Rashid MH, Islam MS (2007) Adaptation to climate change for sustainable development of Bangladesh agriculture. Bangladesh Country Paper, APCAEM.
24. Rosenzweig C, Parry ML (1994)Potential impact of climate change on world food supply. Nature367: 133-38.
25. Deressa TT, Hassan RM, Ringler C, Alemu T, Yesuf M (2009)Determinants of farmers' choice of adaptation methods to climate change in the Nile Basin of Ethiopia. Global Environmental Change19: 248-55.
26. Gbetibouo GA (2009) Understanding farmers' perceptions and adaptations to climate change and variability: The case of the Limpopo Basin, South Africa. International Food Policy Research Institute Discussion Paper 00849, Washington, Washington DC.
27. Kropko J (2007) Choosing between Multinomial Logit and Multinomial Probit Models for Analysis of Unordered Choice Data. Pro Quest, Ann Arbor.
28. Islam T, Neelim A (2010) Climate Change in Bangladesh: A Closer Look into Temperature and Rainfall Data. University Press Ltd, Dhaka.
29. Roncoli C, Ingram K, Kirshen P (2002) Reading the rains: Local knowledge and rainfall forecasting in Burkina Faso. Society and Natural Resources15: 409-27.
30. Eakin H (2003)The social vulnerability of irrigated vegetable farming households in central Puebla. The Journal of Environment and Development12: 414-29.
31. Agrawal A (2008)The role of local institutions in adaptation to climate change. International Forestry Research and Institutions Program Working PaperW08I-3.
32. Yip, WC, Wang, H and Liu, Y (1998) Determinants of patient choice of medical provider: a case study in rural China. Health policy and planning13: 311-22.
33. Asfaw A, Admassie A (2004) The role of education on the adoption of chemical fertilizer under different socioeconomic environments in Ethiopia. Agricultural Economics30: 215-28.
34. Norris PE, Batie SS (1987)Virginia farmers' soil conservation decisions: An application of Tobit analysis. Southern Journal of Agricultural Economics 19: 79-90.
35. Lin JY (1991)Education and innovation adoption in agriculture: evidence from hybrid rice in China. American Journal of Agricultural Economics 73: 713-23.
36. Maddison D (2006)The perception of an adaptation to climate change in Africa. University of Pretoria, Pretoria.

CHAPTER 12

Perception-Based Analysis (PBA) of Climate Change Impacts on the Forest and Agricultural Ecosystem of Shropshire, United Kingdom

Chukwunonye Ezeah*, Emmah Simbabure

Faculty of Sciences and Engineering, University of Wolverhampton, City Campus-South, Wolverhampton, UK

ABSTRACT

Agriculture and forestry are vital sectors providing services, food and other environmental benefits that could be most affected by the impact of climate change (CC). This study analysed the impact of CC on forestry and agriculture in a typical UK rural environment. The study interrogates this complex question using the Perception Based Analysis (PBA) methodological approach. Data analysis utilized chi square test and one-way analysis of variance (Anova) in comparing the impact of climate change and human factors on forest and agricultural ecosystems, (significance level $\alpha = 5\%$), calculated $\rho = 0.36 > 0.05$. This non-significant ρ value suggests that the null hypothesis Ho "climate change is responsible for the changes in forest and agricultural ecosystem in the case study area" could be true.

Keywords: Climate Change Impact, Forest and Agricultural Ecosystem, UK Rural Environment

1. INTRODUCTION

The contemporary discourse on climate change CC has often been framed from a complex and rather predictive scientific perspective, based mostly on scenario modelling. Studies indicate that rather than bring greater clarity to the subject, such approach often becomes a vehicle of conflict by promoting the beliefs of different sides to the debate as scientific facts [1] . In addition to this, inherent difficulties associated with many environmental problems also sometimes help to undermine the effectiveness of scientific communication because of the mismatch between “usual modes of understanding” and what is needed to grasp the essence of the problem [2] . Both proponents as well as opponents of climate change have been accused of sometimes deliberately misleading the public [3] . Instead of providing the public with objective basis for making rational decisions on issues, science communication have sometimes been found to be either ineffective or even worsen the situation entirely. In contrast to this trend, this investigation attempts an analysis of the CC phenomenon from a public perception perspective.

There is increasing concern about the variability and impacts of CC on forestry and agricultural production worldwide. These changes have been mostly attributed to anthropogenic influences on the earth’s environment [4] . In particular, IPCC Assessment Report 4 [5] supports the view that climate change will have significant impacts on biodiversity, ecosystems and genetic diversity within species as well as on ecological interactions. DEFRA [6] went further to buttress this point, arguing that while on the one hand CC poses added risks to agriculture due to changes in growing seasons, on the other hand it could present positive outcomes to plants that might adapt to anticipated climatic changes. Furthermore, IPCC Working Group II (Adaptation and Vulnerability) estimates that 20% - 30% of plants and animals species evaluated so far in climate change studies were at risk of extinction following temperature spikes projected to occur towards the end of this century. As a result, it has been predicted that some species may migrate northwards or pole wards to where conditions will be more favourable for their existence. Current research indicates that species failing to adapt may be faced with possible extinction or fragmentation. In certain cases, phenological changes are already evident. It therefore follows that, agriculture and forestry are vital sectors providing services, food and other environmental benefits that could be most affected by climate change. This study analyses the impact of CC on a typical UK rural ecosystem with a view to providing empirical evidence that could underpin future CC policies and programmes.

To adequately understand and mitigate CC impacts, evidence based research is required particularly in the agricultural sector. Perception-

Based Analysis (PBA) has been used as an alternative to more traditional response-based segmentation [7] [8] . Perception based studies on CC impacts on the forest and agricultural sector are currently under reported; it is therefore imperative for a research like this to use as a case study, a typical rural environment such as Shropshire in middle England. Shropshire borders Wales to the west, Cheshire to the north, Staffordshire to the east, Worcestershire to the south east and Herefordshire to the south. Shropshire is divided into North and South with the Shropshire hills being a major land feature at the south western end (see Figure 1). It is estimated that 86% (approximately 128,147 ha) of Shropshire land use is predominantly for agriculture and permanent grasslands [9] . According to DEFRA Statistics [9] , in 2008 there were 7026 farm holdings in Shropshire covering 274,549 ha. A total of 7624 individuals, equivalent to 70.2% are engaged in some form of agricultural activities either in full or part time capacity. Shropshire gov. [10] elaborates further that Shropshire's economy depends largely on agriculture, rural industries and tourism. According to Shropshire RCC [11] , 43% of Shropshire residents live in rural areas; therefore a disruption in agriculture and forest ecosystems would amount to a significant disruption to their livelihood and economy.

Figure 1. Map of the study area.

2. REVIEW OF LITERATURE

An ecosystem is a dynamic and complex system of plant, animal, and microorganism communities and the non-living environment interacting with each other as a functional unit [12] - [14] . Climate exerts a dominant control over the natural distribution of species within ecosystems. There is increasing evidence from literature that the distribution, composition, structure and function of ecosystems do respond to changes in temperature, precipitation and increased CO_2 levels in the environment [15] [16] . Evidence from fossil records [17] [18] as well as from more recently observed trends [19] [20] also shows that changing climate has a profound influence on species' range, expansion and contraction. Although there is relatively limited evidence of current extinctions caused by climate change, studies suggest that CC could surpass habitat destruction as the greatest global threat to biodiversity over the next few decades [21] in [15] . According to [22] , phenology which provides environmental records of leafing, flowering of plants, insect activity (beetles and butterflies) and bird migration are also affected by climate change and global warming [15] .

Recent estimates project that agricultural activities are contributing between 12% - 14% of global anthropogenic greenhouse gas emissions, not including emissions arising from land clearing [23] . The precise magnitude of CO_2 emissions from land-use change is still very unclear [5] . In essence, agriculture is highly sensitive to climate change and climatic variations. As a result, there are not only differences among regions, but also differences caused by interannual variability of production and disruption of ecosystem services within a single region [24] . Climate change impacts on agricultural production are also associated with impacts on human well-being and welfare. Changes in agricultural systems, driven by socioeconomic changes, greenhouse gas emissions, agricultural policies and other factors, are also affecting natural and managed ecosystems [25] . From the fore going, CC impacts on ecosystems are far reaching, and could be divided into two main groups: biophysical and socio-economic. This research focuses on the biophysical impacts and their consequential effects such as droughts and floods, which could cause low crop yields and increased incidence of pests and diseases.

Associated with climate change are several factors that affect forest ecosystems, which can act independently or in combination [26] . The changes in timing of seasonal events and migration patterns can result in mismatches between species such as predator prey host relationships hence an imbalance in the ecosystem. A study of 9650 interspecific systems, including pollinators and parasites, suggested that around 6300 species could disappear following the extinction of their associated species [15] [27] . In addition, for many species, the primary impact of climate change may be mediated through effects on synchrony with species' food and habitat requirements. Climate change is also said to have led to phenological shifts in flowering plants and insect pollinators, causing mismatches between plant and pollinator populations that lead to the extinctions of both the plant and the pollinator with expected consequences on the structure of plant?pollinator networks [15] [27] .

Bellard et al., [15] predicted alarming scenarios of species extinction due to climate change impact in the near future. Indications are that in Europe, with regards to plant diversity, up to 84% plant species could become extinct while as much as 75% of fish depending on river conditions could be wiped out. The study elaborates further that 52 per cent of amphibians could become extinct and 35% of birds could be susceptible to climate change effects and become extinct. These figures are significant and are a cause for concern. Drier conditions and drought could lead to reduced agricultural and timber yields and could also affect woodland conditions [4] . Briner et al. [28] and Lindner et al. [26] argue that the time spent under drought conditions in most crop growing regions of the world will have drastic consequences on yields and tree growth. While Ray et al., [29] , Lindner et al., [26] , Briner et al., [28] , and the IPCC [4] are unanimous in agreement that drier and warmer summers could heighten the risk of abiotic factors such as wild fires. IPCC [30] goes further to predict increased forest fires as a result of prevailing climatic conditions. This goes to support the position that drought which is exacerbated by climate change could have significant negative impact on forestry and agricultural ecosystem. Fantahun [12] suggests that after wildfires the soil is left without plant cover leaving it susceptible to soil erosion and leaching, rendering it unproductive for agricultural purposes and creating desert like conditions.

[29] [31] and [32] predict that other species may dominate as some species become dormant or extinct. Increased drought stress could equally induce a shift in forest species composition. [31] and [29] elaborated further on what they call "forest disturbances" such as fires, drought, alien species, insects, pathogen outbreaks and wind storms, agreeing that these will have impacts on species composition and distribution thereby affecting habitat quality and quantity and even cause disruption to ecosystems services [15] . The Fourth Assessment Report, IPCC [5] , predicts that some of the species and insects like butterflies for example will begin to appear in hitherto unusual habitats while others will move northwards to where conditions are favourable for their survival. This could have significant effect on those ecosystems as well. The consequence of this is that more drought resistant species are predicted to become dominant component of the forest in the future. Ray et al. [29] argue that these new woodlands must be resilient to the impact of climate change. IPCC [30] further argues that there will be increased colonization of alien species in Europe, possibly this could be due

to new opportunities in the agricultural sector and woodland environment. Ray et al. [29] suggest further that in the future, woodlands could be ideal for carbon sequestration, apart from providing wood fiber for timber products.

Climatic change is also predicted to impact on biotic factors such as frequency and consequences of pests and disease outbreaks [26] [31] . Ray et al. [29] emphasized on the influence of warmer weather and warmer winter, especially that they could cause a diversity of pests and pathogens. IPCC [5] supports this analogy and predicts that pests especially beetles outbreak will have fatal effects on pine trees population; Dale et al. [31] also agreed with this position while reporting that in their research they found evidence that pine bark beetle disease destroys forests. It has been noted that weeds and pests thrive under warmer and increased temperatures, wetter climates and enhanced CO_2 levels. Under this scenario, the wider range of weeds and pests are likely to move northwards and different weeds will occupy their space. This could cause increase in pollution as more pesticides and fungicides will be required to control the weeds and pests which will in turn harm water courses.

Bellard et al. [15] stressed the positive aspects of climate change, particularly with regard to increased temperatures and increased CO_2, pointing out that they could result in an accelerated biomass production. Increased precipitation could also benefit some plants and communities depending on how they will respond to extreme rainfall [15] . In a separate study Bellamy and Hulme [33] discussed people's perceptions and touched on the emotional issues as well as cultural perceptions pertaining to climate change. Bellamy and Hulme [33] used terms such as "catastrophic" and "beyond the tipping point", to describe events and scenarios around climate change. Such terms evoke fear in people regarding climate change. Without doubt, there will be noticeable changes both in woodland as well as in forest ecosystems [31] , though the extent of such changes may vary from place to place. Invariably, woodland recreation may be boosted as people seek shelter from the increased temperatures.

3. METHODOLOGY

This research utilizes a multi-pronged PBA methodological approach incorporating, desk top study, participant observation, key informant interview and questionnaire survey. At the desktop study stage, mostly secondary data were collected; in the course of reviewing of current literature on the subject. Thereafter the study area was divided into four zones, Shrewsbury, Telford, Shifnal and Bridgnorth. Following the desktop survey, the authors undertook a number of visits to the site to observe the case study area and interview selected key informants from Shropshire Council, National Farmers Union, and Shrewsbury Botanical Gardens. Finally, a survey questionnaire consisting of 13 questions was designed. The survey focused on ascertaining respondents' perception on the relative impact of climate change on forest and agricultural ecosystems in the case study area. To reduce the time and cost of questionnaire administration, an online format of the questionnaire was adapted using the Survey Monkey sampling software. The online questionnaires were thereafter emailed to 250 individuals who live and work in Bridgnorth, Telford, Shrewsbury and Shifnal during the summer of 2014, using the stratified random sampling approach. Figure 2 is an indicative map with questionnaire sampling and interview areas inset. A total of 100 responses were returned equivalent to about 40% response rate. This rate is well above the norm (20% - 30%) for postal questionnaires [34] - [36] . Most of the data collected were ordinal in nature, although there were some nominal data as well. Data analysis was carried out using the Statistical Programme for Social Sciences (SPSS) for Windows 17.0.

Statistical Data Analysis

Define Statistical analysis utilized chi square test and analysis of variance (ANOVA) to analyse some of the numerical data. A null hypothesis was propounded that "climate change is responsible for the changes in forest and agricultural ecosystem in the case study area". Questions 3, 5, 6, 7 and 11 from the questionnaire were subjected to chi square test, represented by the formula:

$$X^2 = \sum \frac{(\text{observed} - \text{expected})^2}{\text{expected}} \tag{1}$$

where:

o = Observed Frequency in each category,

e = Expected Frequency in the corresponding category,

d_o = degree of freedom (n – 1),

n = 100,

X^2 = Chi Square.

In addition, we used one-way comparison analyses of variance (Anova) in predicting any variations in CC indicators such as rainfall and temperature over a defined period.

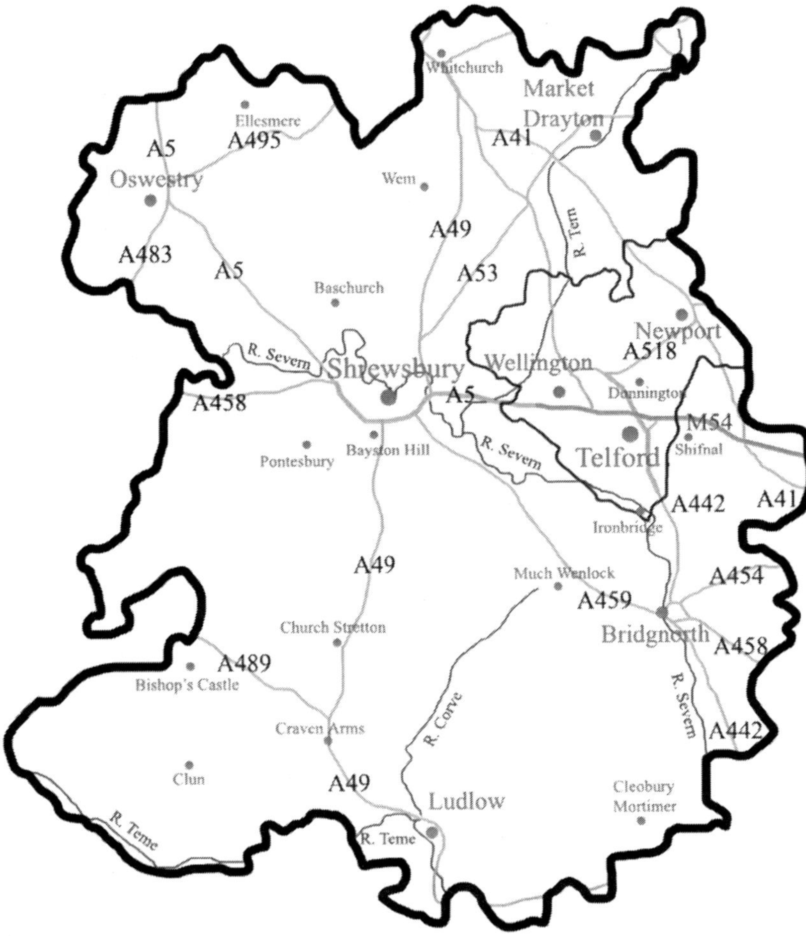

Figure 2. Key interview areas [★] andpaper based questionnaire distribution areas [★].

4. RESULTS

Table 1 is an overview of respondents' occupation groups. Table 2 is an outline of respondents' perception on the causes and impacts of CC. One way analyses of variance (Anova) was used in comparing climate change and human factors. The significance level α = 5%, calculated ρ = 0.36 > 0.05. Table 3 represents results for test of homogeneitythat the population variances are equal. Based on Anova results, the null hypothesis H_0 that "climate change is responsible for changes in forest and agricultural ecosystem in the case study area" is accepted.

Table 1. Respondents' occupation group.

Response	Occupation Group				
	C	M	T	O	Total
Agree	14	1	2	12	**29**
Neutral	6	3	4	5	**18**
Disagree	6	10	12	16	**44**
Total	**26**	**14**	**18**	**33**	**91**

C = Carers, M = Managers, T = Technical, O = Others.

Table 2. Causes and impact of climate change.

	Percentage of Respondents						
Human factors are causing climate change	3.30	3.30	24.10	52.70	16.40	100	3.76
Climate change has had significant impact on businesses in Shropshire	3.26	5.43	52.17	35.87	3.26	100	3.3
	Strongly disagree	Disagree	Neither agree nor disagree	Agree	Strongly agree	Total	Average rating

5. DISCUSSION

Primary data were collected using the stratified random sampling technique across four districts of Shropshire and Telford councils: Bridgnorth, Telford, Shrewsbury and Shifnal. Bellamy and Hulme [33] have previously studied public perceptions on CC and noted that the subject was often associated with a culture of fear, best reflected by some of the words used to describe scenarios and events surrounding climate change, such as "catastrophic" and "point of no return". This investigation was designed to avoid such elements of fear and subjectivity in analyzing people's perception of climate change. Like Bellamy and Hulme [33] , the research employed both qualitative and quantitative approaches but differed slightly from earlier PBA studies because it did not touch on emotions and the myth or the abruptness of climate change.

5.1. Relative Impact of Climate Change on Ecosystems

In response to questions 5 and 6 designed to find out whether "Climate change is causing floods, rainfall variability and warmer temperatures" 63.33% of respondents agreed while 13.53% strongly agreed (see Figure 3 and Figure 4). It is possible that there could be other factors responsible for the perceived impacts (see Table 2), for instance respondents could have been influenced by recent memories of the floods that inundated Shropshire's Abbey Foregate road and other places in 2014. Elaborating further on this theme, National Farmers Union (NFU) in Cork [37] suggested that Shropshire "has periodically been affected by floods that threaten livestock; in particular, farmers around River Severn and Shrewsbury have been the most impacted with entire farmlands sometimes underwater, this was the case following late arrival of spring in 2013. While Blessington and Shields [38] were of the opinion that it is too early to conclude that the changes in the ecosystems were solely due to CC, IPCC [5] suggests that CC could cause significant changes in ecosystem composition if increases in global average temperature exceed 1.5°C - 2.5°C in which case, they predict that as much as 20 - 30 percent of species will be at risk of extinction.

Drier conditions and drought are other key impacts of CC that could lead to reduced agricultural and timber yields and also affect woodland conditions [4] . Briner et al. [28] and Lindner et al. [26] argue that the time spent under drought conditions in most crop growing regions of the world will have drastic consequences on yields and tree growth. While Ray et al. [29] , Lindner et al., [26] , Briner et al. [28] , and the IPCC [4] suggest that drier and warmer summers will heighten the risk of abiotic factors such as wild fire; IPCC [30] and Herbeck [39] on the other hand predict increased burnt forest areas in the near future. There appears to be a convergence of opinion that drought, which evidence indicates is exacerbated by climate change, will continue to have negative impacts on forest and agricultural ecosystem, particularly in rural environments. Fantahun [12] went further to point out that after wildfire incidents, the soil is often left bare and becomes more susceptible to soil erosion and leaching. This renders areas of farmlands unproductive for agricultural purposes and creates desert like conditions.

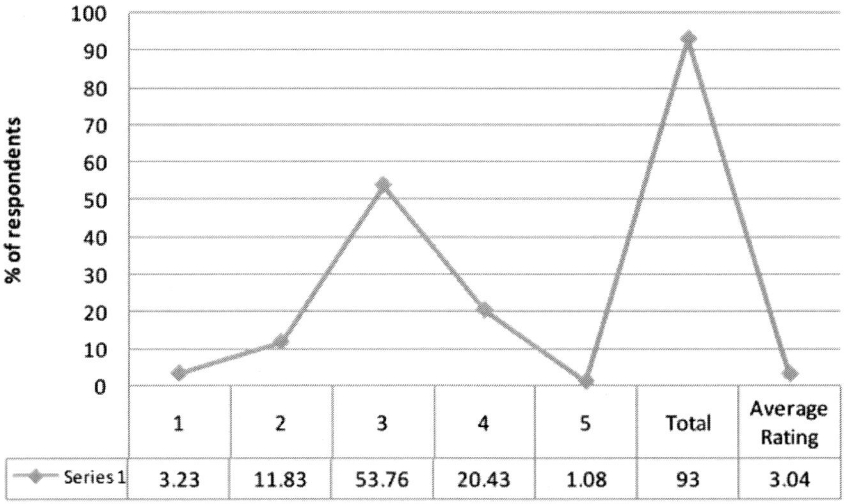

Figure 3. On a sliding scale where 1 is minimal and 5 is maximum impact, do you agree that there have been changes in the forest and agricultural eco- systems in rural Shropshire.

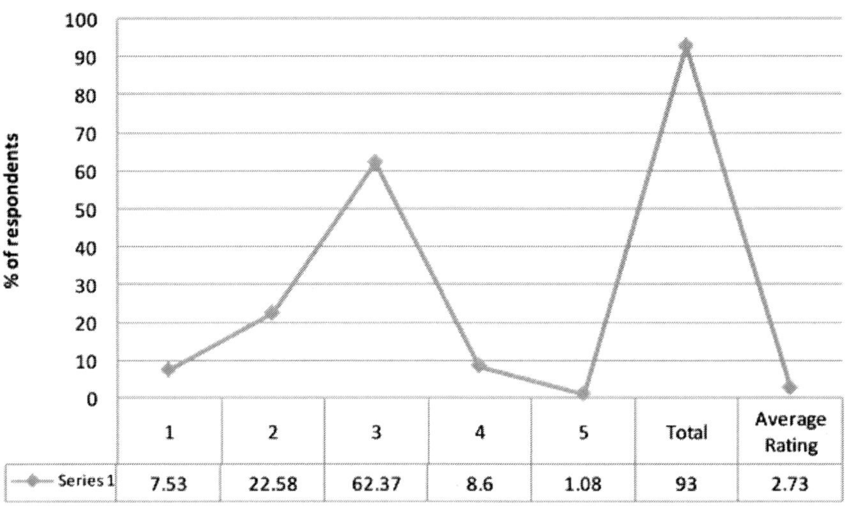

Figure 4. On a sliding scale where 1 is minimal and 5 is maximum impact, do you agree that the changes in the forest and agricultural ecosystems in rural Shropshire can be ascribed mainly to climate change.

Table 3. Test of Homogeneity of variances.

	Levene Statistic	df 1	df 2	Sig.
Climate change	0.653	5	94	0.660
Human factors	4.733	5	93	0.001

Interview results from this investigation provided particular insight on the fate of a number of plant species such as sycamore tree in rural Shropshire. Most respondents agreed with Blessington and Shield [38] on the reason why sycamore trees were doing well in Shropshire and pointed out that Sycamore trees were much more adaptable than other plant species. This is possibly due to their capacity to tolerate significant variations in weather conditions, for instance soil salinity and acidity (pH). It has also been suggested that the sycamore tree is capable of thriving under increased temperatures and flooding [5] as well as heavy precipitation [40] [41] . FAO [42] agreed with the notion that CC could provide opportunities for certain plants species to thrive due to high CO_2 levels and lengthening growing season during the year.

5.2. Livestock

Predictably, it is not only plant species that are impacted on by CC but also livestock. It has been reported that CC has both direct and indirect impact on livestock productivity through changes on the availability of fodder and pastures [43] . IPCC [30] and National Farmers Union [37] elaborated further on the likely impact of floods on livestock in rural areas such as Shropshire, particularly through disease infestation on pastoral herds. A recent global assessment found out that climate change could alter livestock numbers at the "local breed" level [44] . Herbeck [39] reported that in most regions of the world local breeds make up two thirds of all breeds, equivalent to 5067 out of a total of 7600 breeds recorded by FAO. FAO [44] predicts the possible extinction of local livestock breeds in the near future while pointing out that around 1500 breeds have already become extinct in the last 15 years. Though results from the this research could not validate findings in the FAO reports, IPCC [30] however agrees that beef production is likely to decline and sheep and goat production increase in the case study area, possibly as a result of changes in growing conditions. Large-scale commercial beef cattle farmers are most vulnerable to climate change, particularly since they are less likely to have diversified. Shropshire has good numbers of sheep but this research could not provide evidence to indicate a decline of beef cattle in the area as a result of CC impact.

5.3. Fish

A separate species study cited by Bellard et al. [15] predicted alarming scenarios of species extinction due to climate change impact. In Europe in particular, estimates indicate that up to 75% of fish population could become extinct, depending on river habitat conditions and severity of CC impact. The study also indicates that 52% amphibians and 35% of birds could be susceptible to climate change adverse impacts and become extinct. These figures are significant and are a cause for concern. This research did not generate any evidence to back the position of literature on species extinction but public opinion seems to point to the fact that species are declining in rural Shropshire [38] . It has been further argued that CC impacts, such as warming of oceans, rivers and lakes and changes in precipitation, water salinity and ocean acidity as well as the increases in extreme weather events, will increase uncertainties in the supply of fish from capture fisheries and aquaculture [44] . Severe weather events could also upset the balance of supply [30] . Results from this investigation agree that CC will impact on fish numbers, production and jobs. On the other hand, FAO [42] agrees that fish numbers will be affected but argue that it would be due to water scarcity as there will be competition for water usage.

5.4. Infrastructure

Cork [37] reported that farm sheds where broken down when Shropshire was hit by extreme weather in 2013. This is similar to the situation described by [5] and [30] where it was reported that extreme weather events caused by CC will affect food security by damaging infrastructure for food supply. Post-harvest aspects of agriculture i.e. on-farm storage and commercial activities, handling and transport, will also be affected by changes in temperature, rainfall, humidity, and by other weather dependant factors. As a result, a significant numbers of adaptation projects have been reported by literature. The IPCC [5] and Herbeck [39] are of the opinion that biodiversity of food crops might suffer from repeated extreme weather events and monoculture crops may not be able to adapt to repeated flooding [5] [37] [40] [45] or drought [4] [26] [28] [41] .

5.5. Business

Evaluating the results from question No. 11, "Climate has had significant impact on businesses in Shropshire" (Table 2), 36% of respondents agreed that climate change is having an impact on businesses in rural Shropshire and 52% were neutral. The rationale behind the high proportion of neutral answer is unclear. The UK derives more than 70 percent of its timber imports from EU27. In addition, about 18 percent is supplied domestically [46] . The IPCC [30] argues that, "Change in economic values will impact rural communities with the linkages between biodiversity, tourism and rural livelihoods and rural landscapes being an established one both for developing and developed countries". From the personal observation of the authors, there appears to be noticeable changes in woodland recreation [31] . Business may also be boosted by "environmental tourism" as well as by people seeking shelter from the increased temperatures and other adverse CC impacts [46] .

5.6. Mitigation and Adaptation

IPCC [30] and Allen et al. [47] reported significant increases in the number of planned adaptation responses at the local level in rural and urban communities. Local governments play a central role in addressing the challenges of adaptation planning and implementation [30] [45] . On the other hand, a number of scholars have stressed the important role partnerships between public and private sectors could play towards CC adaptation and mitigation. In addition, it has been stated that social networks can influence vulnerability in complex ways [48] . Well networked neighborhoods have been shown to respond better in emergency situations and social isolation can increase vulnerability. A study in Sweden found out that limited co-operation between local sector organizations, lack of local co-ordination, and an absence of methods and traditions to build institutional knowledge, present barriers to managing vulnerability [30] . In addition, the study also found out that communication is vital for mitigation and adaptation to work effectively.

It might be possible for rural people in some cases to adapt to climate change by using their own knowledge, resources and networks [30] . In other cases governments and other outside actors will have to assist rural people, or plan and execute adaptation on a scale that individual rural households and communities cannot. Brisley et al. [49] and Laukkonen et al. [50] argue that most local authorities do not consider adequately the vulnerable population of the community with regard to adaptation to the impact. Examples of rural adaptations observed during this investigation includes modifying farming and fishing practices, introducing new species, varieties and production techniques, managing water in different ways, diversification of livelihoods, modifying infrastructure, and using or establishing risk sharing mechanisms, both formal and informal. Adaptation also includes changes in institutional and governance practices [30] .

6. CONCLUSION

Evidence indicates that climate change could cause significant changes in ecosystem composition if increases in global average temperature continue unabated. As much as 30 percent of species in forest and agricultural ecosystems could be at risk of extinction. Apart from CC other factors such as land development practices, government policies as well as unsustainable agricultural practices are also contributory in causing fragmentation and extinction of species. Results from this study indicate that climate change will bring increased frequency and severity of extreme weather events to Shropshire region. As a result, increased resilience and adaptation to these climatic threats is urgently required at local (Shropshire), regional and national levels. As in other UK counties, Shropshire has recently taken a proactive step by developing a Climate Change Strategy but now faces additional challenges associated with implementation.

CITE THIS PAPER

Chukwunonye Ezeah, Emmah Simbabure (2015) Perception-Based Analysis (PBA) of Climate Change Impacts on the Forest and Agricultural Ecosystem of Shropshire, United Kingdom. *Journal of Geoscience and Environment Protection*,**03**,111-121. doi: 10.4236/gep.2015.35013

REFERENCES

1. Hoffman, A.J. (2011) Talking past Each Other? Cultural Framing of Skeptical and Convinced Logics in the Climate Change Debate. Organization and Environment, 24, 3-33.
2. Weber, E.U. and Stern, P.C. (2011) Public Understanding of Climate Change in the United States. American Psychologist, 66, 315-328.
3. Oreskes, N. and Conway, E.M. (2010) Merchants of Doubt: How a Handful of Scientists Obscured the Truth on Issues from Tobacco Smoke to Global Warming. Bloomsbury Press, New York.
4. Intergovernmental Panel for Climate Change IPCC (2013) 1st Report on the Physical Science Basis of Climate Change. IPCC.
5. Intergovernmental Panel for Climate Change IPCC (2007) Climate Change 2007: Mitigation. Contribution of Working Group III to the Fourth Assessment Report of the Intergovernmental Panel for Climate Change. Contribution of Working Group III, 4th Assessment Report.
6. Department of Environment, Food and Rural Affairs DEFRA (2009) Adapting to Climate Change UK Climate Projections.
7. Mazanec, J.A. and Strasser, H. (2001) A Nonparametric Approach to Perception-Based Market Segmentation Foundation. Springer-Verlang, Berlin.

8. Mazanec, J.A. and Strasser, H. (2007) Perception-Based Analysis of Tourism Products and Service Providers. Journal of Travel Research, 45, 387-401.
9. Department of Environment, Food and Rural Affairs DEFRA (2013) Faming Statistics. https://www.gov.uk/government/collections/structure -of-the-agricultural-industry#2
013-publications
10. Pearson, R.G. and Dawson, T.P. (2012) Predicting the Impacts of Climate Change on the Distribution of Species: Are Bioclimate Envelope Models Useful. Global Ecology & Biogeography, 12, 361-371.
11. Shropshire Rural Community Council (Shropshire RCC) (2014) Annual Report. Shropshire Rural Community Council, Shrewsbury.
12. Fantahun, A. (2013) Impacts of Climate Change on Plant Growth, Ecosystems Services, Biodiversity and Potential Adaptation Measure. Master's Thesis, Department of Biological and Environmental Sciences, University of Gothenburg, Gothenburg. http://bioenv.gu.se/digitalAssets/1432/1432197_fantahun.pdf
13. Lymn, N., Glasener, K.M. and Wagester, T. (2014) Agriculture, Biology and Conservation: The A, B, C's of How Nature Serves the Nation. American Society of Agronomy Crop Science, Society of America, USA.
14. Environmental Protection Agency EPA (2014) Climate Impact on Ecosystems. http://www.epa.gov/climatechange/impacts-adaptation/ecosystems.html
15. Bellard, C., Bertelsmeier, C., Leadley, P., Thuiller, W. and Courchamp, F. (2012) Impacts of Climate Change on the Future of Biodiversity. Ecology Letters, 15, 365-377.
16. Pearson, R.G., Dawson, T.P. and Liu, C. (2004) Modelling Species Distributions in Britain: A Hierarchical Integration of Climate and Land-Cover Data. Ecography, 27, 285-298.
17. Huntley, B. (1999) Species Distribution and Environmental Change: Considerations from the Site to the Landscape Scale. In: Maltby, E., Holdgate, M., Acreman, M. and Weir, A., Eds., Ecosystem Management: Questions for Science and Society, Royal Holloway Institute for Environmental Research, Virginia Water, 115-130.
18. Davis, M.B. and Shaw, R.G. (2001) Range Shifts and Adaptive Responses to Quaternary Climate Change. Science, 292, 673-679.
19. McCarty, J.P. (2001) Ecological Consequences of Recent Climate Change. Conservation Biology, 15, 320-331.
20. Walther, G.R., Post, E., Convey, P., Menzel, A., Parmesan, C., Beebee, T.J.C., Fromentin, J.M., Hoegh-Guldberg, O. and Bairlein, F. (2002) Ecological Responses to Recent Climate Change. Nature, 416, 389-395.
21. Leadley, P., Pereira, H.M., Alkemade, R., Fernandez-Manjarres, J.F., Proenca, V. and Scharlemann, J.P.W. (2010) Biodiversity Scenarios: Projections of 21st Century Change in Biodiversity and Associated Ecosystem Services. In: Bellard, C., Bertelsmeier, C., Leadley, P., Thuiller, W. and Courchamp, F. (2012) Impacts of Climate Change on the Future of Biodiversity. Ecology Letters, 15, 365-377.
22. Ray, D. (2013) Forestry Commission Research Note. Forestry Commission, Edinburgh.
23. Power, A.G. (2010) Ecosystem Services and Agriculture: Tradeoffs and Synergies. Philosophical Transactions of the Royal Society B—Biological Sciences, 365, 2959-2971.
24. Howden, S.M., Soussana, J.F., Tubiello, F.N., Chhetri, N., Dunlop, M. and Meinke, H. (2007) Adapting Agriculture to Climate Change. Proceedings of the National Academy of Sciences of the United States of America, 104, 19691-19696.
25. Zaehle, S., Bondeau, A., Carter, R.T., Cramer, W., Erhard, M., Prentice, C., Reginster, I., Rounsevell, M.A.D., Sitch, S., Smith, B., Smith, P.C. and Sykes, M. (2007) Projected Changes in Terrestrial Carbon Storage in Europe under Climate and Land-Use Change, 1990-2100. Ecosystems, 10, 380-401.
26. Lindner, M., Maroschek, M., Netherer, S., Kremer, A., Barbati, A., Garcia-Gonzalo, J., Seidl, R., Delzon, S., Corona, P. and Kolstrom, M. (2010) Climate Change Impacts,

Adaptive Capacity, and Vulnerability of European Forest Ecosystems. Forest Ecology and Management, 259, 698-709.
27. Bishaw, B., Henry, N., Jermias, M., Abdu, A., Jonathan, M., Gemedo, D., Tewodros, A., Habtemariam, K., Ian, K. and Eike, L. (2013) Farmer's Strategies for Adapting to and Mitigating Climate Variability and Change through Agroforestry in Ethiopia and Kenya. In: Davis, C.M., Bernart, B. and Dmitriev, A., Eds., Forestry Communications Group, Oregon State University, Oregon.
28. Briner, S., Elkin, C. and Huber, R. (2013) Evaluating the Relative Impact of Climate and Economic Changes on Forest and Agricultural Ecosystem Services in Mountain Regions. Journal of Environmental Management, 129, 414-422.
29. Ray, D., Morison, J. and Broadmeadow, M. (2010) Climate Change: Impacts and Adaptation in England's Woodlands. Forestry Commission Research Note 201. Forestry Commission, Edinburgh.
30. Intergovernmental Panel for Climate Change (IPCC) (2014) Climate Change 2014: Impacts, Adaptation and Vulnerability Contribution of the Working Group, Group 11 AR5. Cambridge University Press, Cambridge, United Kingdom, IPCC, New York. [Citation Time(s):13]
31. Dale, V.H., Tharp, M.L., Lannom, K.O. and Hodges, D.G. (2010) Modeling Transient Response of Forests to Climate Change. Science of the Total Environment, 408, 1888-1901.
32. Briner, S., Elkin, C., Huber, R. and Grêt-Regamey, A. (2012) Assessing the Impacts of Economic and Climate Changes on Land-Use in Mountain Regions: A Spatial Dynamic Modeling Approach. Agriculture, Ecosystems & Environment, 149, 50-63.
33. Bellamy, R. and Hulme, M. (2011) Beyond the Tipping Point: Understanding Perceptions of Abrupt Climate Change and Their Implications. Weather, Climate, and Society, 3, 48-60.
34. Black, C., Akintiye, A. and Fitzgerald, E. (2000) An Analysis of Success Factors and Benefits of Partnering in Construction. International Journal of Project Management, 18, 423-434.
35. Babbie, E.R. (1990) Survey Research Methods. Wadsworth Publishing Co., Belmont.
36. Creswell, J.W. (2003) Research Design: Qualitative, Quantitative, and Mixed Method Approaches. Sage Publications, Thousand Oaks.
37. Cork, H. (2014) Agriculture and Climate Change Interview, 1. National Farmers' Union Telford Shropshire.
38. Blessington, J. and Shields, P. (2014) Oral Interview: Climate Change and Forestry. Shropshire Council, Shrewsbury.
39. Herbeck, J. (2008) Biodiversity and Climate Change: 1st and 2nd Order Effects in Agriculture, Forestry and Fisheries. (2170788).
40. Eliasch Review (2008) Climate Change: Financial Global Forest. The Stationery Office Limited on Behalf of the Controller of Her Majesty's Stationery Office. https://www.gov.uk/government/uploads/system/uploads/attachment_data/file/228833/9780108507632.pdf
41. Petr, M., Boerboom, L.G., van der Veen, A. and Ray, D. (2014) A Spatial and Temporal Drought Risk Assessment of Three Major Tree Species in Britain Using Probabilistic Climate Change Projections. Climatic Change, 124, 791-803.
42. Food and Agriculture Organization (FAO) (2013) Climate Change Impacts, Adaptation and Mitigation in REU Regions.
43. Department of Environment, Food and Rural Affairs (DEFRA) (2012) Climate Change Risk Assessment for the Agriculture Sector. file://prs-store2.unv.wlv.ac.uk/home2$/in0587/home/Profile/Downloads/CCRAfortheAgricultureSector%20(1).pdf
44. Food and Agriculture Organization (FAO) (2012) Climate Change, State of the World Forests. http://www.fao.org/docrep/016/i3010e/i3010e.pdf
45. Food and Agriculture Organization (FAO) (2012) Forest Management and Climate Change: A Literature Review. Forests and Climate Change Working Paper10. http://www.fao.org/docrep/015/md012e/md012e00.pdf

46. Read, D.J., Freer-Smith, P.H., Morison, J.I.L., Hanley, N., West, C.C. and Snowdon, P. (2009) Combating Climate Change—A Role for UK Forests. An Assessment of the Potential of the UK's Trees and Woodlands to Mitigate and Adapt to Climate Change. http://www.forestry.gov.uk/pdf/SynthesisUKAssessmentfinal.pdf/$FILE/SynthesisUKAssessmentfinal.pdf
47. Allen, P., Blake, L., Harper, P., Hooker-Stroud, A., James, P. and Kellner, T. (2013) Zero Carbon Britain, Rethinking the Future. http://library.uniteddiversity.coop/Climate_Change/Zero_Carbon_Britain_2013.pdf
48. Preston, I., Banks, N., Hargreaves, K., Kazmierczak, A., Lucas, K., Mayne, R., Downing C. and Street, R. (2014) Climate Change and Social Justice: An Evidence Review. http://www.jrf.org.uk/sites/files/jrf/climate-change-social-justice-full.pdf
49. Brisley, R., Welstead, J., Hindle, R. and Paavola, J. (2012) Socially Just Adaptation to Climate Change. http://www.jrf.org.uk/sites/files/jrf/climate-change-adaptation-full_0.pdf
50. Laukkonen, J., Blanco, P., Lenhart, J., Keiner, M., Cavric, B. and Kinuthia-Njenga, C. (2009) Combining Climate Change Adaptation and Mitigation Measures at the Local Level. Habitat International, 33, 287-292.

CHAPTER 13

DMPP is Ineffective at Mitigating N_2O Emissions From Sheep Urine Patches in a UK Grassland under Summer Conditions

Karina A. Marsden , Davey L. Jones, David R. Chadwick

School of Environment, Natural Resources and Geography, Bangor University, Bangor, Gwynedd, LL57 2UW, UK

ABSTRACT

Nitrification inhibitors are a potential technology to mitigate N_2O emissions from the urine patches of grazing animals. At present, there is limited information regarding the efficacy of the nitrification inhibitor 3,4-dimethylpyrazole phosphate (DMPP) in reducing N_2O emissions from ruminant urine patches, as opposed to the well-studied nitrification inhibitor, dicyandiamide. In practical terms, urine patches would be deposited to soil at various times following the application of a nitrification inhibitor to soil. We hypothesised that the effectiveness of DMPP in reducing cumulative N_2O emissions would decrease the longer the time since DMPP application. This study utilised an automated closed chamber technique, to monitor fluxes of N_2O from sheep urine patches (725 kg N ha^{-1}; 150 ml; 300 cm^2) deposited to a Eutric Cambisol, where DMPP was applied (1 kg ha^{-1}) on the same day, 2 weeks before and 4 weeks before urine application. Fluxes were monitored continuously from 4 weeks before, to 9 weeks after, urine application. DMPP was found to be ineffective at reducing cumulative N_2O emissions and 9-week urine-N_2O emission factors when applied at the same time as the sheep urine, although a low number of replicates were used in this study. Some effect of DMPP in delaying the accumulation of soil NO_3^- was observed, with effects being greater the shorter the time since DMPP application. The temporal dynamics of N_2O fluxes were also altered where DMPP was applied on the same day as the urine. Heterogeneity in soil conditions were deemed responsible for the large spatial variability of N_2O emissions observed in this study. The use of Rhizon samplers were useful for detecting spatial variability within the soil solution directly beneath the urine patches (within the flux chambers), which may not have been detected within duplicate urine patches (outside the chambers), where soil sampling was conducted. Further work is required to determine the loading rate and duration of efficacy of DMPP to reduce emissions from urine patches under temperate summer conditions.

Keywords: Diurnal; Grassland; Nitrous oxide; Nitrogen cycle; Mitigation strategy

1. INTRODUCTION

The sustainable management of livestock excreta will form an important component of reducing agricultural greenhouse gas (GHG) emissions, especially under projected increases in global meat consumption (Davidson, 2009). Urine patches deposited to pastures are known hotspots for N cycling, where microbial transformations (e.g. nitrification and denitrification) can result in elevated N_2O emissions (De Klein and van Logtestijn, 1994; Carter, 2007). Excreta (dung and urine) deposited to pasture soils account for ca. 40% of the global N_2O emissions arising from animal production systems (Oenema et al., 2005).

The use of nitrification inhibitors within grazed pastures offers a potential means of reducing N losses whilst promoting nitrogen use efficiency (Di and Cameron, 2002; Di and Cameron, 2007; Cardenas *et al.*, 2016). By delaying the first and rate-limiting step of nitrification (the oxidation of NH_4^+ to NO_2^-; Chaves *et al.*, 2006; Fiencke and Bock, 2006; Benckiser *et al.*, 2013), nitrification inhibitors enable N to persist in the soil in the ammoniacal form for longer. This increases opportunity for plant acquisition of applied N, reduces NO_3^- leaching and can reduce N_2O emissions from both nitrification and denitrification processes. Two of the most widely researched nitrification inhibitors are dicyandiamide (DCD) and 3,4-dimethylpyrazole phosphate (DMPP; Liu et al., 2013), where the latter is a newer product, developed by BASF (Ludwigshafen, Germany; Barth *et al.*, 2001 ; Zerulla *et al.*, 2001). Reviews on the efficacy of nitrification inhibitors have revealed a greater effectiveness of these products in reducing N_2O emissions from grasslands, in comparison to other land use types (Akiyama *et al.*, 2010; Abalos *et al.*, 2014 ; Gilsanz *et al.*, 2016).

There have been a greater number of studies exploring the effect of DCD in reducing N_2O emissions from urine patches compared to DMPP (e.g. Di and Cameron, 2003; De Klein *et al.*, 2011; Misselbrook *et al.*, 2014 ; Bell *et al.*, 2015). DMPP has been suggested for use in pasture soils as it has a lower phytotoxicity and it has slower degradation rates in soil in comparison to DCD (Weiske *et al.*, 2001 ; Zerulla *et al.*, 2001). DMPP has been shown to be effective in reducing N_2O emissions from fertilisers (Linzmeier *et al.*, 2001; Menéndez *et al.*, 2012 ; Bell *et al.*, 2015) and livestock slurry (Dittert *et al.*, 2001; Hatch *et al.*, 2005 ; Merino *et al.*, 2005), but there are few studies investigating DMPP efficacy where ruminant urine is the nitrogen source (e.g. Di and Cameron, 2011; Di and Cameron, 2012 ; Gilsanz *et al.*, 2016). Therefore, further information is required to determine the suitability of DMPP as a means of reducing N_2O emissions from grazed grasslands. When applied at the same time as a cattle urine application (1000 kg N ha^{-1}), a liquid DMPP (5 kg ha^{-1}) application reduced N_2O emissions by 66% (Di and Cameron, 2012). Misselbrook et al. (2014) and Barneze et al. (2015) investigated the N_2O reduction efficacy of pyrazole derivatives (which are structurally similar to DMPP; highest application rate of 3.76 kg ha^{-1}) applied to cattle urine under UK summer conditions, however, no significant reduction in N_2O emissions were observed. The nil effect may have been related to higher temperatures under summer conditions, or an effect of the loading rate of the applied compounds. DMPP performance has been found to be best under cold and wet conditions (Menéndez et al., 2012), therefore, whether DMPP is effective under warmer soil conditions is still unclear.

This study was carried out to determine whether DMPP is effective at reducing N_2O emissions arising from pasture soils influenced by sheep urine, under UK summer conditions. Given that sheep will deposit urine at various times following an application of DMPP, and that cost and labour associated with application will increase with the need for repeated applications, we wanted to assess the duration of DMPP efficacy in reducing N_2O emissions. We used a high frequency automated GHG monitoring system to evaluate the efficacy of a liquid DMPP application in reducing N_2O emissions from sheep urine applied immediately after, 2 weeks after and 4 weeks after DMPP application. We hypothesised that the efficacy of DMPP in reducing N_2O emissions would decrease with the greater the time-since application, due to degradation and immobilisation of the nitrification inhibitor within the soil. The experimental design also allowed i) an assessment of different approaches to measuring N and C dynamics within the urine patch, both inside (Rhizon samplers) and outside (standard soil extractions) of the chambers, and ii) an insight into spatial and temporal variability of soil derived N_2O emissions.

2. MATERIALS AND METHODS

2.1. Field site

The study site was established on a lowland grazed grassland at Henfaes Research station, Abergwyngregyn, North Wales (53°14′N, 4°01′W). The soil at the site is classified as a Eutric Cambisol, and is of glacial till (deposited ca. 10 000 years ago) in origin. The recent history includes moderate fertiliser applications (60–80 kg N ha^{-1} since 2002, and 120 kg N ha^{-1} from 1990 to 2002), light grazing by Welsh Mountain ewes (2–3 livestock units ha^{-1}) and reseeding in 1990 with a *Lolium perenne* L. and *Trifolium repens* L. mix. To prevent the effects of recent livestock excretal depositions on monitored gas fluxes, livestocks were excluded from the study site five months prior to treatment application.

2.2. Soil sampling and analysis

To provide the background characteristics (Table 1) of the study site, soil was sampled (0–10 cm, 1.5 cm diameter) in triplicate, at the block level ($n = 3$). Soil bulk density was measured by inserting a 100 cm^3 metal ring into the ground (0–5 cm), oven-drying the removed soil core (105 °C; 24 h), and sieving (<2 mm) the dry soil to remove and weigh stones. Gravimetric moisture content was determined by oven drying soil (105 °C; 24 h) and organic matter was determined by loss-on-ignition (450 °C; 16 h; Ball, 1964). Soil pH and electrical conductivity (EC) were measured with standard electrodes, submerged in 1:2.5 (w/v) soil-to-distilled water suspensions. The total soil C and N content of oven-dried and ground samples were determined using a TruSpec® Analyzer (Leco Corp., St. Joseph, MI); pasture foliar C and N contents, before and during the study, were also analysed using the same instrument. Dissolved (1:5; w/v; soil-to-0.5 M K_2SO_4) total C and N concentrations in soil were determined by the method of Jones and Willett (2006), using a multi N/C 2100S analyser (AnalytikJena, Jena, Germany). The $CHCl_3$-fumigation-extraction method of Voroney et al. (2008), was used to determine microbial biomass C and N, where the resulting C and N in the extracts were analysed as described previously, using K_{EC} and K_{EN} correction factors of 0.35 and 0.5, respectively. Concentrations of P, NO_3^- and NH_4^+ were determined in 0.5 M K_2SO_4 soil (0–5 cm) extracts using colorimetric methods described in Murphy and Riley (1962), Miranda et al. (2001) and Mulvaney (1996), respectively. Exchangeable cations (Na, K, Ca) were measured using a Sherwood Model 410 flame photometer (Sherwood Scientific Ltd., Cambridge, UK) in 1 M NH_4Cl extracts (1:5, w/v, soil-to-1 M NH_4Cl).

Table 1. Properties of the Eutric Cambisol at the field site receiving urine and DMPP applications, at the time of the first DMPP Application. Values represent means ± SEM (n = 3) and results are reported on a dry weight basis.

Eutric Cambisol properties (0–10 cm)	
Texture	Sandy Clay Loam
Bulk density (g cm^{-3})	1.08 ± 0.05
Gravimetric moisture content (%)	21.5 ± 0.9
Organic matter (%)	9.99 ± 0.32
pH	5.91 ± 0.03
EC (μS cm^{-1})	32.7 ± 0.9
Total C (%)	3.99 ± 0.01
Total N (%)	0.34 ± 0.01
C:N ratio	11.6 ± 0.2
Dissolved organic C (mg C kg^{-1})	170 ± 25
Total dissolved N (mg N kg^{-1})	23.1 ± 2.8
Microbial biomass C (g C kg^{-1})	2.44 ± 0.30
Microbial biomass N (mg N kg^{-1})	171 ± 10
Extractable NO_3^- (mg N kg^{-1})	3.97 ± 0.45
Extractable NH_4^+ (mg N kg^{-1})	3.23 ± 0.99
Extractable P (mg P kg^{-1})	3.37 ± 0.44
Exchangeable Na (mg kg^{-1})	43.8 ± 1.1
Exchangeable K (mg kg^{-1})	140 ± 12
Exchangeable Ca (g kg^{-1})	1.07 ± 0.09

2.3. Sheep urine collection and analysis

Welsh Mountain ewes (n = 6) were housed in individual pens, containing slatted plastic flooring specifically designed for sheep (Rimco Ltd., Yorkshire, UK). The pen flooring was slightly raised above the ground, allowing room for urine collection trays. The sheep were allowed free access to water and provided with a diet (ad libitum) of freshly cut pasture similar in composition to the field site. During the urine collection period individual urine volumes were recorded and the samples were filtered (0.45 μm) and frozen prior to use. The urine collected from all sheep was bulked (to provide sufficient volume of homogenous composition) and thoroughly mixed before application to the soil. Total dissolved organic C, total N, P, NO_3^-, NH_4^+ and cations in the sheep urine were analysed as described for the soil extracts, and the urea content was determined via the method of Orsonneau et al. (1992).

2.4. Experimental design, treatments and application dates

The study was laid out in a randomised block design, consisting of five treatments (n = 3) as follows: 1) no urine application (control), 2) sheep urine only, 3) sheep urine plus DMPP applied at the same time, 4) sheep urine plus DMPP applied 2 weeks before urine application, and 5) sheep urine + DMPP applied 4 weeks before urine application, hereafter referred to as NU, SU, SU + DMPP, SU + DMPP 2, and SU + DMPP 4, respectively. For each treatment, urine patches were duplicated next to each flux chamber, allowing one patch for monitoring greenhouse gases from undisturbed soils and another for soil sampling. The pooled sheep urine (150 ml at 14.5 g N l^{-1}) was poured onto respective plots using a fixed template as an area guide (300 cm^2), to represent an average sheep urine volume and patch size (Doak, 1952). This resulted in a urine-N loading rate of 725 kg N ha^{-1} and a total of 2.17 g urine-N applied to the chambers receiving urine applications. Liquid applications of DMPP (250 ml; 0.1 g DMPP l^{-1}) were sprayed (a normal nitrification inhibitor application method for grazed grasslands; Di and Cameron, 2011) by hand onto respective plots (0.25 m^2, across the chamber basal area), at a frequently used equivalent rate of 1 kg DMPP ha^{-1} (Zerulla et al., 2001). The first DMPP treatment application to respective plots (4 weeks before urine application; SU + DMPP 4) took place on 29/6/2015, the second DMPP treatment application (2 weeks before urine application; SU + DMPP 2) was applied on 13/7/2015, and the final DMPP treatment application (SU + DMPP) took place on 27/7/2015. Following this, sheep urine was applied to all plots except for the control on the same date (27/7/2015) as the final DMPP application.

2.5. Greenhouse gas emission monitoring

An automated high frequency (8 flux measurements per day/night cycle) greenhouse gas monitoring system (Queensland University of Technology,

Institute for Future Environments, Brisbane, Australia), as described in Scheer et al. (2014), was used to measure fluxes of N_2O from the urine-treated soils. As the automated system consisted of 12 chambers only (i.e. enough for 4 treatments, where $n = 3$), fluxes of N_2O from the NU treatment were measured manually using the conventional static chamber technique. Here, 1 flux measurement was made as close to daily as possible, which on the majority of sample dates (>86%) was taken between 10 am and 12 noon as suggested in Rochette et al. (2015), with the remainder being sampled between 12 noon and 2 pm. Fluxes of N_2O were expected to be lower and less temporally variable from the NU treatment, compared to urine treatments, as no N was applied. We, therefore, assumed the difference in measurement techniques to cause minimal bias in the results.

For the automated chambers, stainless steel chamber bases (0.25 m^2 basal area) were inserted into the soil (10 cm depth), and chambers (50 cm × 50 cm × 15 cm) were clamped to the bases. The chambers opened and closed via pneumatic actuators, where one block of chambers would close sequentially every 1 h. While one block of chambers were closed, the other two blocks of chambers were open, allowing ambient conditions to be restored within the chambers. Chamber headspace gas samples were pumped (ca. 200 ml min^{-1}) to a sampling unit, through Teflon tubing. The sampling unit housed a LI-COR LI-820 non-dispersive infrared gas analyser (LI-COR, St Joseph, MI, USA) to measure CO_2. Samples were then passed through an Ascarite (sodium hydroxide coated silica) filter before being pumped to a gas chromatograph (SRI 8610C, Torrance, USA), equipped with a ^{63}Ni electron capture detector (ECD) and flame ionization detector (FID) to measure N_2O and CH_4 concentrations, respectively. Over the 1 h chamber closure period, each chamber ($n = 4$) was sampled at 3 min intervals followed by a calibration standard (500 ppb N_2O; 880 ppm CO_2; 3 ppm CH_4; ±2% of the certified value; BOC Gases Ltd., Liverpool, UK). For each chamber, this results in headspace concentration measurements once every 15 min, over the course of 1 h. A full cycle takes 3 h to complete, resulting in (where uninterrupted measurement occurs) 8 flux measurements per 24 h.

The manual gas samples from the NU treatment were taken by placing circular polypropylene chambers (ca. 26 cm in height) onto collars (26 cm diameter), inserted to a depth of 10 cm. The chambers were fitted with a re-sealable vent to allow pressure equalisation when placing chambers onto bases, and were fitted with Suba-Seals®(Sigma, Gillingham, UK) to allow headspace gas sampling. Headspace samples were taken with a syringe every 15 min over the period of 1 h, to match the automated system. Samples were stored in pre evacuated 20 ml glass vials, before being analysed on a PerkinElmer 580 Gas Chromatograph served by a TurboMatrix 110 auto sampler (PerkinElmer, CT, USA). The system contained two Elite-Q PLOT megabore capillary columns, with split injection. One column was linked to an ECD (375 °C), and the other to an FID (350 °C) with methaniser, and the oven temperature was 50 °C. Fluxes from the automated system were measured for a total of 13 weeks, (beginning 4 weeks before urine application and ending 9 weeks after urine application), and fluxes which were sampled manually were measured for 9 weeks following urine application. Prior to the start of the experiment, standard gases were analysed by both GCs (laboratory and field system) and shown to be not significantly different (data not shown).

2.6. Ancillary measurements

To support the emission measurements, several ancillary measurements were made. Soil cores (0–5 cm) were taken from replicated urine patches on several dates throughout the experiment. Gravimetric moisture content measurement and 1:5 (w/v) soil-to-0.5 M K_2SO_4 extractions were conducted on the resulting soil samples. The extracts were analysed for NO_3^-, NH_4^+, extractable dissolved organic C and total N, as described previously. In order to allow sampling of the soil solution from within the chambers in a non-destructive manner, Rhizon® suction samplers (2.5 mm diameter, 5 cm porous part, 12 cm length tubing; Rhizosphere Research Products, Wageningen, Netherlands) were inserted into the centre of the chamber plots, at a 45° angle in relation to the soil surface, and were in situ four weeks before urine application. 1 ml soil solution samples were collected periodically from the Rhizons through a needle inserted into an evacuated 9 ml plastic container, when soil conditions were moist enough for successful sample collection. The soil solution samples were also analysed for NO_3^-, NH_4^+, dissolved organic C and dissolved total N, as described previously. Soil moisture sensors (Acclima SDI-12 digital TDT® sensors; Acclima Inc., ID, USA) were inserted diagonally through the urine patch (but were placed in situ several weeks prior to urine application) within the chambers, in order to measure soil moisture non-destructively. The pasture foliar N content was measured within the chambers at block level prior to urine application, and pasture biomass and foliar N content was measured in each chamber at 3, 6 and 9 weeks following urine application. The grass was also cut in the duplicated plots at the same time as the chambers to ensure consistency. Air temperature was monitored inside and outside the chambers using Thermochron iButtons® (iButtonLink, LLC, WI, USA) logging temperature every 1 h. Rainfall, soil (0–10 cm) and air temperature were monitored hourly at a weather station near to the field site.

2.7. Data processing and statistical analysis

Fluxes of N_2O were calculated from each chamber as described in Scheer et al. (2014). The linear slope of N_2O concentration over time was comprised of either three or four data points. Cumulative N_2O emissions were calculated by trapezoidal integration (from the point of urine application onwards). As urine was applied to only a 0.03 m^2 area of soil, out of a possible 0.25 m^2 enclosed by the chamber, cumulative emissions were corrected in a similar fashion to Marsden et al. (2016). Briefly, the results for all the treatments with a urine application were corrected for the area unaffected by urine within the chamber (0.22 m^2), by deducting the cumulative N_2O emissions arising from the NU treatment over 0.22 m^2. This provided emissions arising directly from the urine patch area (0.03 m^2), where results were then up-scaled to express the direct urine patch cumulative N_2O emissions over 1 m^2. These corrected values were then used to calculate the emission factors, using the following equation:

equation(1)

$$EF = (\text{treatment } N_2O\text{-}N - \text{control } N_2O\text{-}N)/\text{Total N applied} \times 100\%$$

Differences in cumulative N_2O emissions were analysed by a one way ANOVA, with Tukey's post-hoc test, after assessing the normality (Shapiro Wilk test) and homogeneity of variance (Levene's test) assumptions. The ANOVA was conducted between treatments NU, SU and SU + DMPP only, due to the extreme variability observed in the SU + DMPP 2 and SU + DMPP 4 treatments. For the same reasons, the urine patch 9-week N_2O emission factors were only compared statistically between SU and SU + DMPP, by the use of a *t*-test.

An analysis was conducted to determine the best time within a 24 h period to sample chambers in a manual sampling campaign, in order to best represent emissions calculated from the automated system. Days without a full 8 flux measurements were removed from the dataset and cumulative N_2O emissions were then calculated for the automated sampling dataset alongside eight different sampling-time scenarios (00:00–03:00, 03:00–06:00, 06:00–09:00, 09:00–12:00, 12:00–15:00, 15:00–18:00, 18:00–21:00 and 21:00–24:00), across all chambers ($n = 12$). The cumulative emissions from the simulated sampling campaigns were compared to that arising from the automated chamber dataset, to determine how accurate emission measurements would have been if taken within the sampling time windows tested.

ANOVA followed by Tukey's post-hoc test was conducted for extractable NH_4^+, NO_3^-, N and dissolved organic C at each sampling point, following log transformation of the data, in order to satisfy the normality and homogeneity of variance assumptions (same tests used as above). The soil solution NH_4^+, NO_3^-, N and dissolved organic C had a variable *n*, ranging from 1 to 3, due to successful sample collection being dependent on soil conditions being moist enough. Therefore, numerical trends in the soil solution NH_4^+, NO_3^-, total N and dissolved organic C data were reported rather than statistical differences. The soil solution mineral N data allowed determination of whether N transformations were occurring at similar rates in the chambers and the duplicated plots used for the soil coring, as the Rhizon samplers were located directly inside the chambers. Pasture biomass and foliar N content were also analysed by ANOVA, as above, on each of the three biomass cutting dates. All statistical analyses were conducted in Minitab 17.0 (Minitab Inc., State College, PA, USA).

3. RESULTS

3.1. Urine characteristics

The composite sheep urine sample contained 14.5 g N l^{-1} and 28.7 g C l^{-1}, which resulted in an equivalent N application rate of 725 kg N ha^{-1} in all applied urine patches. The N content of the composite urine sample was high, although within reported ranges for sheep (Hoogendoorn et al., 2010). The high N content within the urine may have been a consequence of the sheep being fed ad libitum, whilst not pregnant/lactating. The urine contained 12.4 g urea-N l^{-1}, 0.4 mg NO_3^--N l^{-1}, 211 mg NH_4^+-N l^{-1}, 5.5 mg P l^{-1}, 1.85 g Na l^{-1}, 13.9 g K l^{-1} and 284 mg Ca l^{-1}. The urine had a pH and EC of 8.48 and 25.3 mS cm^{-1}, respectively. Out of a total of 40 urine events between the 6 sheep, the mean and range of individual urine volumes were 104 (18–397) ml per urination event.

3.2. Weather data and soil moisture

The hourly rainfall, soil temperature (0–10 cm) and water-filled pore space (WFPS) data collected throughout the study period can be seen in Figure 1. The DMPP application on the 29/06/15 for treatment SU + DMPP 4, was applied under warm and dry conditions. The DMPP application on the 13/07/15, for treatment SU + DMPP 2, was applied during a period of rainfall. The DMPP and urine applications (27/07/15) were also under periods of rainfall. The total rainfall over the course of the study was 137 mm, and the average soil temperature (0–10 cm) was 16 °C, ranging from 11 to 20 °C. The average WFPS over the course of the study was 27%, and ranged from 5 to 69%, with minimum values at the beginning of the experiment following an extended dry period, and maximum values occurring after rainfall events.

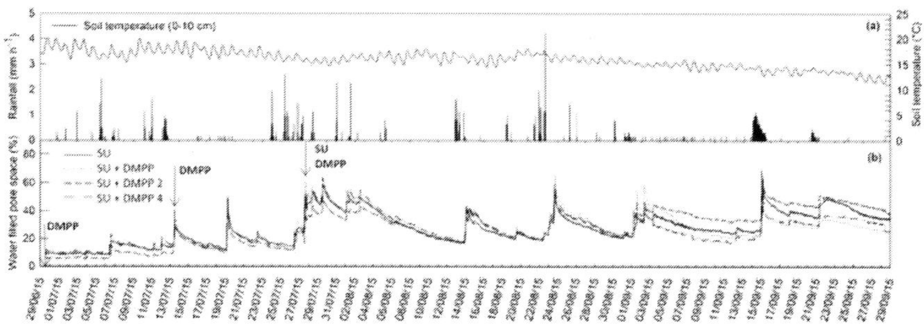

Figure 1. Rainfall, soil temperature and soil water-filled pore space during the course of the field trial. Hourly rainfall and soil (0–10 cm) temperature are shown in panel a) and water-filled pore space within the automated chambers are displayed in panel b). Arrows indicate the timing of either sheep urine (SU) and DMPP applications, the date of measurements are displayed in dd/mm/yy format and legends apply to each respective panel only.

3.3. Nitrous oxide emissions

The fluxes of N_2O from all treatments can be seen in Figure 2. When considering the whole automated chamber data set (8420 flux measurements), 79% of the headspace concentration data had an $R^2 > 0.95$, 88% had an $R^2 > 0.90$ and 97% of the data had an $R^2 > 0.80$, indicating the increase/decrease in N_2O within the chamber headspace over time fitted well to the assumption of linearity. In the vast majority of cases three, rather than four, data points of increase in chamber headspace concentration over time resulted in a higher R^2. As can be seen in Figure 2b, the main N_2O peak from the SU treatment occurred within the first nine days following urine application, with a maxima of 264 ± 179 μg N_2O-N m^{-2} h^{-1}. DMPP was effective in reducing this initial N_2O peak when applied at the same time as the urine (Figure 2c), however, a broader, shallower peak began around the 30/08/15, which was not evident in the SU treatment. In both treatments, SU + DMPP 2 and SU + DMPP 4, this initial peak was similar in size (Figure 2d and e, respectively) with maximum peaks in the first nine days of 192 ± 151 and 231 ± 89 μg N_2O-N m^{-2} h^{-1}, respectively. Notably the variation in N_2O emissions is higher in Figure 2d and e compared to Figure 2b and c.

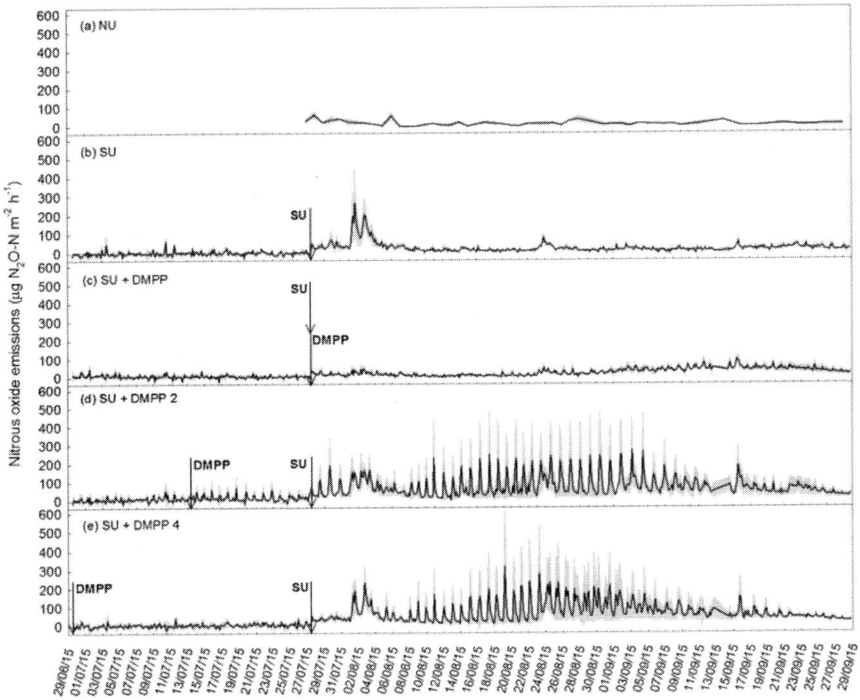

Figure 2. N_2O fluxes following sheep urine and DMPP applications to a Eutric Cambisol, where panel a) displays fluxes from the NU treatment as measured by manual static chambers, panels b), c), d) and e) display fluxes from the automated chambers for treatments SU, SU + DMPP, SU + DMPP 2 and SU + DMPP 4, respectively. In each panel the black line represents the mean flux ($n = 3$) and the grey shaded area represents the upper and lower bounds of the SEM. Arrows indicate the timing of either sheep urine (SU) or DMPP applications and the dates along the x-axis are in dd/mm/yy format.

3.4. Cumulative N_2O emissions and emission factors

Cumulative N_2O emissions were calculated from the point of urine application onwards, and can be seen in Figure 3. Mean cumulative emissions over 9 weeks ranged from 0.06–6.49 g N_2O-N m^{-2} across all treatments. Particularly high emissions were observed within one replicate from both the SU + DMPP 2 and the SU + DMPP 4 treatments, where cumulative N_2O emissions were ca. 5 and 7 times larger than the mean of the remaining two replicates, respectively. Due to the observed extreme variability of cumulative N_2O emissions from SU + DMPP 2 and SU + DMPP 4, these treatments were not included in the ANOVA, due to large violations of the equality of variance assumptions; nevertheless, the results are shown in Figure 3, as the data could shed light on spatial variability observed in soil N_2O emissions. In Figure 3, it can be seen that the SU and SU + DMPP treatments emitted significantly more ($p < 0.01$) N_2O in comparison to NU. In contrast to our hypotheses, DMPP did not reduce the cumulative N_2O emissions from sheep urine patches when applied at the same time as the urine. Although the SU + DMPP 2 and SU + DMPP 4 data show large variability and a relatively low number of replicates, based on the mean values, DMPP applied 2 and 4 weeks before urine application also appeared to be ineffective in reducing N_2O emissions from the urine patch (Figure 3).

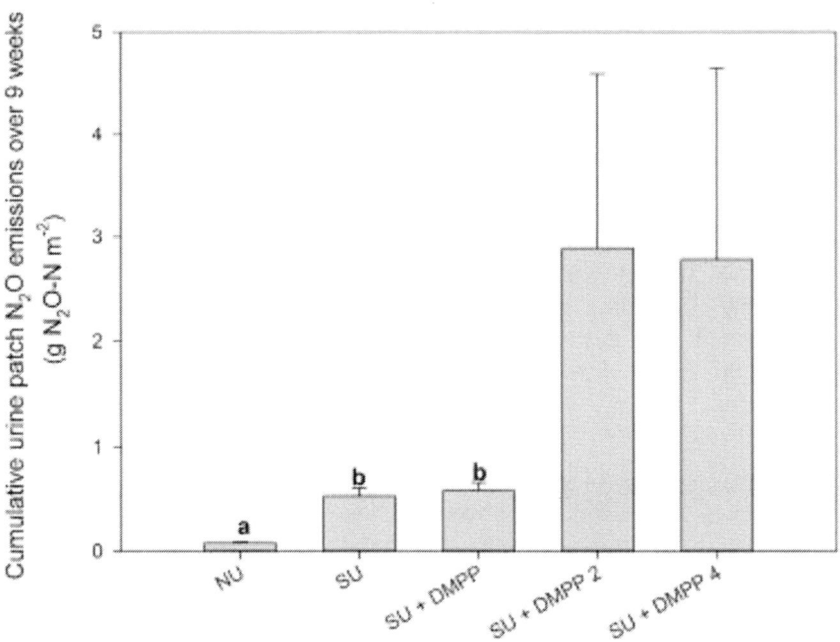

Figure 3. Cumulative nitrous oxide emissions (g N_2O-N m^{-2}) over 9 weeks post urine and DMPP applications to a Eutric Cambisol. Bars represent arithmetic means ($n = 3$) and error bars denote SEM. Letters represent significant differences (one-way ANOVA with Tukey's post-hoc test; $p < 0.05$), where SU + DMPP 2 and SU + DMPP 4 were excluded from the analysis due to extremely high variability between replicates.

The urine patch N_2O emission factors over 9 weeks were 0.63 ± 0.10, 0.70 ± 0.10, 3.87 ± 2.36 and 3.73 ± 2.57% of the applied N for SU, SU + DMPP, SU + DMPP 2 and SU + DMPP 4 treatments, respectively. No significant differences ($p > 0.05$) were found between the amount of N lost as N_2O from the SU and SU + DMPP treatments. Emission factor values within the 'high-emitting' replicates within SU + DMPP 2 and SU + DMPP 4 treatments were as high as 8.20 and 8.84% of the N applied, respectively.

3.5. Diurnal nature of N_2O emissions

An analysis of the best three hour window (within a 24 h period) to sample fluxes from the chambers in this study was conducted, to determine the best time of day to sample chambers within a manual gas sampling campaign. As can be seen in Figure S1 (Supplementary information), sampling between 09:00 in the morning and 12:00 noon most closely approximated the cumulative flux as measured by the automated system, where daily sampling within this time window would have underestimated the cumulative emissions by 0.22 ± 3.51% across all chambers ($n = 12$). Sampling between 18:00 and 21:00 was the next best time window, where emissions would have been overestimated by 1.60 ± 3.83%. Sampling between 21:00 in the evening and 09:00 in the morning would have underestimated cumulative fluxes by a similar amount (cumulative fluxes underestimated by 7.49 ± 3.63%, 7.10 ± 4.66%, 8.09 ± 4.71% and 5.51 ± 3.28% at 00:00–03:00, 03:00–06:00, 06:00–09:00 and 21:00–24:00, respectively). Sampling between the hours of 12:00–18:00 would have overestimated cumulative fluxes (22.46 ± 5.05% and 16.14 ± 4.83% at 12:00–15:00 and 15:00–18:00, respectively). These data show that N_2O fluxes across the whole experimental period were generally highest between the hours of 12.00 and 15.00 and lowest between 06:00 and 09:00.

In some chambers (e.g. 'high emitting' replicate from SU + DMPP 2 treatment), however, emissions were highest between ca. 00:30 and 06:30 (hh:mm), with minimum emissions in the diurnal cycle occurred between ca. 15:30 and 21:30 (hh:mm). An example of this is shown in Figure S2 (Supplementary information) between 08/08/2015 and 14/08/15, where N_2O emissions appear to be out of phase with soil temperature. Within this individual chamber both N and labile C were present within the soil solution and assumed to not be limiting N_2O production. The maximum soil temperature over the period of observation shown in Figure S2 was 26 °C, and the minimum soil temperature was 12 °C. The highest soil temperatures generally occurred at around 13:30 (hh:mm), however the minimum temperature occurred anywhere between 00:00 and 06:00 (hh:mm).

3.6. Mineral N dynamics

The extractable NH_4^+ and NO_3^- from soil cores in replicated plots can be seen in Figure 4a and b, respectively, and the soil solution NH_4^+ and NO_3^- from Rhizon samples can be seen in Figure 4c and d, respectively. In order to make the conversion between the mg kg^{-1}as expressed for the soil extract data, into mg l^{-1} of soil solution, a conversion factor of multiplying by ca. 3 is required (the exact conversion factor depends on the soil moisture at the time of sampling, therefore, conversion factors for each soil sampling date are supplied in Supplementary information: Table S1.0). Prior to urine application, no significant differences ($p > 0.05$) in extractable NH_4^+ concentrations were observed between treatments. On the day of urine application (27/07/15), all extractable NH_4^+concentrations were greater than the control ($p < 0.01$), with no further differences between treatments. On the second day following urine application, extractable $NH_4^+$was only higher ($p < 0.05$) in the SU treatment in comparison to the control. Four days following urine application the extractable NH_4^+ was greater in all DMPP containing treatments ($p < 0.05$) compared to the control, but the SU treatment had returned to control levels ($p > 0.05$), indicating DMPP applications were delaying nitrification. One week following urine application the SU and SU + DMPP 4 treatments had returned to control levels ($p > 0.05$), however, the SU + DMPP and SU + DMPP 2 treatments had higher extractable NH_4^+ concentrations in comparison to the control ($p < 0.05$). This indicates that the effect of DMPP was shorter lived in the SU + DMPP 4, compared to SU + DMPP 2 and SU + DMPP treatments. No differences were observed in extractable NH_4^+ between treatments 10 days after urine application. Two weeks following urine application SU + DMPP had a higher extractable NH_4^+ content compared to the control ($p < 0.05$), and after 3 weeks extractable NH_4^+ was greater in SU + DMPP and the SU + DMPP 2 compared to the control ($p < 0.05$). No further differences in extractable NH_4^+ concentrations were observed beyond three weeks following urine application ($p > 0.05$).

No differences were observed in extractable NO_3^- concentrations between treatments prior to, on the day of urine application, or two and four days following urine application ($p > 0.05$). After one week, however, all urine containing treatments had a higher extractable NO_3^- content in comparison to NU ($p < 0.001$). Ten days following urine application the extractable NO_3^- concentrations were still greater than NU in all urine containing treatments. Here, the SU had a higher extractable NO_3^- concentration than the SU + DMPP treatment, but not the SU + DMPP 2 or SU + DMPP 4 treatments, indicating a delay in nitrification when DMPP was applied at the same time as the urine only. Two weeks after urine application, extractable NO_3^- was greater in all urine containing treatments compared to NU, but no other differences were detected beyond this date ($p > 0.05$).

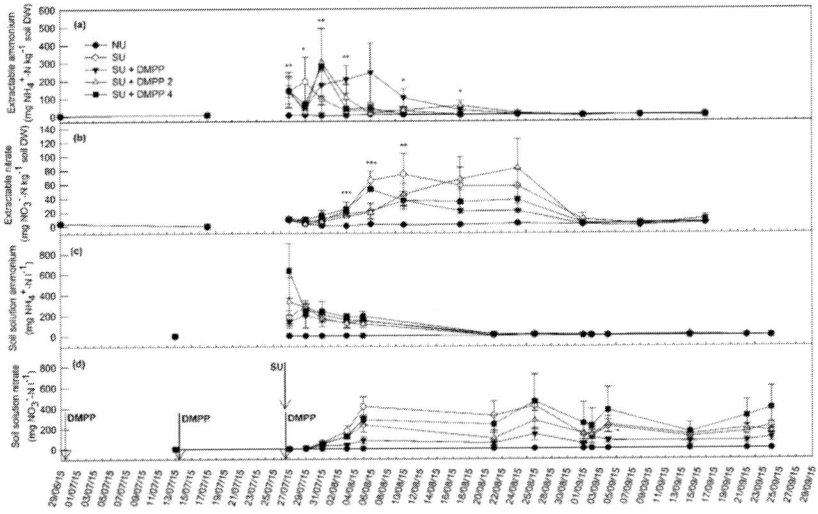

Figure 4. Mineral nitrogen dynamics during the course of the field trial. Extractable ammonium-N can be seen in panel a) and extractable nitrate-N in panel b), which were measured on extracted soils from replicated plots (outside chambers). The soil solution ammonium-N is shown in panel c) and soil solution nitrate-N in panel d), which were measured from Rhizon samples located within the chambers. The figure legend applies to all panels, symbols represent means ($n = 3$), and error bars denote SEM. Arrows in panel d) represent timings of sheep urine (SU) and/or DMPP applications to the respective treatments. Asterisks above symbols represent level of significant differences (ANOVA) and was conducted for data in panels a) and b) only. The date format along the x-axis is dd/mm/yy.

Rhizon soil solution samples were only obtained when soil conditions were moist enough, therefore, *n* varied on each sample date, depending on whether sufficient sample was obtained for analysis. Soil solution NH_4^+ concentrations in the treatments containing sheep urine appeared to return to control values at a similar time to the soil extracts. The soil solution NO_3^- concentrations increased steadily during the 9 d following urine application. At the point of 9 d following urine application, the soil solution NO_3^- followed the numerical trend NU < SU + DMPP < SU + DMPP 2 < SU + DMPP 4 < SU, which indicates DMPP delayed nitrification rates compared to sheep urine alone, but was less effective the longer the time since DMPP application. Notably, the soil solution NO_3^- concentrations were numerically higher than NU for a longer period in comparison to the extractable NO_3^- data, indicating a difference in N processing rates between duplicated plots and the chambers.

3.7. Dissolved organic C and N dynamics

The total extractable N and dissolved organic C during the course of the field trial can be seen in Figure 5a and b, respectively; the total N and dissolved organic C measured in the soil solution samples can be seen in Figure 5c and d, respectively. Differences between total extractable N were only found at one sampling point following urine application to the plots, this was at two weeks following urine application where the SU + DMPP extractable N was greater than the control ($p < 0.05$). Similarly, not many differences in extractable dissolved organic C were found between the treatments at the different sampling points. Here, the only differences found were two days following urine application, where SU and SU + DMPP 2 had higher levels of extractable dissolved organic C compared to the control ($p < 0.05$).

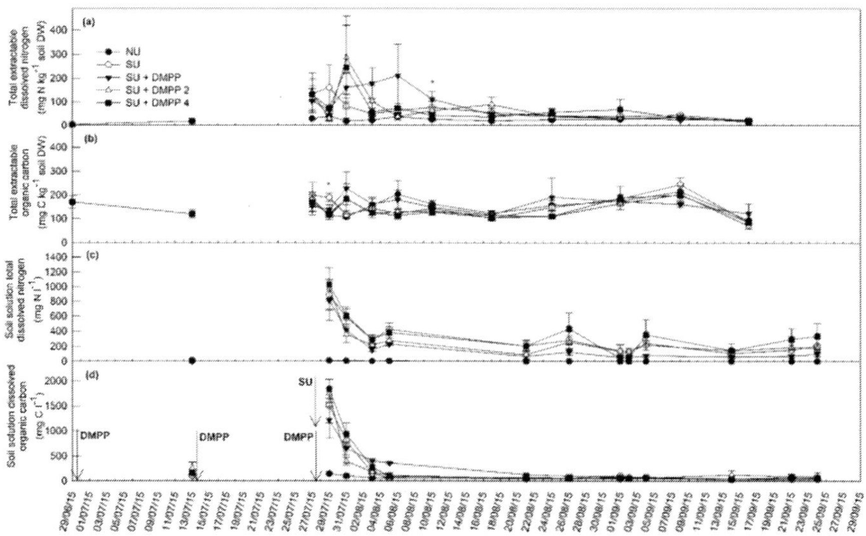

Figure 5. Total extractable and soil solution dissolved N and C dynamics. Total extractable dissolved N is displayed in panel a) and total extractable dissolved organic C is displayed in panel b), which were measured on extracted soils from replicated plots (outside chambers). Soil solution dissolved N is displayed in panel c) and soil solution dissolved organic C is displayed in panel d), which were measured using Rhizon samplers located within the chambers. The figure legend applies to all panels, symbols represent means ($n = 3$), and error bars denote SEM. Arrows in panel d) represent timings of sheep urine (SU) and/or DMPP applications to the respective treatments. Asterisks above symbols represent level of significant differences (ANOVA) and was conducted for data in panels a) and b) only. The date along the x-axis is expressed in dd/mm/yy.

The soil solution N increased following urine application, and remained numerically higher than the control until the end of the study, which reflected the remaining NO_3^- in the soil solution of the chamber plots. In the two chambers which emitted the highest amounts of N_2O, it was found that the soil solution dissolved organic C concentration was ca. 2 and 4 times higher than the other samples collected across the plots, prior to urine application. The soil solution dissolved organic C concentration increased following urine application and treatments appeared to decline at similar rates. No observable differences in soil solution dissolved organic C were found between treatments after ca. 3 weeks following urine deposition.

3.8. Pasture biomass and foliar N content

The foliar N content measured at block level ($n = 3$), was 3.5 ± 0.5% prior to urine application. No significant differences ($p > 0.05$) in foliar N content were observed between treatments 3 weeks following urine application, where the mean foliar N content across all treatments was 3.4 ± 0.2%. The urine-containing treatments all had a significantly greater pasture foliar N content in comparison to the control ($p < 0.01$) at 6 and 9 weeks following urine application, where the mean foliar N content of all the urine-containing treatments was 4.3 ± 0.1% and 4.6 ± 0.2%, compared to the control value of 3.2 ± 0.2% and 3.0 ± 0.4%, respectively. No significant differences ($p > 0.05$) were found for pasture biomass in the plots at 3, 6 or 9 weeks following urine application, where the mean pasture biomass across all samples were 0.47 ± 0.2, 0.51 ± 0.03 and 0.29 ± 0.04 t DM ha^{-1}, respectively.

4. DISCUSSION

4.1. Effect of DMPP on sheep urine patch N_2O emissions and mineral N dynamics

In this study, a liquid DMPP application (1 kg ha^{-1}) did not reduce cumulative N_2O emissions from sheep urine patches in a temperate grassland under summer conditions, when applied at the same time as the urine. Whilst no significant effect of DMPP was found for cumulative N_2O emissions, there was an observable effect of the nitrification inhibitor based on temporal trends in the N_2O fluxes and mineral N data. When considering the SU + DMPP treatment, the initial N_2O peak following urine application was much reduced in comparison to the SU-only treatment, however, a broader, shallower peak towards the end of the measurement period was observed. In addition, the extractable NH_4^+ concentrations were higher in the SU + DMPP and SU + DMPP 2 treatments up to three weeks following urine application. This suggests that nitrification rates were delayed, with the effect being greater the shorter the time since DMPP application. Whilst some effect of DMPP was observable, this tended to delay, rather than reduce overall N_2O emissions in this study. The N application in the sheep urine patches was relatively high at 725 kg N ha^{-1}, yet Di and Cameron (2012) found DMPP reduced N_2O emissions by 66% from cattle urine N applications as high as 1000 kg N ha^{-1}, although a higher DMPP application of 5 kg ha^{-1} was used in this study.

There could be several reasons for not observing an effect of DMPP on cumulative N_2O emissions. It could be due to the DMPP load applied (1 kg ha^{-1}), which was low compared to other studies e.g. 5 kg ha^{-1} application in Di and Cameron (2012) and 10 kg ha^{-1} application in the laboratory study of Di and Cameron (2011). It could also be due to high soil temperatures, which may have resulted in rapid DMPP degradation by microbial activity. Menéndez et al. (2012) and Vasquez et al. (unpubl.), however, found no significant effect of increasing temperature between 10 and 20 °C on DMPP degradation, which represents the typical range of soil temperatures observed in this field trial. The soil water content has also been identified as an important parameter influencing the persistence of this molecule in soil, where a lower water content (and hence greater oxygen content) may increase oxidation of the molecule (Chen *et al.*, 2010 ; Menéndez *et al.*, 2012). In this study, the maximum observed water-filled pore space was 69%, but more often it remained between 20 and 40%, which suggests depletion of DMPP via oxidation may have occurred. Further depletion of DMPP from the soil solution could have occurred due to immobilisation within non-target microbial biomass, sorption to the soil's solid phase and root surfaces or canopy interception of the applied inhibitor.

Another possible reason for the lack of effect of DMPP on cumulative N_2O emissions could be inefficient mixing with the soil solution and urine-N. Given that the DMPP was applied under rainfall conditions in the treatments SU + DMPP and SU + DMPP 2, and that there was an observable effect of DMPP on nitrification rates and the temporal nature of N_2O fluxes in SU + DMPP treatment, it would suggest that the inhibitor at least partially mixed with the urine derived N. The DMPP applied 4 weeks before urine application, however, may have experienced greater degradation or pasture canopy interception during application, as it was not applied during a rainfall event. It cannot be excluded that weather conditions during the time of DMPP applications may have confounded observations of a treatment effect. Improvements to the experimental design could have been made via the use of rain shelters and/or irrigation systems to maintain standard conditions throughout the study period. Monitoring N_2O emissions for 9 weeks following urine application may have also not been a long enough measurement period for treatment effects to be observed.

Further research based on different DMPP application methods is required, e.g. investigating whether DMPP would have been more effective if mixed with the urine prior to application to soil. Comparatively more research has been conducted on different application methods of DCD in comparison to DMPP (e.g. Ledgard *et al.*, 2008; Minet *et al.*, 2013; Wakelin *et al.*, 2013 ; Luo *et al.*, 2015). This may be due to the original commercial development of DMPP as a fertiliser coating (Pasda *et al.*, 2001 ; Zerulla *et al.*, 2001), rather than for its use on urine-influenced soils.

4.2. N_2O emission factors

The 9-week emission factor for the SU treatment was 0.63 ± 0.01% of the applied N, which is lower than the default IPCC emission factor for sheep excreta of 1% (IPCC, 2007). This is consistent with reported emission factors being lower under spring-summer conditions in comparison to autumn-winter (Allen et al., 1996). No significant differences were found for N_2O emission factors between SU and SU + DMPP. The mean 9 week N_2O emission factor across all urine receiving treatments (two extreme values removed as values were >1.5 times the interquartile range; $n = 10$) was 0.98 ± 0.30% of the applied N. This is remarkably similar to the default N_2O emission factor proposed by the IPCC of 1%. The highest emission factors occurred in the SU + DMPP 2 and SU + DMPP 4 treatments (8.20 and 8.84%, respectively). Whilst these are remarkably high values, results of a similar magnitude have been reported in De Klein and van Logtestijn (1994). It is suggested that the overlapping of nutrient hotspots greatly increased the emission factors for sheep urine. As the study was fenced off from grazing livestock, we suggest the nutrient hotspot may have come from a source other than a urine patch (see Supplementary information S1.0, for additional discussion).

4.3. Spatial and temporal variability in N_2O fluxes

N_2O emissions from agricultural soils are notoriously variable, both in time and space (Hénault et al., 2012), and this study proved to be no exception. Notably the SU and the SU + DMPP treatments did not have as high a variability in N_2O emissions as the SU + DMPP 2 and SU + DMPP 4 treatments (Figure 2 ; Figure 3). Due to this we can conclude with more confidence that DMPP had no effect on cumulative N_2O emissions when applied at the same time as the urine, under these experimental conditions. Bearing this in mind, it is unlikely that the DMPP would have had an appreciable effect on reducing N_2O emissions when DMPP was applied weeks before the urine, even if the data had not suffered from the large spatial variability.

Whilst a high temporal variability was observed, the automated chamber system captured this variability, thereby reducing bias in emission calculations due to sampling time and frequency, which can be an artefact of manually sampled static chamber based campaigns (e.g. Yamulki *et al.*, 2001 ; Scheer *et al.*, 2014). The N_2O emissions in this study were also, however, highly spatially variable. Here, one replicate in each of the treatments SU + DMPP 2 and SU + DMPP 4 demonstrated a marked difference to their respective replicates. Future studies should consider maximising the number of replicates in order to capture spatial variability in N_2O emissions, when using an automated chamber based system. This will potentially result in a having to select fewer treatments for investigation, however, would be appropriate where emission factors are to be measured for inventory purposes.

4.4. Diurnal nature of N_2O fluxes

The existence of diurnal trends in N_2O flux data have been identified in several studies (e.g. Thomson *et al.*, 1997; Williams *et al.*, 1999; Yamulki *et al.*, 2000; Yamulki *et al.*, 2001; Hyde *et al.*, 2005a; Hyde *et al.*, 2005b; Hatch *et al.*, 2005; Alves *et al.*, 2012; Scheer *et al.*, 2014; Shurpali *et al.*, 2016 ; Wei *et al.*, 2016), yet contrasting timings of daily minimum and maximum N_2O flux are reported between studies. When considering the current study, the best time of day to manually sample chambers in order to best represent emissions from the automated system, would have been between 09:00 and 12:00. This is in agreement with Rochette et al. (2015) and similar to the recommended sampling time of 09:00–10:00 as found in Alves et al. (2012). Scheer et al. (2014) found the diurnal nature of N_2O fluxes to be greatest when the mineral N and moisture were non-limiting. This appeared to be true in our study, and the diurnal trend was also strongest in the chambers where urine was applied to a dissolved organic C hotspot.

Large overestimations of the cumulative flux would have occurred if sampling was conducted between 12:00–15:00 and 15:00–18:00, suggesting maximum diurnal emissions generally occurred within these time frames over the study period. This was not the case for all chambers at all times though, and in some chambers maximum diurnal N_2O fluxes occurred at night time and early morning. The existence of these contrasting diurnal trends, even within the same study, have also been shown by Shurpali et al. (2016), where N_2O emissions had maximum peaks during both the day and the night at different times. In Shurapli et al.

(2016) night-time diurnal peaks occurred when N was limiting and the magnitude of N_2O flux was low, however, in the current study emissions were greater at night when N and C were presumably non-limiting, following the urine application.

Some studies show N_2O emissions to be closely in phase with soil temperature (e.g. Williams *et al.*, 1999 ; Hatch *et al.*, 2005), whilst others observe a greater lag time between soil temperature and N_2O production. The observed variation in timings of minimum and maximum daily flux could be due to the fact that N_2O emissions are dependent on several factors (not just soil temperature), which may interact in different ways to produce different timings of minimum and maximum flux. The solubility and microbial consumption of O_2 increases alongside temperature, which may lead to reductions in N_2O due to the occurrence of complete denitrification (Yamulki et al., 2000). The pattern of N_2O production and consumption across soil depth may also vary with temperature, where maximum and minimum temperatures would occur later at soil depth. Another possible reason is that greater compaction could result in slower gas diffusivity (Ball *et al.*, 1999 ; Sitaula *et al.*, 2000) from the source of N_2O to the chamber headspace i.e. a shorter lag between maximum temperature and maximum N_2O flux was found in the sieved and repacked soils of Hatch et al. (2005), as well as soils which had been subjected to rotary hoeing during incorporation of vegetable residues in Scheer et al. (2014). This area of study requires further research as it has implications for recommendations of the best time to sample N_2O emissions over a 24 h period to reflect the average daily fluxes, and to more accurately calculate N_2O losses from soils (Shurpali et al., 2016). It also supports the use of methods with a high temporal resolution for monitoring N_2O emissions, where possible.

4.5. Destructive vs. non-destructive soil sampling for measuring soil mineral N pools

In this study, differences were found between the mineral N dynamics as measured by taking soil cores from duplicate urine patches (destructive) and as measured in the soil solution of the chambers (non-destructive). The soil solution NO_3^- concentrations in the urine treated chambers remained higher than the control for a longer period than the extractable NO_3^-, therefore, further N_2O emissions could have occurred beyond the duration of this study. The differences in N processing found between the duplicated urine patches may reflect artefacts of coring the duplicated urine patches. This may have influenced the aeration and water infiltration dynamics in these areas. Conversely, the chambers may have caused more shading and/or interception of rainfall in comparison to the duplicated plots. The difference in N processing rates highlights the need for the development and use of non-destructive sampling techniques for use inside chambers. Rhizon samplers were useful in this study for determining differences, which would not have been observed from only taking soil cores from duplicated urine patches. Limitations of the Rhizons were that successful sample collection depended on sufficient soil moisture, which sometimes resulted in a variable number of replicates on different sample dates. Deploying multiple Rhizons within a urine patch could potentially overcome the issue of the occasional small sample volume. Nevertheless, Rhizon samplers would be ineffective under extended dry periods.

5. CONCLUSIONS

A recent review by Gilsanz et al. (2016), highlights the paucity of studies regarding DMPP efficacy in reducing N_2O emissions, in comparison to the well-researched inhibitor DCD. This is especially true of urine-influenced soils, where this study contributes new information for DMPP efficacy in sheep urine patches, under temperate summer conditions. In this study, DMPP applied at a rate of 1 kg ha^{-1} influenced the dynamics of N processing rates, but had no effect on overall cumulative N_2O emissions or emission factors. Care should be taken when interpreting these results, due to a low number of treatment replicates within this study and the short N_2O measurement period. In order to be able to assess the practicality and cost-effectiveness of DMPP use in temperate grazed grasslands, further information is required on the loading rates of DMPP required to reduce N_2O emissions from urine patches under summer conditions and whether different application methods alter the efficacy of this nitrification inhibitor. Whilst the duration of DMPP efficacy remained unestablished in this study, this is an important factor to determine, due to the potential of applying inhibitors to urine patches of varying ages.

ACKNOWLEDGEMENTS

KAM would like to thank the School of Environment, Natural Resources and Geography, Bangor University, for sponsoring this study. This work was supported by the UK Natural Environment Research Council, under grant award NE/M015351/1/. In addition, thanks go to Neil Donovan at Rothamsted Research for analysing gas samples, Sarah Chesworth, Terry Cox, Rory Shaw, Jonathan Roberts, Gordon Turner, Mark Hughes, Llinos Hughes and Jairo Cardoso Abreu at Bangor University for all the assistance with setting up the field trial and sample processing.

REFERENCES

1. Abalos, D., Jeffery, S., Sanz-Cobena, A., Guardia, G., Vallejo, A., 2014. Meta-analysis of the effect of urease and nitrification inhibitors on crop productivity and nitrogen use efficiency. Agric. Ecosyst. Environ. 189, 136–144.
2. Akiyama, H., Yan, X., Yagi, K., 2010. Evaluation of effectiveness of enhanced-efficiency fertilizers as mitigation options for N2O and NO emissions from agricultural soils: meta-analysis. Global Change Biol. 16, 1837–1846.
3. Allen, A.G., Jarvis, S.C., Headon, D.M., 1996. Nitrous oxide emissions from soils due to inputs of nitrogen from excreta return by livestock on grazed grassland in the UK. Soil Biol. Biochem. 28, 597–607.
4. Alves, B.J.R., Smith, K.A., Flores, R.A., Cardoso, A.S., Olivieria, W.R.D., Jantalia, C.P., Urquiaga, S., Boddey, R.M., 2012. Selection of the most suitable sampling time for static chambers for the estimation of daily mean N2O flux from soils. Soil Biol. Biochem. 46, 129–135.
5. Ball, B.C., Scott, A., Parker, J.P., 1999. Field N2O, CO2 and CH4 fluxes in relation to tillage, compaction and soil quality in Scotland. Soil Till. Res. 53, 29–39.
6. Ball, D.F., 1964. Loss-on-ignition as an estimate of organic matter and organic carbon in non-calcareous soils. J. Soil Sci. 15, 84–92.
7. Barneze, A.S., Minet, E.P., Cerri, C.C., Misselbrook, T., 2015. The effect of nitrification inhibitors on nitrous oxide emissions from cattle urine depositions to grassland under summer conditions in the UK. Chemosphere 119, 122–129.
8. Barth, G., von Tucher, S., Schmidhalter, U., 2001. Influence of soil parameters on the effect of 3,4-dimethylpyrazole-phosphate as a nitrification inhibitor. Biol. Fertil. Soils 34, 98–102.
9. Bell, M.J., Rees, R.M., Cloy, J.M., Topp, C.F.E., Bagnall, A., Chadwick, D.R., 2015. Nitrous oxide emissions from cattle excreta applied to a Scottish grassland: effects of soil and climatic conditions and a nitrification inhibitor. Sci. Total Environ. 508, 343–353.
10. Benckiser, G., Christ, E., Herbert, T., Weiske, A., Blome, J., Hardt, M., 2013. The nitrification inhibitor 3,4-dimethylpyrazole-phosphate (DMPP) –quantification and effects on soil metabolism. Plant Soil 371, 257–266.
11. Cardenas, L.M., Misselbrook, T.M., Hodgson, C., Donovan, N., Gilhespy, S., Smith, K.A., Dhanoa, M.S., Chadwick, D., 2016. Effect of the application of cattle urine with or without the nitrification inhibitor DCD, and dung on greenhouse gas emissions from a UK grassland soil. Agric. Ecosyst. Environ. 235, 229–241.
12. Carter, M.S., 2007. Contribution of nitrification and denitrification to N2O emissions from urine patches. Soil Biol. Biochem. 39, 2091–2102.
13. Chaves, B., Opoku, A., De Neve, S., Boeckx, P., Van Cleemput, O., Hofman, G., 2006. Influence of DCD and DMPP on soil N dynamics after incorporation of vegetable crop residues. Biol. Fertil. Soils 43, 62–68.
14. Chen, D., Suter, H.C., Islam, A., Edis, R., 2010. Influence of nitrification inhibitors on nitrification and nitrous oxide (N2O) emissions from a clay loam soil fertilized with urea. Soil Biol. Biochem. 42, 660–664.
15. Davidson, E.A., 2009. The contribution of manure and fertilizer nitrogen to atmospheric nitrous oxide since 1860. Nat. Geosci. 2, 659–662.
16. De Klein, C.A.M., van Logtestijn, R.S.P., 1994. Denitrification and N2O emission from urine- affected grassland soil. Plant Soil 163, 235–242.
17. De Klein, C.A.M., Cameron, K.C., Di, H.J., Rys, G., Monaghan, R.M., Sherlock, R.R., 2011. Repeated annual use of the nitrification inhibitor dicyandiamide (DCD) does not alter its effectiveness in reducing N2O emissions from cow urine. Anim. Feed Sci. Technol. 166, 480–491.

18. Di, H., Cameron, K.C., 2002. The use of a nitrification inhibitor, dicyandiamide (DCD), to decrease nitrate leaching and nitrous oxide emission in a simulated grazed and irrigated grassland. Soil Use Manage. 4, 395–403.
19. Di, H.J., Cameron, K.C., 2003. Mitigation of nitrous oxide emissions in a spray-irrigated grazed grassland by treating the soil with dicyandiamide, a nitrification inhibitor. Soil Use Manage. 19, 284–290.
20. Di, H., Cameron, K.C., 2007. Effects of the nitrification inhibitor dicyandiamide on soil mineral N, pasture yield, nutrient uptake and pasture quality in a grazed pasture system. Soil Use Manage. 23, 111–120.
21. Di, H.J., Cameron, K.C., 2011. Inhibition of ammonium oxidation by a liquid formulation of 3,4-dimethylpyrazole phosphate (DMPP) compared with dicyandiamide (DCD) in six New Zealand grazed grassland soils. J. Soils Sediments 11, 1032–1039.
22. Di, H.J., Cameron, K.C., 2012. How does the application of different nitrification inhibitors affect nitrous oxide emissions and nitrate leaching from cow urine in grazed pastures? Soil Use Manage. 28, 54–61.
23. Dittert, K., Bol, R., Chadwick, D., Hatch, D., 2001. Nitrous oxide emissions from injected 15N-labelled cattle slurry into grassland soil as affected by DMPP nitrification inhibitor. Plant Nutrition. Springer, Netherlands, pp. 768–769.
24. Doak, B.W., 1952. Some chemical changes in the nitrogenous constituents of urine when voided on pasture. J. Agric. Sci. 42, 162–171.
25. Fiencke, C., Bock, E., 2006. Immunocytochemical localization of membrane-bound ammonia monooxygenase in cells of ammonia oxidizing bacteria. Arch. Microbiol. 185, 99–106.
26. Gilsanz, C., Báez, D., Misselbrook, T.H., Dhanoa, M.S., Cárdenas, L.M., 2016. Development of emission factors and efficiency of two nitrification inhibitors, DCD and DMPP. Agric. Ecosyst. Environ. 216, 1–8.
27. Hénault, C., Grossel, A., Mary, B., Roussel, M., Léonard, J., 2012. Nitrous oxide emission by agricultural soils: a review of spatial and temporal variability for mitigation. Pedosphere 22, 426–433.
28. Hatch, D., Trindade, H., Cardenas, L., Carneiro, J., Hawkins, J., Scholefield, D., Chadwick, D., 2005. Laboratory study of the effects of two nitrification inhibitors on greenhouse gas emissions from a slurry-treated arable soil: impact of diurnal temperature cycle. Biol. Fertil. Soils 41, 225–232.
29. Hoogendoorn, C.J., Betteridge, K., Costall, D.A., Ledgard, S.F., 2010. Nitrogen concentration in the urine of cattle, sheep and deer grazing a common ryegrass/cocksfoot/white clover pasture. New Zeal. J. Agric. Res. 53, 235–243.
30. Hyde, B., Ryan, M., Hawkins, M., Connolly, J., Carton, O.T., 2005a. Nitrous Oxide Emissions. Teagasc Environment Research Centre, Wexford, Ireland.
31. Hyde, B.P., Fanning, A.F., Ryan, M., Carton, O.T., 2005b. An examination of the diurnal variability in nitrous oxide emissions. In: O'Mara, F.P. (Ed.), 20th International Grassland Congress. Wageningen Academic Pub..
32. IPCC,, 2007. Climate Change 2007: The Physical Science Basis. Contribution of Working Group I to the Fourth Assessment Report of the Intergovernmental Panel on Climate Change. In: Solomon, S., Qin, D., Manning, M., Chen, Z., Marquis, M., Averyt, K.B., Tignor, M., Miller, H.L. (Eds.), Cambridge University Press, Cambridge, UK and New York, USA. J
33. ones, D.L., Willett, V.B., 2006. Experimental evaluation methods to quantify dissolved organic nitrogen (DON) and dissolved organic carbon (DOC) in soil. Soil Biol. Biochem. 38, 991–999.
34. Ledgard, S.F., Menneer, J.C., Dexter, M.M., Kear, M.J., Lindsey, S., Peters, J.S., Pacheco, D., 2008. A novel concept to reduce nitrogen losses from grazed pastures by administering soil nitrogen process inhibitors to ruminant animals: a study with sheep. Agric. Ecosyst. Environ. 125, 148–158.

35. Linzmeier, W., Gutser, R., Schmidhalter, U., 2001. Nitrous oxide emission form soil and from a nitrogen-15-labelled fertilizer with the new nitrification inhibitor 3,4-dimethylpyrazole phosphate (DMPP). Biol. Fertil. Soils 34, 103–108.
36. Liu, C., Wang, K., Zheng, X., 2013. Effects of nitrification inhibitors (DCD and DMPP) on nitrous oxide emission, crop yield and nitrogen uptake in a wheat-maize cropping system. Biogeosci. Discuss. 10, 711–737.
37. Luo, J., Ledgard, S., Wise, B., Welten, B., Lindsey, S., Judge, A., Sprosen, M., 2015. Effect of dicyandiamide (DCD) delivery method, application rate and season on pasture urine patch nitrous oxide emissions. Biol. Fertil. Soils 51, 453–464.
38. Marsden, K.A., Jones, D.L., Chadwick, D.R., 2016. Disentangling the effect of sheep urine patch size and nitrogen loading rate on cumulative N2O emissions. Anim. Prod. Sci. 56, 265–275.
39. Menéndez, S., Barrena, I., Setien, I., González-Murua, C., Estavillo, J.M., 2012. Efficiency of nitrification inhibitor DMPP to reduce nitrous oxide emissions under different temperature and moisture conditions. Soil Biol. Biochem. 53, 82–89.
40. Merino, P., Menéndez, S., Pinto, M., González-Murua, C., Estavillo, J.M., 2005. 3,4-Dimethylpyrazole phosphate reduces nitrous oxide emissions from grassland after slurry application. Soil Use Manage. 21, 53–57.
41. Minet, E.P., O'Carroll, C., Rooney, D., Breslin, C., McCarthy, C.P., Gallagher, L., Richards, K.G., 2013. Slow delivery of a nitrification inhibitor (dicyandiamide) to soil using a biodegradable hydrogel of chitosan. Chemosphere 93, 2854–2858.
42. Miranda, K.M., Epsey, M.G., Wink, D.A., 2001. A rapid, simple, spectrophotometric method for simultaneous detection of nitrate and nitrite. Nitric Oxide 5, 62–71.
43. Misselbrook, T.H., Cardenas, L.M., Camp, V., Thorman, R.E., Williams, J.R., Rollett, A.J., Chambers, B.J., 2014. An assessment of nitrification inhibitors to reduce nitrous oxide emissions form UK agriculture. Environ. Res. Lett. 9, 11506.
44. Mulvaney, R.L., 1996. Nitrogen –inorganic forms. In: Sparks, D.L. (Ed.), Methods of Soil Analysis. Part 3. Soil Science Society of America Inc., Madison WI, USA, pp. 1123–1184.
45. Murphy, J., Riley, J.P., 1962. A modified single solution method for determination of phosphate in natural waters. Anal. Chim. Acta 27, 31–36.
46. Oenema, O., Wrage, N., Velthof, G.L., van Groenigen, J.W., Dolfing, J., Kuikman, P.J., 2005. Trends in global nitrous oxide emissions from animal production systems. Nutr. Cycl. Agroecosys. 72, 51–65.
47. Orsonneau, J.-L., Massoubre, C., Cabanes, M., Lustenberger, P., 1992. Simple and sensitive determination of urea in serum and urine. Clin. Chem. 38, 619–623.
48. Pasda, G., Hähndel, R., Zerulla, W., 2001. Effect of fertilizers with the new nitrification inhibitor DMPP (3,4-dimethylpyrazole phosphate) on yield and quality of agricultural and horticultural crops. Biol. Fertil. Soils 34, 85–97.
49. Rochette, P., Chadwick, D.R., de Klein, C.A.M., Cameron, K., 2015. Deployment protocol. In: de Klein, C.A.M., Harvey, M. (Eds.), Nitrous Oxide Chamber Methodology Guidelines. Ministry for Primary Industries, Wellington, New Zealand, pp. 34–55.
50. Scheer, C., Rowlings, D.W., Firrel, M., Deuter, P., Morris, S., Grace, P.R., 2014. Impact of nitrification inhibitor (DMPP) on soil nitrous oxide emissions from an intensive broccoli production system in sub-tropical Australia. Soil Biol. Biochem. 77, 243–251.
51. Shurpali, N.J., Rannik, Ü, Jokinen, S., Lind, S., Biasi, C., Mammarella, I., Peltola, O., Pihlatie, M., Hyvönen, N., Räty, M., Haapanala, S., Zahniser, M., Virkajärvi, P., Vesala, T., Martikainen, P.J., 2016. Neglecting diurnal variations leads to uncertainties in terrestrial nitrous oxide emissions. Sci. Rep. 6, 25739.
52. Sitaula, B.K., Hansen, S., Sitaula, J.I.B., Bakken, L.R., 2000. Effects of soil compaction on N2O emission in agricultural soil. Chemosphere Global Change Sci. 2, 367–371.
53. Thomson, P.E., Parker, J.P., Arah, J.R., Clayton, H., Smith, K.A., 1997. Automated soil monolith-flux chamber system for the study of trace gas fluxes. Soil Sci. Soc. Am. J. 61, 1323–1330.

54. Voroney, R.P., Brookes, P.C., Beyaert, R.P., 2008. Soil microbial biomass C, N, P and S. In: Carter, M.R., Gregorich, E.G. (Eds.), Soil Sampling and Methods of Analysis, 2nd edn. CRC Press, Boca Raton, FL, USA, pp. 637–651.
55. Wakelin, S.A., Clough, T.J., Gerard, E.M., O'Callaghan, G.M., 2013. Impact of shortinterval, repeat application of dicyandiamide on soil N transformation in urine patches. Agric. Ecosyst. Environ. 167, 60–70.
56. Wei, Q., Xu, J., Yang, S., Li, X., Twecan, D., 2016. The sampling strategy for the estimation of daily average soil N2O flux by static chamber method. Fresen. Environ. Bull. 25 1962-1700.
57. Weiske, A., Benckiser, G., Herbert, T., Ottow, J.G.C., 2001. Influence of the nitrification inhibitor 3,4-dimethylpyrazole phosphate (DMPP) in comparison to dicyandiamide (DCD) on nitrous oxide emissions, carbon dioxide fluxes and methane oxidation during three years of repeated application in field experiments. Biol. Fert. Soils 34, 109–117.
58. Williams, D. Ll., Ineson, P., Coward, P.A., 1999. Temporal variations in nitrous oxide fluxes from urine-affected grassland. Soil Biol. Biochem. 31, 779–788.
59. Yamulki, S., Wolf, I., Bol, R., Grant, B., Brumme, R., Veldkamp, E., Jarvis, S.C., 2000. Effects of dung and urine amendments on the isotopic content of N2O released from grasslands. Rapid Commun. Mass Spec. 14, 1356–1360.
60. Yamulki, S., Toyoda, S., Yoshida, N., Veldkamp, E., Grant, B., Bol, R., 2001. Diurnal fluxes and the isotopomer ratios of N2O in a temperate grassland following urine amendment. Rapid Commun. Mass Spec. 15, 1263–1269.
61. Zerulla, W., Barth, T., Dressel, J., Erhardt, K., von Locquenghien, K.H., Pasda, G., Rädle, M., Wissemeier, A.H., 2001. 3,4-Dimethylpyrazole phosphate (DMPP) –a new nitrification inhibitor for agriculture and horticulture. Biol. Fertil. Soils 34, 79–84.

CHAPTER 14

Environmental and Ecological Aspects in the Overall Assessment of Bioeconomy

Andra´s Sze´ka´cs[1]

[1].Agro-Environmental Research InstituteNational Agricultural Research and Innovation CentreBudapestHungary

ABSTRACT

Bioeconomy solutions potentially reduce the utilization demand of natural resources, and therefore, represent steps towards circular economy, but are not per se equivalent to sustainability. Thus, production may remain to be achieved against losses in natural resources or at other environmental costs, and materials produced by bioeconomy are not necessarily biodegradable. As a consequence, the assumption that emerging bioeconomy by itself provides an environmentally sustainable economy is not justified, as technologies do not necessarily become sustainable merely through their conversion to using renewable resources for their production. A source of the above assumption is that the utility of bioeconomy is mostly assessed in interaction between technology developers and economists, resulting in biased assessment with private commercial technology benefits being included, but environmental costs, especially longer term ones, not being sufficiently considered in the economic models. A possible solution to this conceptual contradiction may come from bioethics, as a strong concept in environmental ethics is that no technological intervention can be imposed on nature beyond its receptive capacity. To achieve a better balanced analysis of bioeconomy, environmental and ecological, as well as non-economic social aspects, need to be included in the overall assessment.

Keywords: Bioeconomy Circular economy Ecological aspects Natural resources Bioethics

INTRODUCTION

There are complex tensions—sometimes contradictions—between two central policy commitments of most modern democratic regimes, namely the emerging bioeconomy and sustainable development. While the latter is endorsed in for example the UN Convention on Biodiversity as well as the EU Lisbon Treaty as a constitutional principle for all relevant legislative Directives, developing the bioeconomy is also a universal driving concern for policy, R&D, and innovation. For reconciling emerging bioeconomy developments with genuine sustainable development however, a more detailed knowledge is needed about what comprises the bioeconomy, what is its contribution to economic growth, including possible negative consequences which may have been (knowingly or inadvertently) externalized. For example, the major current commitment in global agriculture to genetically modified (GM) crops, as a bioeconomic commercial scale innovation from the 1990s, has also built the need for chemical herbicides into its main crop-innovation biotechnologies, and these chemicals are controversial in terms of health and environmental impacts, e.g. the recent case of the herbicide active ingredient glyphosate and its formulating agent polyethoxylated tallowamine (POEA). How all of these factors can be reliably measured is also an important issue for documenting the contribution and for assessing the impacts of supporting policies as well as policies responding to citizens' concerns. An equally important and possibly even more urgent issue is, however, whether currently dominant bioeconomy solutions do indeed represent a step towards the ultimate sustainable development goal of a circular economy, i.e. ecological "zero waste" technology (Stahel 2016), or towards truly sustainable ecocycles (Nemethy and Komives 2016). From the disciplinary perspectives of economic analysis and policy driven strategies, the market potential, profitability and some (though selective) societal aspects of bioeconomy have been analyzed. Much less concern has been shown, and even less implemented, to reveal environmental and ecological costs, and therefore, in spite of the achievements realized so far, bioeconomy still operates on the basis of natural resource utilization, conversion of natural assets into more "useful" forms (i.e. economically measurable, but neglecting the costs of natural resources-depletion), while creating less "useful" by-products. For a real transition a true conversion to the principles of ecological economics (Costanza et al. 1997, 2015; Daly and Farley 2011; Baveye et al. 2013), reliance on biomimicry to support ecological innovations instead of exploitative technological approaches (Blok and Gremmen 2016), and the abandonment of the economic growth concept (El-Chichakli et al. 2016) is needed. Without a conceptually improved, ecology-based assessment and implementation, bioeconomy will remain a substantially improved, yet fundamentally equivalent version to unsustainable resource-intensive chemical technologies. 'Bio-'as a preface does not automatically mean ecologically sound, and without the transformations indicated above, and discussed further below, public policy and debate could mislead itself into "talking the talk, but not walking the walk", of sustainability.

BIOECONOMY AS A CONCEPT

The Millennium Ecosystem Assessment (UNEP-WRI 2005) by the United Nations Environment Programme evidenced the effects of anthropogenic activities on the ecosystems and services they provide, such as food, water, disease management, climate regulation, spiritual fulfilment and aesthetic enjoyment. It also further emphasized the value of research at the interfaces between natural and social sciences, and humanities (Reid and Mooney 2016); and it urged efforts to conserve these more complex to measure thus too-often neglected economic assets in order to achieve sustainability. In an attempt towards such a sustainable management practice, bioeconomy (EC 2012a) (formerly bio-based economy—Langeveld et al. 2010) aims for the production and utilization of renewable biological resources in agribusiness. Bio-based products have been specified as one of the six areas selected for the Lead Market Initiative for the EU (EC 2008). The European Union defines bioeconomy as "the production of renewable biological resources and their conversion into food, feed, bio-based products and bioenergy" (EC 2012a, b), often produced in systemic, both materially and financially interconnected networks on the basis of the cascade principle (de Besi and McCormick 2015). Bioeconomy is being implied in various segments of the industrial and agribusiness sectors, including agriculture, forestry, fisheries, food, pulp and paper production, parts of chemical, biotechnological and energy industries. These areas do not develop as insular entities, but influence each other, demanding an integrated systems approach in their regulation, currently occasionally fragmented into risk assessment or management in artificially isolated components.

The main objectives of growth of the bioeconomy in Europe are: (a) ensuring food security; (b) managing limited and depleting natural resources sustainably; (c) reducing dependence on non-renewable resources; (d) mitigating and adapting to climate change; as well as (e) creating jobs and maintaining European competitiveness (EC 2012b). The opportunities (or hopes) regarding bioeconomy include its potential to move production technologies towards a renewable resource base; to reduce pollution; to improve and enhance food security; and to accelerate adaptation and mitigation of climate change. These hopes are aimed to be achieved by broadening novel science-based applications, yet scientific literature surveys indicate that the concept of bioeconomy is conceived rather differently from various stakeholder perspectives. The industrial biotechnology vision focuses on the rapid utilization and commercialization of biotechnology research in various sectors of the economy; the bio-resource vision emphasizes sustainable utilization of biological raw materials; while the more recent, and yet less prominent, bio-ecology vision promotes maintenance or improvement of biodiversity and ecosystem services, as well as the avoidance of monocultures and soil degradation (Bugge et al. 2016). Due to the complexity of the issues, the topic has been evaluated in detail by the Standing Committee of Agricultural Research (SCAR) of the European Commission (EC 2011, 2015). Thus, the forecasted broadening of technology development is faced with serious challenges from at least three main directions, such as knowledge-based technology itself, the prevailing natural resources base, and the policy environment. All of these are also influenced, including in terms of continuing concentration of ownership over such as intellectual property rights (IPRs), by political economy. To enhance a properly open and more diverse knowledge base for sustainability (Leach et al. 2010), effectively targeted research, development and innovation (RDI) need to be promoted; access to knowledge has to be enhanced; skilled workforces need to be developed; and environmental ethics need to be taken into more pronounced consideration to discriminate between technology being integrated in or being imposed on nature. In this context, targeted RDI means focusing on true sustainability instead of global business interests, monopoly control, IPRs or distribution (see later, under 'The policy environment'). In spite of all potential achievements in technology, however, natural resources remain the key limiting factor, as long as their utilization rate exceeds their natural renewal rate. It may be unsustainable therefore, if a new crop is innovated which may produce more yield but only on condition that more energy, chemicals and other resources are used, and which reduces both agricultural and natural biodiversity. Therefore, it should be the main role of a safe and efficient regulatory—and innovation—system that it is formulated so as to harmonize the production goals with the real background resource capacity.

Bioeconomy and Economic Growth

By producing biofuels and bio-based chemicals on the basis of renewable biological resources, bioeconomy currently has substantial economic growth potential. However, this growth is limited by two main factors. The initial expansion of the sector will eventually need to lessen as bioeconomy solutions gradually replace fossil fuels-based chemical technologies. As bioeconomy reaches full capacity, the external limiting factor is the renewal rate of the bio-resources used. Should bioeconomy go beyond that limit, it would fail to achieve an equilibrium state, and would fall into the unsustainability trap of fossil fuels-based chemical technologies. The renewal rate should not be underestimated, as it allows substantial steady-state operational capacity, but obviously it cannot be considered limitless, either.

Bioeconomy and Circular Economy

From an ecological aspect, a circular economy in its objective to produce no waste or pollution is a concept to organize industrial economy on the basis of stable ecosystems, where the output of every technological process serves as an input for another process or processes. Thus, the concept is a 21st century manifestation of the pioneering approaches in the early seventies of the last century by Barry Commoner (Commoner 1971) and the Club of Rome (Meadows et al. 1972). Ecological innovation, including biomimicry, can provide solutions that are better embedded and more in harmony with natural ecosystems (Blok and Gremmen 2016), and thus provide steps towards a circular economy. Bioeconomy in its particular form of technological solutions represents a step towards the principle of a circular economy; but it is not equivalent with it. Bioeconomy does not accomplish circular economy, but aims "to pave the way to a more innovative, resource efficient and competitive society that reconciles food security with the sustainable use of renewable resources for industrial purposes, while ensuring environmental protection" (EC 2012a). The ambiguous idea of a "competitive society" in this statement would also need some collective reflection. Is this "competitive" in the sense which dominates and drives current, unsustainable, policy and society? Or might it mean a form of society which is more "competitive with" and more humanly civilized and attractive than conventional unsustainable societal habits and "consumption-as-therapy" drivers of demand for commodities which extend—and shape—production systems far beyond what would realistically be called meeting social *needs*? This is part of the general point that sustainability-governance would need not only to manage *supply* of goods and services within any economy we manage. It would also have to manage *demand* to reasonable and sustainable levels, as has been extensively analyzed and debated (Wynne 2011; Owen et al. 2013) for example for energy and food.

Cascading use of Biomass

Besides the use and regeneration of renewable resources, a key concept in bioeconomy to achieve resource efficiency is the so-called cascade principle. The bio-resource aspect of bioeconomy emphasizes cascading use of biomass and raw bio-materials in recycling, to maximize the efficiency of biomass use. Yet the various and often long-term, indirect environmental and economic impacts of increased industrial biomass-uses need to be better understood (Keegan et al. 2013; Bugge et al. 2016).

The cascade principle is mentioned in particular cases, referring to innovation cascades, i.e. series of innovative technologies built on each-other. The two types of cascades—cascades of material utilization and of technology development—are mutually related. The development of cascading processing of new materials requires novel technologies, and vice versa, these emerging technologies may produce new by-products needed to be cycled into the processing cascade. Moreover, as the number of technologies increases, the possibilities for their recombination also rises (Arthur 2009), making innovation cascades possible.

KNOWLEDGE-BASED TECHNOLOGY

Bioeconomy approaches aim to replace the functions of conventional synthetic, usually fossil fuels-based industrial chemicals by new technologies based on biological processes, natural or GM organisms, fermentation, biotechnology and molecular biology. This has led to a rapid expansion of the bioeconomy market, realizing an annual turnover reported by Eurostat to be 2,1 trillion EUR for the EU (EU-28) in 2013 (Piotrowski et al. 2016). With food, beverages and tobacco products excluded from this, the remaining bio-based sector represents an annual turnover of 1 trillion EUR, biofuels and bioenergy accounting for 8%, agriculture and forestry for 43%.

Within bioeconomy, particular emphasis has been given to GM crops, which has generated a politically heated international debate. Scientific opinions claiming the safety of current (first generation) agricultural biotechnology (GM) products and urging their application (Barrows et al. 2014; Domingo 2016) confront others claiming lack of evidence on their safety (Hilbeck et al. 2015; Krimsky 2015), and exaggeration of what risk assessment can demonstrate as so-called "lack of harm". Both positions were considered in the latest evaluation by the US National Academy of Sciences (US NAS 2016). GM crops developed by agricultural biotechnology are classified into several (four) generations on the basis of the type of genetic transformation applied and intended utilization. Following its mandate (EC 2002), the European Food Safety Authority (EFSA) is assigned to carry out environmental and food safety risk assessment of these GM crops if used for food and feed purposes. First generation GM crops are modified for their agronomic traits, their main representatives containing transgenic traits that result in resistance to insect pests or tolerance to particular herbicides. These GM crops rely on pesticide applications: the crop either produces an insecticide, or it is designed for actual herbicides (e.g. glyphosate along with its formulnt POEA) being sprayed on it. Second generation GM crops contain genetic transformations intended to improve/modify their product quality (composition, tolerance to environmental conditions like drought, etc.), while third generation GM crops are intended to express industrial products and pharmaceutical drugs. Somewhat distant from the first three, fourth generation GM crops are being produced with new methods in molecular biology, also termed emerging technologies (Lusser et al. 2011). Although these emerging technologies alter plant genomes so as to create new plant varieties, it is debated whether all result in GM organisms according to the legal definition set in the EU (EC 2001). Industrial and pro-GM scientific bodies are promoting deregulation of such new technologies or their products on the claim that they are not GM. This has extended and elaborated on the long-standing vigorous societal and scientific debates at the UN Convention on Biodiversity and in Europe, regarding the environmental safety and regulatory questions surrounding all four generations of GM crops. In this context, it has been stated that the EU, unlike the US, is less biotech oriented, or even negative about biotechnology. A correction to this opinion is that the EU is not less biotech oriented, but less oriented to open biotechnologies (see below)—closed system biotechnologies (e.g. the pharmaceutical industry's use of GM in insulin production) are not subject to greater limitation in the EU than in the US.

In spite of the spectacular boom in the bioeconomy sector, several controversies remain unresolved, considering technology, societal and assessment-related issues. Besides resource utilization, a major issue regarding the environmental aspect of bioeconomy technologies is the question mentioned above and discussed in detail in the next section. This question is whether the technology is isolated from the environment, or open to it. In fact, this aspect is not unique to bioeconomy technologies, but has to be considered for all technologies—should technologies be fully open to the environment, as in agriculture, fisheries or forestry; isolated by physical means and waste management methods, such as chemical industry technologies; or fully isolated, such as technologies used in isotope techniques.

A means to overcome these conflicts and consequent public distrust has been proposed through the approach of responsible research and innovation (RRI) (Asveld et al. 2015). If implemented universally RRI should provide transparency and responsiveness, and is also aimed to enhance public trustworthiness in socially sensitive issues involving bioeconomy and other technologies.

Another burning issue relates to the accelerating rate of innovation occurring. It is often questionable, whether health and environmental risk assessment is able to keep pace with, or rather to resist, the powerful demand for rapid commercial development seen e.g. for GM organisms [GMOs, also termed living modified organisms, LMOs in the UN Cartagena Biosafety Protocol (CBD 2000)], by the Precautionary Principle declared by the implemented in over 160 signatory countries and assessed in and compared among nine of them, Australia, Brazil, Canada, China, Cuba, Germany, Japan, South Africa and the USA (Flint et al. 2012).

NATURAL RESOURCES

Socio-economic and Earth System trends between 1750 and 2010 on the basis of analyses of resource utilization evidence that anthropogenic activities have resulted in more extensive transformations in our environment during the last half-century than at any period in the Earth's known history (UNEP-WRI 2005; Steffen et al. 2015). Data indicate that while population growth has been seen in the non-OECD world since 1950, the vast majority of the world's economy (measured in gross domestic product, GDP) and consumption is realized by OECD countries. Economic growth has been achieved against growing costs in natural resources and ecosystem services, and has caused the largest human imprint on the planet, the Anthropocene Era (Crutzen 2002). This rate of reduction in fossil raw materials and ecosystem services cannot subsist as planetary boundaries are claimed to have been reached, or perhaps exceeded, with the greatest emphasis on the extreme rate of global biodiversity loss. Yet, environmental issues related to bioeconomy are mentioned mostly from a human population aspect, with the strongest focus on climate change, much less from the aspect of the state and adequacy of natural resources. Moreover, the decreasing availability of natural resources is even mentioned as a factor in favour of bioeconomy (bio-based vs. fossil products).

The Earth System aspects of bioeconomy have to be assessed by both environmental and ecological perspectives. Thus, environmental (pollution, resource utilization) and ecological (biodiversity) aspects should both be an integral part of the assessment of bioeconomy innovations aiming to achieve a circular economy.

Environmental aspects cover the use of natural resources, including the balance (use minus renewal) of given resources, such as plants, minerals, nutrients or fossil fuel, and also evaluation of alternative uses of arable land (land used for cultivation of food/feed or energy crops). Ecological aspects cover biodiversity of the utilized areas as habitats, maintaining ecosystem services (UNEP-WRI 2005) (supporting or habitat, provisioning, regulating and cultural services) provided by various participants of the given ecosystems. Careful assessment of the changes in the natural resources and biodiversity of the ecosystem due to different technologies should be an integral part of bioeconomy assessment, and therefore, environmental science experts and ecologists should be included in such analysis.

Environmental pollution remains high in various regions of the world, justifying the need for programs to re-establish natural ecosystems, to reduce water and air pollution, to serve soil protection, and to protect animal/plant species and their habitats. This applies to the Pannonian Biogeographical Region (see below) of the EU, where indices indicate deterioration (Kelemen et al. 2014), including river natural floodplain ecosystems (e.g. the Danube) which need to be restored.

Major environmental constraints to technologies (including bioeconomy) are the availability/exploitation of non-renewable natural resources (e.g. fossil fuels, phosphate, nitrogen and carbon dioxide); the limited capacity/renewal rate of renewable natural resources; as well as the availability of water (fresh water and seas/oceans) and land (crop- and shrub-lands, pastures, forests, even urban areas). Soil is also a natural resource. Although soil fertility is renewable by character (its renewability is the basis or our existence), its renewal capacity is limited: intensive cultivation, either for food of for bioeconomy purposes, causes chemical pressure on soil ecosystems and removes energy represented by soil nutrients and components. Intensive soil utilization leaves less energy resources to maintain soil ecosystem functions, which jeopardizes soil fertility. Such degradation of ecosystem functions due to unsustainable industrial agricultural production needs to be avoided, not only from an ecological, but even from a pragmatic, productivity-oriented aspect, as compromized ecosystem functions eventually can maintain neither habitat ecosystem services, nor agricultural production, as the latter is also tightly interlinked with natural systems (Miko and Storch 2015).

As stated in the EU Nature Directives i.e. the Habitats Directive (EEC 1992) and the Birds Directive (EC 2009a), Member States of the EU are legally bound to preserve their natural ecosystems in their existing state. To support compliance with this requirement, the EU is divided into nine terrestrial and five marine biogeographical regions by their ecological conditions in the Natura 2000 network (ten Brink et al. 2011). Of the nine terrestrial regions, the Pannonian Biogeographical Region is one with outstandingly high biodiversity (EC 2009c), but over 50% of its habitats are assessed as 'unfavourable—bad', exceeding the average of the other biogeographical regions (EEA 2010).

An essential environmental issue regarding bioeconomy technologies, is not actually related to features specific to bioeconomy, but is derived from the isolation characteristics of these (and other) technologies. Areas of applied biotechnology are often described with colours: white, red, blue and green biotechnology referring to industrial, health, marine and agricultural applications. Of these, white biotechnology solutions are typically closed systems, operated in closed reactors and subject to strict waste and pollution management. In contrast, blue and particularly green biotechnologies are typically open to the environment throughout their entire process, although also implying waste management practices. This is of particular importance in the application of living organisms, both in cases of invasive alien species to given regions and of GM organisms, since these can reproduce of their own accord once released. The highest concern relates to microorganisms, where similar isolation criteria as those set for closed systems should apply, depending on the type and potential severity of effects of the microorganisms. In the EU these are set by legal measures (EEC 1989), updated several times later (EC 2000).

Biofuels

Increasing utilization of renewable energy resources instead of fossil fuels is certainly an advantage, a step towards a circular economy. Thus, the use of renewable resources for energy production is promoted in the EU by policy (EC 2009b) and by biofuel certification to favour environmental and social sustainability standards (Pols 2015). In fact, not a single, but several steps, as biofuel technologies have been developed by now in several generations on the basis of the biological material the technology is based on (Aro 2016). (Coincidentally biofuel technologies are also classified into four generations, like GM crops, but the two classifications do not relate to each other.) First generation biofuels are produced from sugar, lipid or starch extracted from food crops (EASAC 2012). Thus, first generation biofuels pose sustainability challenges and represent a direct competition between food and bioenergy use of crops and land (Naik et al. 2010; Mohr and Raman 2013; Rulli et al. 2016), and pose the same environmental risks associated with intensive agriculture: biodiversity threats from the use of crop monocultures, environmental contamination from pesticide use, water use for irrigation, soil acidification and erosion, increased carbon emissions from ploughing, indirect fossil fuel use and nitrogen oxide emissions from industrial fertilizers. To avoid this controversy, later generations of biofuels, termed advanced biofuels have also been developed (EASAC 2012). Second generation biofuels are based on cellulose, hemicellulose, lignin or pectin from non-food plant materials, wood, organic waste, food crop waste or specific biomass crops (Sims et al. 2010); third generation biofuel technologies are based on biomass from algae or other aquatic autotrophic organisms (Alaswad et al. 2015); while fourth generation technologies combine biofuel production with CO_2-capture and storage (CCS) in deep geological formations, e.g. old oil and gas fields or saline aquifers.

Practical utility of advanced biofuels is seen in the production of cellulosic chemicals (bioethanol), biokerosene, green diesel, bio-based marine diesel and other biofuels. With the achievements of emerging bioeconomy, the proportion of such biofuels is increasing, and industrially developed countries commit themselves to cover a certain proportion of their energy needs with biofuels. However, the production of these biofuels has to be managed in a sustainable way to fulfil these commitments. The assessment of the environmental and socioeconomic impacts of biofuels in a more coherent and policy-relevant manner has been urged (Lovett et al. 2011; Gasparatos et al. 2013), including environmental sustainability indicators (McBride et al. 2011). Advanced biofuel technologies have been advocated for reducing the social and environmental risks associated with biofuel production and usage (UNCTAD 2014). Yet such 'advanced' biotechnologies also have their environmental risks, e.g. those related to ecological consequences of cultivation or biomass processing; environmental release of GM plants (should GM energy crops be utilized); or CO_2-leakage in CCS, ocean acidification and its consequences on ecosystem functions (Phelps et al. 2015; Wang et al. 2016). Moreover, the biofuel sector is subject to a current market paradox, as explained below: expansion in biofuel utilization increases overall fuel supplies, which in turn, results in price reductions of fossil fuels as well.

Bio-based Products

Bio-based technologies do not produce solely biofuels, but give rise to numerous bio-based chemicals, often difficult to produce by the conventional chemical industry. These bio-based compounds are often produced in cascades, employing chemical conversion, e.g. solvent or supercritical water extraction, hydrolysis of the biomass, resulting in valuable bio-based substances from technological batches from biofuel or food production technologies (Naik et al. 2010; Nattrass et al. 2016; Snyder 2016). This is supported by the European Bioeconomy Panel and the SCAR Strategic Working Group of the European Commission (EC 2014), and is best exemplified by the cascade use of biomass in the wood industry (Scarlat et al. 2015; Hagemann et al. 2016).

Chemicals/plastics and pharmaceuticals so far represent only 10% of the annual turnover of the bio-based sector (Piotrowski et al. 2016), yet this may be considered the most promising segment within the sector, as bio-based chemicals are being produced in closed system technologies, representing the lowest hazard to the environment.

Bio-based chemicals include biomaterials (such as natural fibres, cellulose, starch, sugars, as well as synthesis gases and oils, e.g. plant and animal oils), of which further derived products (such as glycerol or CO_2), as well as fuels (such as hydrogen, methane or ethanol) are produced. From these intermediates, various organic building blocks (such as alkanes and alkenes, furans and ketones, organic acids and alcohols) are (bio)synthesized and converted into a wide range of chemical products. Leading examples in the bio-based industry include essential amino acids (methionine, lysine) as chemical intermediary compounds and feed additives; organic acids, such as lactic acid for bio-based polymer polylactic acid (PLA) used in 3D printing; succinic acid as raw material for various bioplastics, plasticizers and biosolvents; lauric acid from biooils, and levulinic acid from sugar production; castor oil and related bio-based polyamides; 1,4-butanediol/butandiene and other biomonomers, biopolymers, biorubbers, resins, plasticizers, solvents (ethyl acetate, dimethyl succinate, etc.) and biosurfactants; new fibres from cellulosic chemicals, as well as lignin and bran. Yet, the example with the highest public recognition is probably that of polyethylene furanoate (PEF) bottles, built by the plant carbohydrate-based, so-called YXY-technology, capable of building C_6-sugars from plants through catalytic dehydration, catalytic oxidation followed be catalytic polymerization into 100% bio-based PEF, leading to PlantBottle™ plastic, intended to reach the market by 2020 and replace the current polyethylene terephthalate (PET) bottles. New substances built from raw materials of plant origin by innovative technologies may be advantageous by using renewable resources instead of fossil assets, yet their end product, PEF is not biodegradable, similarly to PET. Thus the approach is reducing resource-intensity, but it cannot realize a circular economy and therefore, true sustainability, with respect to its waste utilization or recycling.

THE POLICY ENVIRONMENT

To assess the potential of the emerging bioeconomy, the Organisation for Economic Co-operation and Development compiled a broad-based analysis of the future development perspectives in three sectors where biotechnology has the greatest potential impact: agriculture, health and industry (OECD 2009). Although the bioeconomy/circular economy concept is based on environmental-ecological dimensions; the main driving forces required to underpin its policy and strategy effectiveness are societal/economic, including trade, finance, political economy of RDI, and knowledge transfer between RDI and industry (de Besi and McCormick 2015). Currently, and on a long-term basis, whatever its ecological claims or aspirations, bioeconomy is mostly assessed in interaction between technology developers and economists. This results in biased assessment misnamed as scientific risk assessment for policy, where commercial technology benefits are included, but environmental and social costs not sufficiently considered in the economic and scientific models used.

A societal concern regarding investment and innovation structures is that through current RDI financing, which increasingly prioritizes industrial applications, society is becoming an investor in technology development. This, on the one hand, may be considered contradictory, as the role of the society should be the assurance of public interests, not of business prospects; yet on the other hand, it holds a certain advantage potential, as society may have a better position in assuring issues of sustainability over usual business interests, e.g. IPRs, monopoly or distribution control. As for IPRs, patenting no longer suits bioeconomy, just as other scientific fields with dynamic and interactive complexity, e.g. systems biology or synthetic biology (Cavert 2008). In turn, democratisation of science is becoming of increasing importance (Jasanoff 2006, 2011; Jasanoff et al. 2015), increasing the role of assessment science as impure science in which facts are intermingled with values and judgments (Jasanoff 2015). Whether society as a business investor in bioeconomy protects or advances public welfare depends upon two broad questions. One concerns "protection", and whether risks of harm to health or the environment are properly researched and controlled; the other concerns, whether the directions of innovation which result from such large public investments, such as the EU's Horizon 2020 programme 75 billion EURO funding over 7 years, reflect genuine public priorities and concerns, or private commercial competitive ones. On neither of these broad questions can confidence be justified in Europe (Felt and Wynne 2007). Nor in such controversial domains with enormous commercial interests bearing down on regulatory science, can illegitimate conflicts of interest be assumed to be absent (Guillemaud et al. 2016).

The need of a physical land use balance is expressed in terms of technological changes in cropland use and yields achieved (Kuemmerle et al. 2013; Engström et al. 2016) and as limits of bioenergy production (Scarlat et al. 2015). A global land use assessment by UNEP indicated a need for reduction in land use intensity in the EU, so as to reach a global land use target based on the safe operating space (UNEP 2014). In fact, sustainable landscape management has been assumed to be a proper base for sustainable regional development strategies, as natural resources management, biodiversity, environmental protection, ecosystem services, socio-economic sustainability and cultural heritage are considered as its inherent elements (Nemethy and Komives 2016). Yet, the competition between food and non-food applications in bioeconomy for arable land from an ecological aspect is insufficiently recognized (O'Brien et al. 2015), even though land-availability and soil-quality, as well as water-scarcity, are being emphasized as limiting factors.

Bioeconomy modelling and regulatory policies emphasize the environmental issue mostly as a societal need, but in practice they focus rather on business, economy, R&D and consumer issues (food price, choice, and safety), and bioeconomy is mostly presented from the aspect of development and business opportunities in the food and feed, biomaterials and bioenergy (biomass) sectors. Attempts to overcome this deficiency are still coming from economic analyses, with all their recognized inadequacies. Computable general equilibrium (CGE) models have been proposed to analyse the consequences of bioeconomy policies (Francois et al. 2005; van Vuuren et al. 2016), yet dynamic CGE models mostly remain focused only on economic effects in global production and trade, with intermediate linkages between sectors; to scale economies and imperfect competition; and to assess trade impacts on capital stocks through investment effects. All of this ignores the environmental, ecological and social factors (Laurenti et al. 2016) discussed before, and is thus wholly unable to provide a realistic basis for sustainable innovation.

Traditional economic tools may fail to assess the efficacy of bioeconomy: evaluating the bioeconomy sector by measuring its share in the GDP does not give any useful knowledge, can even be harmful, e.g. leading to the market aspect paradox that the use of bio-based products as market competitors of fossil products leads to a decrease in fossil prices, which in turn, stimulates fossil demand and acts against the desired reduction in fossil usage. This trend, although not to a major scale, is documented in the oil sector: by reducing oil prices in oil-importing countries, the introduction of biofuels contributed to a current increase in fossil fuel consumption (Hochman et al. 2010). Official regulatory limitations on fossil fuel consumption may not be unambiguously beneficial either due to their anticipated stimulatory effect on biofuel consumption to secure GDP, that, if unregulated, would increase the demand for land and water resource utilization (Pols 2015). In lieu of GDP as an economic measure, the concept of consumer surplus has been proposed (Zilberman et al. 2013). This, by itself is still a purely economic approach, but a nonstandard definition of consumer surplus in environmental economy allows possible inclusion of the economic evaluation of ecosystem services (Banzhaf and Boyd 2012).

As long as bioeconomy practices, just as other technologies, are evaluated on the basis of the economic growth they allow, their promise to achieve circular economy cannot be accomplished. To achieve a better balanced analysis of bioeconomy, environmental and ecological—and indeed non-economic social—aspects need to be included in the overall assessment.

CONCLUSION

In spite of the initial ecological approach, altogether, the current business model of the emerging bioeconomy does not appear to be fundamentally different from that of traditional chemical industry, and in reality it focuses on business potential, economic growth and profitability. Therefore, it cannot meet sustainability needs, claims, or aspirations. It can only mitigate unsustainable resource utilization, but does not (yet) hold a promise for a true circular economy. Circular economy cannot be achieved until all products and by-products of technologies gain utility in some other bioeconomy technologies; in other words, no waste is being produced in these processes. Unless they do achieve this, they should not be approved in regulatory processes, but refused until better innovations are developed. Until then, human activity will remain operating on the basis of converting natural resources from their more "useful" forms into their less "useful" forms (Commoner 1971). It is of utmost environmental ethical importance and in turn, practical significance that the ongoing second, bio-based (also termed biomimetic) industrial revolution (Blok and Gremmen 2016) should be transformed from its present inability to enframe its inventions within the copious but demanding limits and processes of natural ecosystems.

In summary, the full and transparent, long-term environmental and ecological assessment of bioeconomy initiatives is urged, in a global context, in all identified biogeographical regions worldwide, and within the EU. This is necessary to ensure their true sustainability, and to press towards development of a circular economy. Assessment should include the environmental status of organic microcontaminants in environmental matrices (including surface water and soil), as well as effects on protected species and habitats, and through them on biodiversity and ecosystem services.

PUBLIC INTEREST STATEMENT

Bioeconomy, as an emerging sector in the national economies of industrial and developing countries, is advocated as a way towards circular economy by the production and utilization of renewable biological resources in agribusiness, bio-based products and bioenergy. Yet in spite of the bio-based approach, bioeconomy solutions cannot provide true sustainability, if they use renewable resources beyond their renewal rate, if they produce non-biodegradable products or wastes resulting in environmental pollution, or if they pose threat to biodiversity. Therefore, as discussed in the present paper, the assessment of bioeconomy innovations on the basis of their current profitability and economic growth is inadequate and misleading, and must consider environmental and ecological aspects rigorously and comprehensively, and not only in appearance.

ACKNOWLEDGEMENTS

The Author expresses his sincere appreciation to Prof. Brian Wynne (Director, Centre for the Study of Environmental Change, University of Lancaster, Lancaster, UK) for his valuable comments and suggestions. This work was supported by Project K109865 of the Hungarian Scientific Research Fund (OTKA). The Author is member of the Scientific Advisory Board of the Co-operative Research Programmes (CRP) of the Organisation for Economic Co-operation and Development (OECD) and of the Management Board of the European Food Safety Authority (EFSA). This publication reflects the views of the Author, and the OECD or EFSA cannot be held responsible for any use which may be made of the information contained therein.

ETHICAL STANDARD

The Author declares to have received no funding that might cause a conflict of interest with regard to this research.

REFERENCES

1. Alaswad, A., Dassisti, M., Prescott, T., & Olabi, A. G. (2015). Technologies and developments of third generation biofuel production. *Renewable and Sustainable Energy Reviews, 51,* 1446–1460.
2. Aro, E.-M. (2016). From first generation biofuels to advanced solar biofuels. *Ambio, 45*(Suppl. 1), S24–S31.
3. Asveld, L., Ganzevles, J., & Osseweijer, P. (2015). Trustworthiness and responsible research and innovation: The case of the bio-economy. *Journal of Agricultural and Environmental Ethics, 28,* 571–588.
4. Athur, W. B. (2009). *The nature of technology: What it is and how it evolves.* New York, USA: The Free Press. http://tuvalu.santafe.edu/~wbarthur /thenatureoftechnology.htm.
5. Banzhaf, H. S., & Boyd, J. (2012). The architecture and measurement of an ecosystem services index. *Sustainability, 4,* 430–461.
6. Barrows, G., Sexton, S., & Zilberman, D. (2014). Agricultural biotechnology: The promise and prospects of genetically modified crops. *Journal of Economic Perspectives, 28*(1), 99–120.
7. Baveye, P., Baveye, J., & Gowdy, J. (2013). Monetary evaluation of ecosystem services: Getting the timeline right. *Ecological Economics, 95,* 231–235.
8. Blok, V., & Gremmen, B. (2016). Ecological innovation: Biomimicry as a new way of thinking and acting ecologically. *Journal of Agricultural and Environmental Ethics, 29,* 203–217.
9. Bugge, M. M., Hansen, T., & Klitkou, A. (2016). What is the bioeconomy? *A Review of the Literature. Sustainability, 8,* 691.
10. Cavert, J. (2008). The commodification of emergence: Systems biology, synthetic biology and intellectual property. *BioSocieties, 3*(4), 383.
11. CBD (Convention on Biological Diversity). (2000). *Cartagena Protocol on Biosafety.* Montreal, Canada: Secretariat of the CBD. http://www.cbd.int/doc/legal/cartagena-protocol-en.pdf.
12. Commoner, B. (1971). *The closing circle. Nature, man, and technology* (pp. 6–26). New York, USA: Bantam Books Inc.
13. Costanza, R., d'Arge, R., de Groot, R., Farber, S., Grasso, M., Hannon, B., et al. (1997). The value of the world's ecosystem services and natural capital. *Nature, 387,* 253–260.
14. Costanza, R., Cumberland, J. H., Daly, H., Goodland, R., Norgaard, R. B., Kubiszewski, I., & Franco, C. (2015). *An Introduction to Ecological Economics.* 2nd Ed., Boca Raton, USA: CRC Press. https://www.crcpress.com/An-Introduction-to-Ecological-Economics-Second-Edition/Costanza-Cumberland-Daly-Goodland-Norgaard-Kubiszewski-Franco/p/book/9781566706841.
15. Crutzen, P. J. (2002). Geology of mankind—The Anthropocene. *Nature, 415*(6867), 23.
16. Daly, H. E., & Farley, J. (2011). *Ecological economics.* 2nd Ed., Washington DC, USA: Island Press. http://islandpress.org/book/ecological-economics-second-edition.
17. de Besi, M., & McCormick, K. (2015). Towards a bioeconomy in Europe: National, regional and industrial strategies. *Sustainability, 7,* 10461–10478.
18. Domingo, J. L. (2016). Safety assessment of GM plants: An updated review of the scientific literature. *Food and Chemical Toxicology, 95,* 12–18.
19. EASAC (European Academies Science Advisory Council). (2012). *The current status of biofuels in the European Union, their environmental impacts and future prospects. EASAC Policy Report 19.* Halle, Germany: EASAC. http://www.easac.eu/fileadmin/PDF_s/reports_statements/Easac_12_Biofuels_Complete.pdf.
20. EC (European Commission). (2000). Council directive 2000/54/EC of 18 September 2000 on the protection of workers from risks related to exposure to biological

agents at work (seventh individual directive within the meaning of Article 16(1) of Directive 89/391/EEC). *Official Journal of the European Union, L 262,* 17/10/2000 pp. 1–21.

21. EC (European Commission). (2001). Council Directive 2001/18/EC of 12 March 2001 on the deliberate release into the environment of genetically modified organisms. *Official Journal of the European Union, L 106,* 17/04/2001 pp. 1–39.
22. EC (European Commission). (2002). Council Regulation 178/2002/EC of 28 January 2002 laying down the general principles and requirements of food law, establishing the European Food Safety Authority and laying down procedures in matters of food safety. *Official Journal of the European Union, L 31,* 01/02/2002 pp. 1–24.
23. EC (European Commission). (2008). *Lead market initiative to unlock innovative markets. IP/08/12.* Brussels, Belgium: EC Directorate-General for Research and Innovation. http://europa.eu/rapid/press-release_IP-08-12_en.htm?locale=en.
24. EC (European Commission). (2009a). Directive 2009/147/EC of the European Parliament and of the Council of 30 November 2009 on the conservation of wild birds. *Official Journal of the European Union, L 20,* 26/01/2010 pp. 7–25.
25. EC (European Commission). (2009b). Directive 2009/28/EC of the European Parliament and of the Council of 23 April 2009 on the promotion of the use of energy from renewable sources and amending and subsequently repealing Directives 2001/77/EC and 2003/30/EC. *Official Journal of the European Union, L 140,* 05/06/2009 pp. 16–62.
26. EC (European Commission). (2009c). *Natura 2000 in the Pannonian Region.* EC Directorate-General of the Environment. Luxembourg: Office for Official Publications of the European Communities. doi: 10.2779/79432.
27. EC (European Commission). (2011). *Sustainable food consumption and production in a resource-constrained world.* Brussels, Belgium: Standing Committee on Agricultural Research (SCAR), EC. http://ec.europa.eu/research /agriculture /scar/pdf/scar_feg3_final_report_01_02_2011.pdf.
28. EC (European Commission). (2012a). *Communication from the Commission to the European Parliament, the Council, the European Economic and Social Committee and the Committee of the Regions. Innovating for Sustainable Growth: A Bioeconomy for Europe.* Brussels, Belgium: EC COM (2012) 60, 13.2.2012. http://eur-lex.europa.eu/legal-content/EN/TXT/?uri=COM:2012:0060:FIN.
29. EC (European Commission). (2012b). *Innovating for Sustainable Growth: A Bioeconomy for Europe.* Brussels, Belgium: EC Directorate-General for Research and Innovation. http://ec.europa.eu/research/bioeconomy/pdf/official-strategy_en.pdf.
30. EC (European Commission). (2014). *Where next for the European bioeconomy? The latest thinking from the European Bioeconomy Panel and the Standing Committee on Agricultural Research Strategic Working Group (SCAR).* Brussels, Belgium: EC Directorate-General for Research and Innovation. http://ec.europa.eu/research/bioeconomy/pdf/where-next-for-european-bioeconomy-report-0809102014_en.pdf.
31. EC (European Commission). (2015). *Sustainable agriculture, forestry and fisheries in the bioeconomy—A challenge for Europe.* Brussels, Belgium: Standing Committee on Agricultural Research (SCAR), EC. http://ec.europa.eu/research/scar/pdf/ki-01-15-295-enn.pdf.
32. EEA (European Environment Agency). (2010). *The European environment state and outlook 2010 biodiversity*. Copenhagen, Denmark: EEA.
33. EEC (Council of the European Communities). (1989). Council Directive 89/391/EEC of 12 June 1989 on the introduction of measures to encourage improvements in the safety and health of workers at work. *Official Journal of the European Union, L 183,* 29/06/1989 pp. 1–8.
34. EEC (Council of the European Communities). (1992). Council Directive 92/43/EEC of 21 May 1992 on the conservation of natural habitats and of wild fauna and flora. *Official Journal of the European Union, L 206,* 22/07/1992 pp. 7–50.

35. El-Chichakli, B., von Braun, J., Lang, C., Barben, D., & Philp, J. (2016). Five cornerstones of a global bioeconomy. *Nature, 535*(7611), 221–223.
36. Engström, K., Rounsevell, M. D. A., Murray-Rust, D., Hardacre, C., Alexander, P., Cui, X., et al. (2016). Applying Occam's razor to global agricultural land use change. *Environmental Modelling and Software, 75,* 212–229.
37. Felt, U. & Wynne, B. (2007). *Taking European knowledge seriously.* Report of the expert group on science and governance to the science, economy and society directorate, directorate-general for research, European Commission. Luxembourg: Office for Official Publications of the European Communities. http://ec.europa.eu/research/science-society/document_library/pdf_06/european-knowledge-society_en.pdf.
38. Flint, S., Heidel, T., Loss, S., Osborne, J., Prescott, K., & Smith, D. (2012). *Summary and comparative analysis of nine national approaches to ecological risk assessment of living modified organisms in the context of the Cartagena protocol on biosafety. Annex III.* Montreal, Canada: Secretariat of the Convention of Biological Diversity. http://bch.cbd.int/protocol/cpb_technicalseries/cpb-ts-02-en.pdf.
39. Francois, J., van Meijl, H., & van Tongeren, F. (2005). Trade liberalization in the Doha development round. *Economic Policy, 42*(April), 350–391.
40. Gasparatos, A., Lehtonen, M., & Stromberg, P. (2013). Do we need a unified appraisal framework to synthesize biofuel impacts? *Biomass and Bioenergy, 50,* 75–80.
41. Guillemaud, T., Lambaert, E., & Bourguet, D. (2016). Conflicts of interest in GM Bt crop efficacy and durability studies. *PLoS ONE, 11*(12), e0167777.
42. Hagemann, N., Gawel, E., Purkus, A., Pannicke, N., & Hauck, J. (2016). Possible futures towards a wood-based bioeconomy: A scenario analysis for Germany. *Sustainability, 8,* 98.
43. Hilbeck, A., Binimelis, R., Defarge, N., Steinbrecher, R., Székács, A., Wickson, F., et al. (2015). No scientific consensus on GMO safety. *Environmental Sciences Europe, 27,* 4.
44. Hochman, G., Rajagopal, D., & Zilberman, D. (2010). *The effect of biofuel on the international oil market. CUDARE Working Paper No. 1099.* Berkeley, USA: Department of Agricultural and Resource Economics, UC Berkeley. http://escholarship.org/uc/item/0k93s7zg.
45. Jasanoff, S. (2006). Biotechnology and empire: The global power of seeds and science. *Osiris,21,* 273–292.
46. Jasanoff, S. (Ed.). (2011). *Reframing rights: Bioconstitutionalism in the genetic age.* Cambridge, USA: MIT Press.
47. Jasanoff, S. (2015). The science of assessment and the assessment of science: New frontiers in food safety evaluation. *Shaping the Future of Food Safety, Together. Proceedings of the 2nd EFSA Scientific Conference (Milan, Italy, 14–16 October 2015). EFSA Journal, 13*(10), s1310, p.11.
48. Jasanoff, S., Hurlbut, B. J., & Saha, K. (2015). CRISPR democracy: Gene editing and the need for inclusive deliberation. *Issues in Science and Technology, 32*(1), 25–32. http://www.jstor.org/stable/24727002.
49. Keegan, D., Kretschmer, B., Elbersen, B., & Panoutsou, K. (2013). Cascading use: A systematic approach to biomass beyond the energy sector. *Biofuels, Bioproducts and Biorefining, 7,* 193–206.
50. Kelemen, A., Varbova, V., & Almassy, D. (2014). *Thematic study: Environmental risk management and climate change. Central Europe Programme.* Vienna, Austria: Regional Environmental Center for Central and Eastern Europe, Joint Technical Secretariat.
51. Krimsky, S. (2015). An illusory consensus behind GMO health assessment. *Science, Technology and Human Values, 40,* 1–32.
52. Kuemmerle, T., Erb, K., Meyfroidt, P., Müller, D., Verburg, P. H., Estel, S., et al. (2013). Challenges and opportunities in mapping land use intensity globally. *Current Opinion in Environmental Sustainability, 5*(5), 484–493.
53. Langeveld, H., Sander, J., & Meeusen, M. (Eds.). (2010). *The biobased economy: Biofuels, materials and chemicals in the post-oil era.* Oxon, UK: Earthscan.

54. Laurenti, R., Sinha, R., Singh, J., & Frostell, B. (2016). Towards addressing unintended environmental consequences: A planning framework. *Sustainable Development, 24,* 1–17.
55. Leach, M., Scoones, I., & Stirling, A. (2010). *Dynamic sustainabilities: Technology, environment, social justice.* London, UK: Earthscan.
56. Lovett, J. C., Hard, S., Clancy, J., & Snell, C. (2011). Multiple objectives in biofuels sustainability policy. *Energy & Environmental Science, 4,* 261–288.
57. Lusser, M., Parisi, C., Plan, D., & Rodríguez-Cerezo, E. (2011). *New plant breeding techniques. State-of-the-art and prospects for commercial development.* Report EUR 24760 EN. Seville, Spain: Joint Research Centre of the European Commission (JRC). http://jrc.es/EURdoc/JRC63971.pdf.
58. McBride, A. C., Dale, V. H., Baskaran, L. M., Downing, M. E., Eaton, L. M., Efroymson, R. A., et al. (2011). Indicators to support environmental sustainability of bioenergy systems. *Ecological Indicators, 11*(5), 1277–1289.
59. Meadows, D. H., Meadows, D. L., Randers, J., & Behrens III, W. W. (1972). *The Limits to Growth: A Report for the Club of Rome's Project on the Predicament of Mankind.* New York, USA: Universe Books. http://www.clubofrome.org/report/the-limits-to-growth.
60. Miko, L., & Storch, D. (2015). Biodiversity conservation under energy limitation: Possible consequences of human productivity appropriation for species richness, ecosystem functioning, and food production. *Ecosystem Services, 16,* 146–149.
61. Mohr, A., & Raman, S. (2013). Lessons from first generation biofuels and implications for the sustainability appraisal of second generation biofuels. *Energy Policy, 63,* 114–122.
62. Naik, S. N., Goud, V. V., Rout, P. K., & Dalai, A. K. (2010). Production of first and second generation biofuels: A comprehensive review. *Renewable and Sustainable Energy Reviews, 14,* 578–597.
63. Nattrass, L., Biggs, C., Bauen, A., Parisi, C., Rodríguez-Cerezo, E., & Manuel, Gómez-Barbero, M. (2016). *The EU bio-based industry: Results from a survey. EUR 27736 EN.* Sevilla, Spain: JRC Science Hub, Joint Research Centre, European Commission.
64. Nemethy, S., & Komives, T. (2016). On ecocycles and circular economy. *Ecocycles, 2*(1), 44–46.
65. O'Brien, M., Schütz, H., & Bringezu, S. (2015). The land footprint of the EU bioeconomy. *Land Use Policy, 47,* 235–246.
66. OECD (Organisation for Economic Co-operation and Development). (2009). *The bioeconomy to 2030. Designing a policy agenda.* Paris, France: OECD. https://www.oecd.org/futures/long-termtechnologicalsocietalchallenges/42837897.pdf.
67. Owen, R., Bessant, J., & Heintz, M. (Eds.). (2013). *Responsible innovation. Managing the responsible emergence of science and innovation in society.* Chichester, UK: John Wiley & Sons Ltd.
68. Phelps, J. J. C., Blackford, J. C., Holt, J. T., & Polton, J. A. (2015). Modelling large-scale CO_2leakages in the North Sea. *International Journal of Greenhouse Gas Control, 38,* 210–220.
69. Piotrowski, S., Carus, M., & Carrez, D. (2016). *European Bioeconomy in Figures.* Hürth, Germany: nova-Institute for Ecology and Innovation. http://biconsortium.eu/sites/biconsortium.eu/files/news-image/16-03-02-Bioeconomy-in-figures.pdf.
70. Pols, A. J. K. (2015). The rationality of biofuel certification: A critical examination of EU biofuel policy. *Journal of Agricultural and Environmental Ethics, 28,* 667–681.
71. Reid, W. V., & Mooney, H. A. (2016). The Millennium Ecosystem Assessment: Testing the limits of interdisciplinary and multi-scale science. *Current Opinion in Environmental Sustainability, 19,* 40–46.
72. Rulli, M. C., Bellomi, D., Cazzoli, A., De Carolis, G., & D'Odorico, P. (2016). The water-land-food nexus of first generation biofuels. *Scientific Reports, 6,* 22521.

73. Scarlat, N., Dallemand, J.-F., Monforti-Ferrario, F., & Nita, V. (2015). The role of biomass and bioenergy in a future bioeconomy: Policies and facts. *Environmental Development, 15,* 3–34.
74. Sims, R. E. H., Mabee, W., Saddler, J. N., & Taylord, M. (2010). An overview of second generation biofuel technologies. *Bioresource technology, 101*(6), 1570–1580.
75. Snyder, S. W. (2016). *Commercializing biobased products: Opportunities, challenges, benefits, and risks.* London, UK: Royal Society of Chemistry.
76. Stahel, W. R. (2016). The circular economy. *Nature, 531,* 435–438.
77. Steffen, W., Broadgate, W., Deutsch, L., Gaffney, O., & Ludwig, C. (2015). The trajectory of the Anthropocene: The great acceleration. *The Anthropocene Review, 2*(1), 81–98.
78. ten Brink, P., Badura, T., Bassi, S., Daly, E., Dickie, I., Ding, H., Gantioler, S., Gerdes, H., Hart, K., Kettunen, M., Lago, M., Lang, S., Markandya, A., Mazza, L., Nunes, P. A. L. D., Pieterse, M., Rayment, M., & Tinch, R. (2011). *Estimating the overall economic value of the benefits provided by the Natura 2000 network. Final synthesis report to the European Commission.* EC Directorate-General of the Environment, Contract ENV.B.2/SER/2008/0038. Brussels, Belgium: Institute for European Environmental Policy/GHK/Ecologic. http://ec.europa.eu/environment/nature/natura2000/financing/docs/Economic_Benefits_of_Natura_2000_report.pdf.
79. UNCTAD (United Nations Conference on Trade and Development). (2014). *The state of the biofuels market: Regulatory, trade and development persepectives. UNCTAD/DITC/TED/2013/8.* Geneva, Switzerland: Division of International Trade and Commodities, United Nations. http://unctad.org/en/PublicationsLibrary/ditcted2013d8_en.pdf.
80. UNEP (United Nations Environment Programme). (2014). *Assessing global land use: Balancing consumption with sustainable supply.* Paris, France: UNEP, International Resource Panel. http://www.unep.org/resourcepanel/Portals/50244/publications/Full_Report-Assessing_Global_Land_UseEnglish_(PDF).pdf.
81. UNEP-WRI (United Nations Environment Programme—World Resources Institute). (2005). *Millennium ecosystem assessment. Ecosystems and human well-being: Synthesis.* Washington DC, USA: WRI—Island Press. http://www.millenniumassessment.org/documents/document.356.aspx.pdf.
82. US NAS (United States National Academy of Sciences). (2016). *Genetically engineered crops: Experiences and prospects.* Washington DC, USA: The National Academy Press. http://http://www.nap.edu/download/23395.
83. van Vuuren, D. P., Stehfest, E., Gernaat, D. E. H. J., Doelman, J. C., van den Berg, M., Harmsen, M., et al. (2016). Energy, land-use and greenhouse gas emissions trajectories under a green growth paradigm. *Global Environmental Change.*
84. Wang, Z., Wang, Y., & Yan, C. (2016). Simulating ocean acidification and CO_2 leakages from carbon capture and storage to assess the effects of pH reduction on cladoceran *Moina mongolica* Daday and its progeny. *Chemosphere, 155,* 621–629.
85. Wynne, B. (2011). Lab work goes social, and vice versa: Strategising public engagement processes. *Science and Engineering Ethics, 17*(4), 791–800.
86. Zilberman, D., Kim, E., Kirschmer, S., Kaplan, S., & Reeves, J. (2013). Technology and the future bioeconomy. *Agricultural Economics, 44,* 95–102.

CHAPTER 15

Resolving Conflicts between Agriculture and the Natural Environment

Andrew J. Tanentzap[1] *, Anthony Lamb[2] , Susan Walker[3] , Andrew Farmer[4]

[1] *Ecosystems and Global Change Group, Department of Plant Sciences, University of Cambridge, Cambridge, United Kingdom,*

[2] *Conservation Science Group, Department of Zoology, University of Cambridge, Cambridge, United Kingdom,*

[3] *Landcare Research, Dunedin, New Zealand,*

[4] *Institute for European Environmental Policy, London, United Kingdom*

ABSTRACT

Agriculture dominates the planet. Yet it has many environmental costs that are unsustainable, especially as global food demand rises. Here, we evaluate ways in which different parts of the world are succeeding in their attempts to resolve conflict between agriculture and wild nature. We envision that coordinated global action in conserving land most sensitive to agricultural activities and policies that internalise the environmental costs of agriculture are needed to deliver a more sustainable future.

INTRODUCTION

Agriculture is the most dominant land use on Earth and will remain so as the world's population and global food demand rise [1] (Figure 1). This poses a fundamental challenge to the natural environment, which requires access to many of the same resources as agricultural producers and is harmed by agriculture's external impacts [2,3]. Using land resources for agriculture prioritises crop and livestock production over the many other benefits that natural ecosystems provide and that agriculture displaces, including biodiversity, carbon storage, and water purification [4]. Governments also heavily support agricultural production in ways that are often bad for the environment, such as rewarding output or lowering input prices [5,6]. Very little is spent on actively mitigating the environmental impacts (Figure 1).

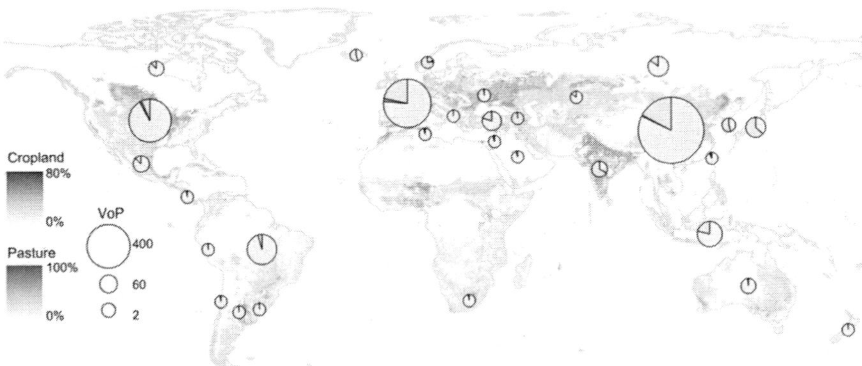

Figure 1. Financial support to farmers from taxpayers and consumers associated with agricultural policies as a proportion of the total value of agricultural production (VoP) at the farm gate.

We distinguished between policies aiming to achieve specific environmental objectives beyond those required by regulation (purple segments) and all other types of support irrespective of their influence on farm production or income (orange segments), e.g., market price support, monetary transfers based on output, input subsidies. Sizes of symbols were scaled to VoP in hundreds of millions of United States dollars. Data are from the most recent year available (2008–2013) and described in S1 Text. Green and brown shading in the background is the percent of cropland and pasture at 5' resolution in the year 2000 [7]. Financial support data were preferentially sourced from the Organisation for Economic Cooperation and Development (OECD) (n = 21 countries), supplemented from the World

Trade Organization (WTO) where possible for additional countries (n = 12). Data used to make this figure are provided in S1 Data.

https://doi.org/10.1371/journal.pbio.1002242.g001

Here, we summarise how agriculture impacts the natural environment and then scrutinise the success of government agricultural policies that we believe offer the most promise for safeguarding nature while feeding a growing human population. We use economic and political theory to help explain why some measures are more prone to succeed than others, recognising that scientific understanding of a problem does not mean it will be solved [8]. Throughout, we touch upon examples from centres of agricultural production.

A CULTIVATED PLANET

Agriculture's demand for land drives conversion of natural habitats, and this is arguably its largest environmental cost. Approximately 40% of ice-free land (4,300 million ha) is already covered by crops or used to raise livestock [7], and an additional 2.7–4.9 million ha of cropland annually, the size of a small European nation, may be required to feed the world's population by 2030 [1,9]. Conversion of land for agriculture is estimated to account for 80% of global deforestation [10], and ca. 53% of terrestrial species assessed as threatened by the International Union for Conservation of Nature (IUCN) [11] are negatively impacted by agriculture (Figure 2). Land conversion also reduces the size of the terrestrial carbon sink. Global simulation models predict that 24% and 10% less carbon is stored in vegetation and soil, respectively, than if present-day landscapes retained their natural vegetation [12].

We counted the total number of species classified as either critically endangered, endangered, or vulnerable by the IUCN Red List in different taxonomic units and the number that were assessed to be threatened by at least one agricultural activity [11]. Agricultural activities were annual and perennial nontimber crops, wood and pulp plantations, livestock farming and ranching, logging and wood harvesting, abstracting of surface water (agricultural use), abstraction of ground water (agricultural use), and agricultural and forestry effluents. Many other species may become threatened with agricultural expansion (S2 Fig).

https://doi.org/10.1371/journal.pbio.1002242.g002

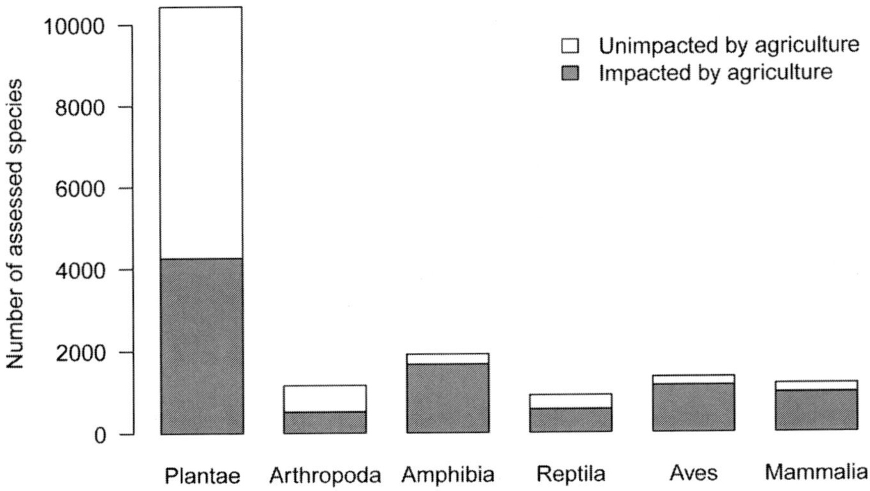

Figure 2. Estimated numbers of threatened species negatively impacted by agriculture.

Agriculture also contributes more to other forms of environmental degradation than any other economic sector. Between 30%–35% of global greenhouse gas (GHG) emissions come from agriculture [3], and crop irrigation accounts for 70% of the world's freshwater withdrawals [13]. The use of synthetic nitrogen fertiliser has increased nearly 21-fold since 1950, and more nitrogen is now added to agricultural soils than from natural processes [14]. Virtually all human-derived nitrogen is lost to the atmosphere or receiving waters [14,15]. This eutrophication suffocates aquatic ecosystems [16]. Phosphorus from fertiliser and livestock manure [14,17], pesticides [18–20], and nanoparticles [21,22] are also exacerbating environmental pollution. Additionally, poor agricultural management causes soil degradation, reducing agricultural productivity and creating further demand for nutrients, water, and land conversion [23,24].

Solutions for limiting agriculture's environmental costs are increasingly well understood. These include reducing land clearance by improving yields of existing crops, minimizing excess nutrient and water use, and shifting production from livestock towards crops [3]. Implementing these practices for only a handful of commodities in a subset of the world can have a large environmental benefit [25]. The success of these practices, however, depends on how well policy puts them into practice.

CHALLENGES TO LIMITING AGRICULTURE'S ENVIRONMENTAL COSTS

Designing and implementing policies to reduce agriculture's environmental costs is difficult. Like other industries, farmers and the businesses reliant upon them, such as input suppliers, processors, and retailers, have a rational self-interest in maximising economic returns and oppose policies that threaten their viability [26,27]. In developed economies, this has historically allowed them to organise into focused lobby groups with considerable advantage over diffuse public interests seeking to safeguard the environment [28,29]. Governments rarely benefit from challenging agri-food lobbies over environmental issues that lack immediate returns for the public good. These lobbies reward government for less demanding environmental policies [30,31], and tailoring policies appropriate for the range of physical and social conditions that farms occupy is costly [26,27]. Concerns about foodborne illness (including genetically modified organisms), environmental impacts (e.g., pesticide use), and animal welfare have ceded some power in the policy arena of developed countries from producers to consumers [32], and increasingly to large retailers, but existing governance structures remain highly susceptible to farmer interests despite spatiotemporal variation in power relations [29]. For example, corporate lobbying efforts in the agricultural sector were deemed effective by 52% of respondents in a 2013 survey of nearly 600 European Union (EU) decision makers, and >80% effectiveness was reported in Denmark, Latvia, and the United Kingdom [33]. Policymakers therefore face major challenges in aligning the private interests of the agri-food sector with public interest in environmental goods across the diversity and extent of farm entities.

In addition to political challenges, measures that do succeed in reducing the environmental harms of agriculture in one place can perversely increase them elsewhere. This is because, in a global market, local policies or technologies that cause agricultural production to carry a greater opportunity cost than using land for environmental goods will displace production to regions where these costs are lower [34]. For example, regulations limiting fertilisers can reduce local water pollution and GHG emissions. But because such measures may reduce yields (i.e., tonnage of production per unit area), food demand will be met elsewhere, leading to intensification and land clearance that potentially creates a net environmental loss [35]. Incentivising low-intensity systems, seen as beneficial to biodiversity conservation in regions such as Europe, may similarly limit production and displace environmental impacts while also lacking the full benefits of natural ecosystems [36]. Efforts to limit the environmental costs of agriculture must therefore consider the environmental impacts beyond farm boundaries as well as the consequences for food production.

Potential Resolutions to the Agriculture–Nature Conflict

Agricultural policy with explicit environmental objectives is generally organized around three approaches (Table 1 and expanded upon in S1 Table; see Box 1 for an exception):

1. regulations, such as limits on pesticide use or abstracting water, which can be enforced through penalties and conditions placed on financial support to farmers;
2. community-based approaches, which support farmers and local stakeholders to work collectively in addressing environmental impacts; and
3. economic instruments, which pay farmers directly or create markets for adopting practices that minimise environmental impacts and provide noncommodity outputs beyond those required by existing regulation; tariffs can also be used to internalise environmental costs.

Table 1. Successes associated with different policy approaches and the local[1]obstacles to their implementation.

Approach	Success story	Obstacles
Regulations	Critical Counties programme launched in 2008 by Brazilian government suspended access to credit for farmers in counties with the highest deforestation rates. This stimulated collective actions to conserve native vegetation, resulting in deforestation rates dropping to <20% of previous 10-yr average [40].	Complexity, enforceability, cost of implementation. Where compliance is linked to payments, as in EU's Common Agricultural Policy, clear environmental standards and operational guidance required at farm level [41]. Payments must also incorporate regional and/or sectorial variation in the cost of compliance.
Community-based	Landcare Australia is among the best known partnerships between communities, government, and organisations. Established by a $360 million programme in 1989 and engaging about 30% of farming community at its height [42]. Notable successes improving water quality, reducing soil degradation, restoring habitat [43].	Delivering social and/or short-term economic benefits to individual farmers; lack of these benefits has limited uptake of Landcare practices in Australia [44,45].
Economic instruments	Conservation Reserve Program (CRP) established in 1985 by US government provides farmers with annual rental payments for removing land most sensitive to agricultural impacts from production for ≥10 years. Currently, >10 million ha enrolled at annual cost of ca. US$2 billion [46]. CRP has eliminated millions of tonnes of GHG emissions and fertiliser additions and established new habitat supporting tens of millions of birds and other wildlife [47]. Gains arisen despite conversion of non-cropland into production predicted over area equal to 20% of CRP land [48].	Can perform poorly if farmers paid for actions rather than actual environmental outcomes associated with their land [49]. For example, temporary retirement schemes emerged initially in EU to reduce overproduction. As no environmental outcome was explicitly targeted, farmers tended to retire least profitable land that was often the poorest quality, limiting environmental benefits [50,51].

[1]Globally, the major obstacle for all approaches is to avoid displacing production and causing environmental degradation elsewhere.

Box 1. Abandoning the Agricultural Sector to Market Forces

New Zealand radically altered its agricultural policy in 1984 to embrace minimal government intervention. It now relies exclusively on regulation to minimise the environmental impacts of agriculture while providing among the least support to domestic producers out of all developed countries (ca. 0.5% of the value of production; S1 Text). Consequently, the natural environment has been left vulnerable to the pressures of market forces, which are rapidly accelerating production and land conversion with increasing demand for dairy products in particular. The rate of conversion of indigenous grassland to exotic pasture in the South Island has increased by 67% from the period between 1990–2001 to 2001–2008 [52], and New Zealand leads the world in livestock emissions per capita, exceeding the global total across developed countries by more than 13-fold [53].

The nonprescriptive Resource Management Act (RMA) is the primary national legislation intended to oversee natural and physical resources, with land regulation devolved to elected district councils. Although the RMA tasks councils with "maintaining" biodiversity in balance with resource development, they generally expedite the latter. Fewer than 5% of agricultural sector consents requiring biodiversity or ecosystem services maintenance were recently found to comply with conditions [54]. Limited resources are often blamed for poor enforcement, but political interference also occurs [55]. Other regulations, such as the Wildlife Act and Native Plant Protection Act, are rarely invoked, and the latter does not apply to landowners [55].

New Zealand also has a dedicated judicial system to contest decisions made under the RMA. While this system has delivered several notable "wins" for the environment (e.g., [56]), many local actions escape scrutiny. In practice, biodiversity protection by rural landowners occurs mainly on small, voluntary, set-asides of residual unproductive land (known as covenants) and through patchy predator-trapping and stock fencing activities, supported by limited local and national funds. It appears that the laissez-faire approach to the agricultural sector does little to safeguard the environment.

Farmers will be motivated to accept one or more of these approaches where they incur small socioeconomic costs relative to the benefits that they receive from achieving environmental objectives [37,38]. Eliminating financial support for farmers to maximise agricultural production is important in helping to reduce the socioeconomic costs of adopting environmental practices. Reforms in developed economies since the 1980s are now supporting producer incomes more through direct payments than by subsidising production or protecting domestic prices [2,39] (Box 2).

Even with farmer acceptance, success of any policy approach will ultimately depend on whether farmers maintain or boost yields on existing agricultural lands rather than convert new natural lands to production. Halting local land conversion can be achieved with strict enforced regulations (Table 1), but in a globalized market, changes in the local supply of a given cash crop will lead to changes in demand and supply in other distant locations [59]. Trends in cereal production in 19 countries between 1990 and 2005 clearly show this link between local actions and global consequences. These countries managed to boost cereal yields on existing lands while removing cropland from production, but without fully meeting local demand. Per capita cereal imports consequently increased by 12-fold in these countries compared with those lacking land retirement policies [60], clearly suggesting that production was displaced elsewhere. Where subsistence agriculture is important, limiting its local production may similarly boost reliance on cash crops produced abroad [61]. National governments implementing any of the three main policy approaches need to begin working in a coordinated way to combat land use displacement.

Box 2. Environmental Harms of Agricultural Price Support

Price support policies, such as import tariffs or minimum commodity prices, encourage overproduction in the absence of supply management. This strains natural resources and makes marginal land artificially profitable [57]. In the US, the federal government historically managed the dairy industry by guaranteeing to purchase milk if its market price fell below a set value. Modelling revealed that this would drive abandonment and reversion to natural forest across thousands of hectares of land in the state of Wisconsin if milk prices were decreased by only 5%–15%, with millions saved in water treatment and government expenditure [58]. The US has now adopted a strategy of paying producers should their margins fall beneath national averages, likely with similar environmental consequences.

Finding the Right Mixture of Solutions

Maintaining the status quo with agricultural policy is unsustainable for the environment [3,35,62]. On their own, regulations and community approaches will offer little help if damaging the environment remains financially lucrative because the associated penalties are trivial, governments continue to subsidise production (Figure 1, S1 Fig), and farmers are not accountable for the environmental costs of their actions [63–65] (Box 3). Economic instruments, such as payments for public goods, are also contentious, especially for biodiversity conservation [66–68].

Positive outcomes often arise only from certain actions, such as enhancing specific food resources or nesting habitat [69], and catch-all interventions may avoid targeting rare species [70,71]. Removing production subsidies and making farmers financially responsible for their environmental impacts, such as with markets for public goods, can help local outcomes but must avoid curbing intensification because this will encourage land conversion elsewhere to meet demand for food [35].

Box 3. Balancing Large-Scale Incentives for the Rural Poor with Strict Penalties for Agri-business

China—the world's largest agricultural producer—offers a stark contrast in government policy to that implemented by New Zealand described in Box 2. The country's economy is heavily controlled by centralised government and relatively free from the political influence wielded by farmers lobbies. Government can therefore act largely in its own self-interest, which may align more often with the public good. This has allowed the Chinese government to implement environmental policies on scales that would be virtually impossible elsewhere in the world. The Grain to Green Project is one example, having converted nearly 9.1 million hectares of cropland into forest whilst incentivising farmers to participate with grain and cash subsidies [72]. This permanent land retirement scheme has reduced soil erosion and increased carbon sequestration at landscape scales [73]. China is also adopting a stricter approach to large industrial agriculture. In 2014, the country's top legislature revised the national environmental protection law for the first time in 25 years. The revisions included harsher penalties to deter companies from accepting fines rather than investing in pollution control [74]. Whether enforcement follow suits remains to be seen.

Ruhl [26] argued that successful agricultural policies will target large industrial operations that can be readily regulated while incentivising changes in production practices by small farmers without compromising their livelihoods. Recent developments in China are therefore promising, but the country's agricultural sector still causes major environmental harm, including poor water quality in almost all rivers and lakes [75], and nearly 40% more GHG emissions per tonne of nitrogen fertiliser than in Europe [76]. Subsidies on inputs such as fertilisers and pesticides are largely to blame as they promote overuse. Fertilisers are estimated to be overused by 30%–60% [77], yet nearly CMY¥108,000 million (US$17,000 million) was spent in 2012 on just one scheme to help farmers purchase more of these inputs [78]. Removing harmful financial subsidies would help greatly in aligning the country's agricultural and environmental aims.

We assert that conflict between agriculture and the environment will be best resolved by policies dedicating high-quality habitat towards nature conservation, while encouraging intensive production on existing farmland with stringent limits on environmental impacts [35,36,79–81]. Measures that make farmland itself more benign—so called "land-sharing" approaches—also deliver local environmental and social benefits [82,83], but can reduce agricultural yields and cause land conversion elsewhere, effectively displacing environmental burdens [80]. Intensifying production on existing farmland may thus be a better option because it avoids displacing impacts and can allow other land to be freed for nature conservation and restoration, with no net loss in agricultural production.

Our consideration of policy instruments approaches and the obstacles to their success (Table 1) suggests that our proposed resolution would practically entail a combination of punitive measures, such as taxes that reduce existing overuse of farm inputs [84], with positive incentives linked to measurable outcomes [49], such as for permanently removing land from production. Taxes on farm inputs can improve upon existing payments to limit input use because they do not explicitly cap nutrient, pesticide, and water use. Without input limits, some farmers may find it profitable to intensify existing production, despite the added taxes, while others might profit more from reducing resource consumption. Across the landscape, yields will be maintained, and potentially increased without additional land conversion. For example, water surcharges in Mexico are used to pay landowners to retain forested watersheds. Displaced land conversion from exploitation of communal properties and changing market pressures reduced the environmental benefits of this programme by only ca. 4% [85]. By enabling farms to respond differently to the same policy, input taxes carry an additional advantage over traditional regulations that hold all farmers to the same standard and ignore sectorial diversity [86]. An aligned but more audacious approach would be to replace taxes on income with taxes on environmental consumption. The idea is to shift from taxing the production of private wealth toward taxing the consumption of public environmental goods, thus promoting economic growth from sustainable use rather than depletion of natural capital. Brown et al. [55] have proposed that such a system would most heavily tax intensive land uses that generate the most negative environmental impacts, while areas of intact ecological function would entitle the owners to a rebate. This approach might encourage farmers to work intensively with the most productive land but set aside and restore less productive areas with net gains for the environment.

Evidence at national levels suggests that investments in land retirement and input-limiting schemes, consistent with our suggested policy approach, do deliver local environmental benefits. To illustrate this, we examined trends in two environmental responses, farmland bird populations and GHG emissions from synthetic fertilisers, across countries with similar agricultural patterns, i.e., dominated by livestock and arable crops (see S1 Text for methods). We calculated the annual, per hectare monetary investment in a specific environmental outcome beyond that required by regulation from a detailed inventory of government policies compiled by the OECD (annotated in S1 Data). As expected, because habitat loss is a major threat to farmland and grassland biota, we found that farmland birds benefited as more money was invested in retiring agricultural land (Figure 2). We also found that total emissions of N_2O declined with increasing investment in reducing fertiliser use (Figure 3). While cursory in nature, these results suggest that targeted policies have the potential to deliver positive environmental outcomes.

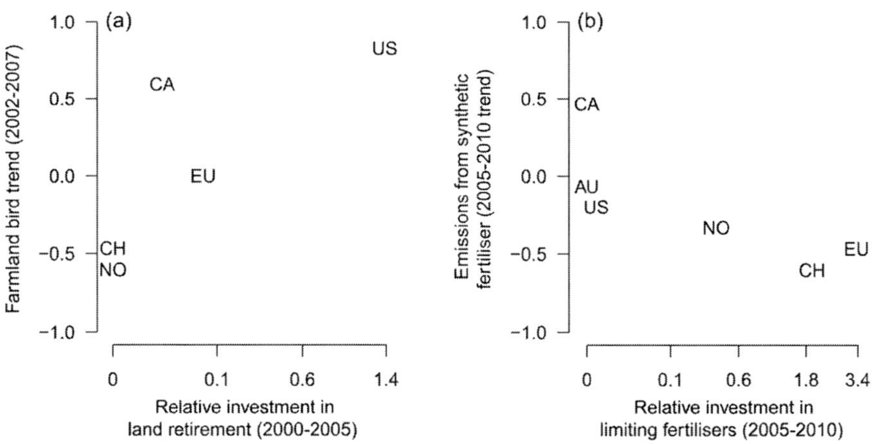

Figure 3. Improvements in environmental performance of agriculture correlate with investments in agri-environment schemes (AES).

Temporal trend in (**A**) farmland bird populations and (**B**) GHG emissions from synthetic fertiliser were calculated over time for each country using Kendall's τ (see S1 Text for full methods). Investments in AES were scaled by the ratio of investments in AES relative to financial transfers to agricultural producers for all other purposes so as to account for the fact that environmental improvements can be offset by policies that maximise production. Values were then summed across 5-y periods with a pre-emptive 2-y lag for land retirement. Habitat quality will take at least one growing season to improve following land retirement and any changes in bird populations from improved breeding success will thus largely be measurable the following year, i.e., two years later [87,88]. Spearman's rank correlation: $\rho = 0.95, p = 0.005$ and $\rho = -0.93, p = 0.008$ for (**A**) and (**B**), respectively. Points are responses in individual countries: AU = Australia [data available for (**B**) only], CA = Canada, EU (aggregated together), NO = Norway, CH = Switzerland, and US. Data used to make this figure are provided in S1 Data.

Measures delivering environmental benefits must also be coupled with mechanisms that promote intensification to avoid displacing production elsewhere, and they must do so with fewer external costs. Research suggests increased yields and better environmental practices are possible by improving nutrient and water management and breeding more efficient crops [3]. Reductions in food waste and human diets with less livestock and dairy consumption can also help meet rising food demands without increasing resource use [89,90]. Dietary change is, however, contentious. Governments may need to consider policies that tax livestock products and/or subsidize the development of livestock-free protein sources, e.g., [91], in order to dislodge entrenched norms.

Here, we have laid out a vision for limiting the environmental impacts of agriculture, yet at least two major challenges remain. First, winning political support for policy change remains a fundamental requirement to any of our proposed reforms. Ultimately, democratically elected governments, while not immune to the lobbying power of agribusiness, will generally respond to electoral pressures [31]. Electorates may express little desire for policy change if they are poorly informed, unmotivated, or misled by industry about limiting the environmental costs of agriculture. Second, implementation of our policy mixture will vary spatially. For example, successful land retirement depends on features of the surrounding landscape, such as habitat connectivity [92], and can reduce yields in regions where the most productive land also has the most valuable environmental assets. Nonetheless, improvements in the management of land resources in developing countries [93], new technologies that optimally allocate farm inputs to reduce overuse [94], advances in crop breeding [95], and tools for better conservation planning [96] give us hope that innovations exist for meeting the need for more sustainable and intensive agriculture.

ACKNOWLEDGMENTS

We thank Andrew Balmford, David Coomes, and staff at the OECD for discussion.

REFERENCES

1. Tilman D, Balzer C, Hill J, Befort BL (2011) Global food demand and the sustainable intensification of agriculture. Proc Natl Acad Sci USA 108: 20260–20264. pmid:22106295

2. Warren J, Lawson C, Belcher K (2007) The agri-environment. Cambridge: Cambridge University Press. 232 p.
3. Foley JA, Ramankutty N, Brauman KA, Cassidy ES, Gerber JS, et al. (2011) Solutions for a cultivated planet. Nature 478: 337–342. pmid:21993620
4. Millennium Ecosystem Assessment. Ecosystems and human well-being: synthesis. Washington, DC: World Resources Institute; 2005. 155 p.
5. James AN, Gaston KJ, Balmford A (1999) Balancing the Earth's accounts. Nature 401: 323–324. pmid:16862091
6. Myers N, Kent J (2001) Perverse subsidies: how tax dollars can undercut the environment and the economy. Washington, DC: Island Press. 277 p.
7. Ramankutty N, Evan AT, Monfreda C, Foley JA. (2008) Farming the planet: 1. Geographic distribution of global agricultural lands in the year 2000. Global Biogeochem Cycles 22: GB1003.
8. Lawton JH (2007) Ecology, politics and policy. J Appl Ecol 44: 465–474.
9. Lambin EF, Meyfroidt P (2011) Global land use change, economic globalization, and the looming land scarcity. Proc Natl Acad Sci USA 108: 3465–3472. pmid:21321211
10. Kissinger G, Herold M, de Sy V (2012) Drivers of deforestation and forest degradation: a synthesis report for REDD+ policymakers. Lexeme Consulting, Vancouver. 46 p.
11. International Union for Conservation of Nature [IUCN] (2014) The IUCN Red List of Threatened Species. Version 2014.3. http://www.iucnredlist.org via the Internet. Accessed 2015 March 14.
12. Bondeau A, Smith PC, Zaehle S, Schaphoff S, Lucht W, et al. (2007) Modelling the role of agriculture for the 20th century global terrestrial carbon balance. Glob Chang Biol 13: 679–706.
13. Food and Agriculture Organization of the United Nations [FAO]. (2015) AQUASTAT database. Available: http://www.fao.org/nr/water/aquastat/data/ via the Internet. Accessed 14 March 2015.
14. Bouwman L, Goldewijk KK, Hoek KWVD, Beusen AHW, Van Vuuren DP, et al. (2013) Exploring global changes in nitrogen and phosphorus cycles in agriculture induced by livestock production over the 1900–2050 period. Proc Natl Acad Sci USA 110: 20882–20887. pmid:21576477
15. Galloway JN, Aber JD, Erisman JW, Seitzinger SP, Howarth RW, et al. (2003) The nitrogen cascade. BioScience 53: 341–356.
16. Diaz RJ, Rosenberg R (2008) Spreading dead zones and consequences for marine ecosystems. Science 321: 926–929. pmid:18703733
17. MacDonald GK, Bennett EM, Potter PA, Ramankutty N (2011) Agronomic phosphorus imbalances across the world's croplands. Proc Natl Acad Sci USA 108: 3086–3091. pmid:21282605
18. Rohr JR, Schotthoefer AM, Raffel TR, Carrick HJ, Halstead N, et al. (2008) Agrochemicals increase trematode infections in a declining amphibian species. Nature 455: 1235–1239 pmid:18972018
19. Henry M, Béguin M, Requier F, Rollin O, Odoux J-F, et al. (2012) A common pesticide decreases foraging success and survival in honey bees. Science 336: 348–350. pmid:22461498
20. Hallmann CA, Foppen RPB, van Turnhout CAM, de Kroon H, Jongejans E (2014) Declines in insectivorous birds are associated with high neonicotinoid concentrations. Nature 511: 341–343. pmid:25030173
21. Priester JH, Ge Y, Mielke RE, Horst AM, Moritz SC, et al. (2012) Soybean susceptibility to manufactured nanomaterials with evidence for food quality and soil fertility interruption. Proc Natl Acad Sci USA 109: E2451–E2456. pmid:22908279
22. Santiago-Martín A, Constantin B, Guesdon G, Kagambega N, Raymond S, et al. (2015) Bioavailability of engineered nanoparticles in soil systems. J. Hazard. Toxic Radioact. Waste in press.

23. Pimentel D, Harvey C, Resosudarmo P, Sinclair K, Kurz D, et al (1995) Environmental and economic costs of soil erosion and conservation benefits. Science 267: 1117–1123. pmid:17789193
24. Lal R (1997) Degradation and resilience of soils. Phil Trans R Soc Lond B 352: 997–1010.
25. West PC, Gerber JS, Engstrom PM, Mueller ND, Brauman KA, Carlson KM, et al. (2014) Leverage points for improving global food security and the environment. Science 345: 325–328. pmid:25035492
26. Ruhl JB (2000) Farms, their environmental harms, and environmental law. Ecol Law Q 27: 263–349.
27. Walker S, Brower AL, Stephens RTT, Lee WG (2009) Why bartering biodiversity fails. Conserv Lett 2: 149–157.
28. Olson M (1965) The logic of collective action. Cambridge: Harvard University Press. 208 p.
29. Daugbjerg C, Swinbank A (2012) An introduction to the "new" politics of agriculture and food. Policy Soc 31: 259–270.
30. Eskridge WN Jr (1988) Politics without romance: implications of public choice theory for statutory interpretation. Va Law Rev 74: 275–338.
31. Bellemare MF, Carnes N (2015) Why do members of congress support agricultural protection? Food Policy 50: 20–34.
32. Grant WP (2012) Can political science contribute to agricultural policy? Policy Soc 31: 271–279.
33. .Burson-Marsteller (2013) A guide to effective lobbying in Europe: the view of policy-makers. Available: http://burson-marsteller.nl/wp-content/uploads/2013/06/Lobbying-survey-Report.pdf. Accessed 1 June 2015.
34. Meyfroidt P, Rudel TK, Lambin EF (2010) Forest transitions, trade, and the global displacement of land use. Proc Natl Acad Sci USA 107: 20917–20922. pmid:21078977
35. Balmford A, Green R, Phalan B (2012) What conservationists need to know about farming. Proc Roy Soc B 279: 2714–2724.
36. Merckx T, Pereira HM (2015) Reshaping agri-environmental subsidies: From marginal farming to large-scale rewilding. Basic and Applied Ecology 16: 95–103.
37. Lastra-Bravo XB, Hubbard C, Garrod G, Tolón-Becerra A (2015) What drives farmers' participation in EU agri-environmental schemes? Results from a qualitative meta-analysis. Environmental Science & Policy 54: 1–9.
38. Wilson GA, Hart K (2000) Financial imperative or conservation concern? EU farmers' motivations for participation in voluntary agri-environmental schemes. Environment and Planning A 32: 2161–2185.
39. Organisation for Economic Co-operation and Development [OECD] (2004) Agriculture and the environment: lessons learned from a decade of OECD work. Paris: OECD. 36 p.
40. Nepstad D, McGrath D, Stickler C, Alencar A, Azevedo A, et al. (2014) Slowing Amazon deforestation through public policy and interventions in beef and soy supply chains. Science 344: 1118–1123. pmid:24904156
41. European Court of Auditors (2008) Is cross compliance an effective policy? Special report no 8/2008. Luxembourg: Office for Official Publications of the European Communities. 57 p.
42. Curtis A, Lockwood M (2000) Landcare and catchment management in Australia: lessons for state-sponsored community participation. Soc Nat Resour 13: 61–73.
43. OECD (2013) Providing agri-environmental public goods through collective action. Paris: OECD. 306 p.
44. Curtis A, Lacy TD (1998) Landcare, stewardship and sustainable agriculture in Australia. Environ Values 7: 59–78.
45. Pannell DJ, Marshall GR, Barr N, Curtis A, Vanclay F, et al. (2006) Understanding and promoting adoption of conservation practices by rural landholders. Aust J Exp Agric 46: 1407–1424.

46. Stubbs M (2014) Conservation Reserve Program (CRP): status and issues. Washington, DC: Congressional Research Service. 24 p.
47. United States Department of Agriculture (2011) CRP benefits. http://www.fsa.usda.gov/Internet/FSA_File/united_states.pdf via the Internet. Accessed 14 March 2015.
48. Wu J (2000) Slippage effects of the conservation reserve program. Am J Agric Econ 82: 979–992.
49. Franks JR, McGloin A (2007) Joint submissions, output related payments and environmental co-operatives: Can the Dutch experience innovate UK agri-environment policy? Journal of Environmental Planning and Management 50: 233–256.
50. Sotherton NW (1998) Land use changes and the decline of farmland wildlife: An appraisal of the set-aside approach. Biol Conserv 83: 259–268.
51. Robinson GM, Lind M (1999) Set-Aside and environment: a case study in southern England. Tijdschr Econ Soc Geogr 90: 296–311.
52. Weeks ES, Walker SF, Dymond JR, Shepherd JD, Clarkson BD (2012) Patterns of past and recent conversion of indigenous grasslands in the South Island, New Zealand. N Z J Ecol 37: 127–138.
53. Caro D, Davis SJ, Bastianoni S, Caldeira K (2014) Global and regional trends in greenhouse gas emissions from livestock. Climatic Change 126: 203–216.
54. Brown MA, Clarkson BD, Stephens RT, Barton BJ (2014) Compensating for ecological harm–the state of play in New Zealand. N Z J Ecol 38: 139–146.
55. Brown MA, Stephens RTT, Peart R, Fedder B (2015) Vanishing nature: facing New Zealand's biodiversity crisis. Auckland: Environmental Defence Society. 208 p.
56. Forest & Bird (2013) 2013 Annual Report. http://www.forestandbird.org.nz/files/file/FB%20Annual%20Report%202013%20V4%20low%20res.pdf via the Internet. Accessed 14 March 2015.
57. Abler D (2004) Multifunctionality, agricultural policy, and environmental policy. Agricultural and Resource Economics Review 33: 8–17.
58. Plantinga AJ (1996) the effect of agricultural policies on land use and environmental quality. Am J Agr Econ 78: 1082–1091.
59. MacDonald GK, Brauman KA, Sun S, Carlson KM, Cassidy ES, Gerber JS, et al. (2015) Rethinking agricultural trade relationships in an era of globalization. BioScience 65: 275–289.
60. Rudel TK, Schneider L, Uriarte M, Turner BL, DeFries R, Lawrence D, et al. (2009) Agricultural intensification and changes in cultivated areas, 1970–2005. Proc Natl Acad Sci USA 106: 20675–20680 pmid:19955435
61. Murray BC (2008) Leakage from an avoided deforestation compensation policy: concepts, empirical evidence, and corrective policy options. Raleigh: Nicholas Institution for Environmental Policy Solutions. 32 p.
62. Pe'er G, Dicks LV, Visconti P, Arlettaz R, Báldi A, et al. (2014) EU agricultural reform fails on biodiversity. Science 344:1090–1092. pmid:24904142
63. Pretty JN, Brett C, Gee D, Hine RE, Mason CF, et al. (2000) An assessment of the total external costs of UK agriculture. Agric Syst 65: 113–136.
64. Pretty J, Brett C, Gee D, Hine RE, Mason CF, et al. (2001) Policy challenges and priorities for internalizing the externalities of modern agriculture. Journal of Environmental Planning and Management 44: 263–283.
65. Balmford A, Bruner A, Cooper P, Costanza R, Farber S, et al. (2002) Economic reasons for conserving wild nature. Science 297: 950–953. pmid:12169718
66. Kleijn D, Sutherland WJ (2003) How effective are European agri-environment schemes in conserving and promoting biodiversity? J Appl Ecol 40: 947–969.
67. Kleijn D, Baquero RA, Clough Y, Díaz M, De Esteban J, et al. (2006) Mixed biodiversity benefits of agri-environment schemes in five European countries. Ecol Lett 9: 243–254. pmid:16958888
68. Whittingham MJ (2007) Will agri-environment schemes deliver substantial biodiversity gain, and if not why not? J Appl Ecol 44: 1–5.

69. Baker DJ, Freeman SN, Grice PV, Siriwardena GM (2012) Landscape-scale responses of birds to agri-environment management: a test of the English Environmental Stewardship scheme. J Appl Ecol 49: 871–882.
70. Kleijn D, Berendse F, Smit R, Gilissen N (2001) Agri-environment schemes do not effectively protect biodiversity in Dutch agricultural landscapes. Nature 413: 723–725. pmid:11607029
71. 71.71.Vickery JA, Bradbury RB, Henderson IG, Eaton MA, Grice PV (2004) The role of agri-environment schemes and farm management practices in reversing the decline of farmland birds in England. Biol Conserv 119: 19–39.
72. State Forestry Administration of China (2014) Forestry in China. http://english.forestry.gov.cn/uploads/Information_Services/Latest_Publication/Forestry_in_China.pdf via the Internet. Accessed 14 March 2015.
73. Liu J, Li S, Ouyang Z, Tam C, Chen X (2008) Ecological and socioeconomic effects of China's policies for ecosystem services. Proc Natl Acad Sci USA 105: 9477–9482 pmid:18621700
74. National People's Congress of the People's Republic of China. (2014) China's legislature adopts revised Environmental Protection Law. http://www.npc.gov.cn/englishnpc/news/Legislation/2014-04/25/content_1861275.htm via the Internet. Accessed 14 March 2015.
75. Sun B, Zhang L, Yang L, Zhang F, Norse D, et al. (2012) Agricultural non-point source pollution in china: causes and mitigation measures. Ambio 41: 370–379. pmid:22311715
76. Zhang W, Dou Z, He P, Ju X-T, Powlson D, et al. (2013) New technologies reduce greenhouse gas emissions from nitrogenous fertilizer in China. Proc Natl Acad Sci USA 110: 8375–8380. pmid:23671096
77. Ju X-T, Xing G-X, Chen X-P, Zhang S-L, Zhang L-J et al. (2009) Reducing environmental risk by improving N management in intensive Chinese agricultural systems. Proc Natl Acad Sci USA 106: 3041–3046. pmid:19223587
78. OECD (2015) Producer and consumer support estimates database. Available http://www.oecd.org via the Internet. Accessed 14 March 2015.
79. Hodgson JA, Kunin WE, Thomas CD, Benton TG, Gabriel D (2010) Comparing organic farming and land sparing: optimizing yield and butterfly populations at a landscape scale. Ecol Lett 13:1358–1367. pmid:20825453
80. Phalan B, Onial M, Balmford A, Green RE (2011) Reconciling food production and biodiversity conservation: land sharing and land sparing compared. Science 333: 1289–1291. pmid:21885781
81. Godfray HCJ, Garnett T (2014) Food security and sustainable intensification. Phil Trans R Soc Lond 369: 20120273.
82. 82.82.Fischer J, Abson DJ, Butsic V, Chappell MJ, Ekroos J, Hanspach J, et al. (2014) Land sparing versus land sharing: moving forward. Conserv Lett 7: 149–157.
83. Frishkoff LO, Karp DS, M'Gonigle LK, Mendenhall CD, Zook J, Kremen C, et al. (2014) Loss of avian phylogenetic diversity in neotropical agricultural systems. Science 345: 1343–1346. pmid:25214627
84. Mueller ND, Gerber JS, Johnston M, Ray DK, Ramankutty N, et al. (2012) Closing yield gaps through nutrient and water management. Nature 490: 254–257. pmid:22932270
85. Alix-Garcia JM, Shapiro EN, Sims KRE (2012) Forest conservation and slippage: evidence from Mexico's national payments for ecosystem services program. Land Econ 88: 613–638.
86. Stavins RN, Whitehead BW (1992) Pollution charges for environmental protection: a policy link between energy and environment. Annual Review of Energy and the Environment 17: 187–210.
87. Newson SE, Johnston A, Renwick AR, Baillie SR, Fuller RJ (2012) Modelling large-scale relationships between changes in woodland deer and bird populations. J Appl Ecol 49: 278–286.

88. Chamberlain D E, Fuller RJ, Bunce RGH, Duckworth JC, Shrubb M (2000) Changes in the abundance of farmland birds in relation to the timing of agricultural intensification in England and Wales. J Appl Ecol 37: 771–788.
89. Bajželj B, Richards KS, Allwood JM, Smith P, Dennis JS, et al. (2014) Importance of food-demand management for climate mitigation. Nature Clim Change 4: 924–929.
90. Tilman D, Clark M (2014) Global diets link environmental sustainability and human health. Nature 515: 518–522. pmid:25383533
91. Pandya R (2014) Milk without the moo. New Sci 222: 28–29.
92. Gilroy JJ, Edwards FA, Uribe CAM, Haugaasen T, Edwards DP (2014) Surrounding habitats mediate the trade-off between land-sharing and land-sparing agriculture in the tropics. J Appl Ecol 51: 1337–1346.
93. Hobbs PR, Sayre K, Gupta R (2008) The role of conservation agriculture in sustainable agriculture. Phil Trans R Soc Lond B 363: 543–555.
94. Chen X-P, Cui Z-L, Vitousek PM, Cassman KG, Matson PA, et al. (2011) Integrated soil–crop system management for food security. Proc Natl Acad Sci USA 108: 6399–6404. pmid:21444818
95. Tester M, Langridge P (2010) Breeding technologies to increase crop production in a changing world. Science 327: 818–822. pmid:20150489
96. Bateman IJ, Harwood AR, Mace GM, Watson RT, Abson DJ, et al. (2013) Bringing ecosystem services into economic decision-making: land use in the United Kingdom. Science 341: 45–50. pmid:23828934
97. Anderson K, Nelgen S (2013) Updated national and global estimates of distortions to agricultural incentives, 1955 to 2011. Available: http://www.worldbank.org/agdistortions via the Internet. Accessed 2015 June 1.
98. Anderson K, Kurzweil M, Martin W, Sandri D, Valenzuela E (2008) Measuring distortions to agricultural incentives, revisited. World Trade Rev 7: 1–30.

CHAPTER 16

Meteorological and Landscape Influences on Pollen Beetle Immigration into Oilseed Rape Crops

Matthew P. Skellern[a], Sue J. Welham[a,b,c] , Nigel P. Watts[a], Samantha M. Cook[a], *

[a]Rothamsted Research, Harpenden, Hertfordshire, AL5 2JQ, UK

[b]VSN International, 2 Amberside House,Wood Lane, Hemel Hempstead, HP2 4TP, UK

[c] Current address: Stats4Biol, Welwyn Garden City, UK

ABSTRACT

Heavy reliance on pesticide inputs to maintain crop yields has been an important aspect of agricultural intensification. Insecticide use has had detrimental impacts on pollinators and natural pest control agents, contributing to a decline in associated ecosystem services, and has also led to resistance development in pest populations. Throughout Europe, in oilseed rape (*Brassica napus* L.) crops, prophylactic use of insecticides against pollen beetles (*Meligethes aeneus* F. also known as *Brassicogethes aeneus*) has led to such issues, and there is an urgent need to develop more sustainable pest management practices for the crop. Although advice is available to oilseed rape growers regarding control thresholds, it may not be adhered to due to the expense of pollen beetle monitoring relative to the inexpensive cost of pyrethroid insecticides. Thus, the key to reducing prophylactic insecticide applications may lie with improved, less labour intensive methods of pollen beetle monitoring. For these to be realized, a better understanding is needed of the effects of agri-landscape features and meteorological conditions on pollen beetle immigration into the crop. In this study, based on data from four years of pollen beetle monitoring on a total of 41 field sites, we model the effects of meteorological (wind speed and direction, rainfall and accumulated temperature) and landscape (areas of woodland, residential gardens, the current and previous seasons' oilseed rape crops, and lengths of hedgerows and treelines) variables on directional sticky trap catches, at both the single trap and field scales. Meteorological variables, particularly accumulated temperature and wind speed were more important than landscape variables in predicting the abundance of pollen beetles immigrating into OSR fields. Sticky traps that were facing downwind caught more beetles than those that were facing across-wind or upwind; this is the first study to show at a landscape-scale, direct evidence for use of upwind anemotaxis by pollen beetles at the point of entry during immigration into the crop. At the field scale, the area of oilseed rape grown in the previous season was found to be positively related to trap catch, but no relationships with other landscape variables were found. Optimally-placed monitoring traps could complement existing decision support systems to reduce pollen beetle monitoring effort and encourage use of insecticides only when control thresholds are breached, thus enhancing the sustainability of oilseed rape production. Knowledge of the area of oilseed rape crops grown during the previous season in the surrounding landscape could contribute to risk assessment of potential pest pressure for individual OSR crops.

Keywords: *Brassicogethes*; Decision support systems; Insecticide resistance; Integrated pest management; *Meligethes aeneus*; Monitoring trap; Sustainable intensification

1. INTRODUCTION

The heavy reliance on pesticide inputs associated with post-WW2 agricultural intensification, along with habitat loss and a reduction in cropping diversity, has contributed to environmental degradation and declining biodiversity within farmed landscapes (Stoate *et al.*, 2001; Robinson and Sutherland, 2002). In particular, the impact of pesticides on pollinators and natural pest control agents has played a role in the deterioration of associated ecosystem services which are vital for sustainable food production (Goulson, 2013; Vanbergen, 2013; Roubos *et al.*, 2014). Furthermore, sustained use or overuse of insecticides has frequently led to development of resistance (Heckel, 2012), threatening the sustainability of existing pest management practices and potentially impacting agricultural productivity. As a result of these issues, and due to increasing public concern, there is an urgent need to reduce pesticide use and develop pest management practices that are more ecologically sustainable.

Throughout Europe, pollen beetles (*Meligethes aeneus* F.; also known as *Brassicogethes aeneus*) are a major pest of oilseed rape (OSR) (*Brassica napus* L.) (Williams, 2010), a crop that has seen significant production increases over recent years (from 19.0 M tonnes produced within the EU in 2008 to 21.7 M tonnes in 2015; Eurostat, 2016a), mostly due to rising demand for biofuel use (Eurostat, 2016b). The beetles overwinter as adults in the leaf litter of hedgerows, woodlands and grassy areas before emerging in early spring (Nilsson, 1988 ; Rusch *et al.*, 2012a), when they feed on pollen from a range of spring flowers before seeking brassicaceous plants for oviposition (Free and Williams, 1978 ; Ouvrard *et al.*, 2016). The beetles use both visual and olfactory cues for host plant location (Williams and Cook, 2010; Cook *et al.*, 2007 ; Cook *et al.*, 2013) and have been observed to use upwind anemotaxis to fly towards their host crops (Evans and Allen-Williams, 1994; Williams *et al.*, 2007 ; Moser *et al.*, 2009). They usually arrive in OSR crops at or around the green bud stage (BBCH growth stage code 51; Lancashire et al., 1991). Feeding damage caused by adults results in yield loss from bud abscission (Nilsson, 1987 ; Alford *et al.*, 2003). The beetle larvae feed within the buds and flowers for around two weeks before dropping to the ground to pupate in the soil. New generation adults emerge during summer and are again polyphagous on several families of flowering plants before overwintering (Alford *et al.*, 2003 ; Williams, 2010).

Although pollen beetle-related yield losses can approach 70% in spring crops (Nilsson, 1987), winter crops have often developed beyond the susceptible bud stages by the time the beetles arrive (Williams, 2010), and control thresholds are breached infrequently (Ellis and Berry, 2012). Despite this, pollen beetles in winter OSR are frequently the target of spring insecticide treatments, usually pyrethroids (Williams, 2010 ; Garthwaite *et al.*, 2011), many of which are applied as prophylactic 'insurance' sprays (Thieme et al., 2010). This overuse has damaging effects on beneficial insects (Mänd *et al.*, 2010 ; Hanson *et al.*, 2015) and has led to increasingly widespread pyrethroid resistance, threatening to limit the arsenal of products remaining active against the pest (Thieme *et al.*, 2010; Zimmer and Nauen, 2011; Nauen *et al.*, 2012 ; Heimbach and Mueller, 2013).

The use of prophylactic treatments for pollen beetle control has been encouraged by the low cost of pyrethroid insecticide products, and application costs are often also minimal as treatments are frequently applied as sprayer tank mixes, at the same time as fungicide programs (Thieme et al., 2010). Although advice on pollen beetle monitoring and control thresholds is available to farmers (e.g. AHDB-HGCA, 2013), its uptake may be discouraged by costly labour intensive monitoring requirements and poor reliability due to variation in spatial distributions of the pest, which are often patchy (Cook *et al.*, 2004; Ferguson *et al.*, 2003 ; Gotlin Culjak *et al.*, 2016). Thus, the key to reducing prophylactic insecticide applications in OSR and improving the sustainability of the crop may lie with improved, less labour intensive methods of pollen beetle monitoring. The recent development of on-line weather-based decision support systems (DSSs) such as proPlant expert (Johnen and von Richthofen, 2013) represents progress towards this goal, by providing regional forecasts of migration risks which reduce monitoring requirements (Ferguson et al., 2016). However, as landscape factors such as woodlands (as overwintering sites) and oilseed crops in the previous and current seasons (as potential sources and alternative sinks, respectively) are known to influence local beetle abundance and crop damage (e.g. Valantin-Morison *et al.*, 2007; Zaller *et al.*, 2008b; Rusch *et al.*, 2012b ; Rusch *et al.*, 2013), the development of models based on both meteorological and landscape factors could be used to further refine such DSSs, and also to help determine optimal placement of traps for monitoring immigration into the crop.

In the current study, based on data from four years of pollen beetle monitoring on a total of 41 field sites, we model meteorological and landscape influences on pollen beetle immigration into the crop as measured by directional sticky trap catches, at both the single trap and field scales. We hypothesize that (i) temperature, rainfall and wind speed will affect trap catches by determining when conditions are optimal for beetle immigration into the crop, (ii) since the beetles may be expected to fly towards the crop using upwind anemotaxis, wind direction will affect the direction from which they enter, and (iii) that landscape features will interact with wind direction to affect trap catches by influencing beetle abundance at source.

2. MATERIALS AND METHODS

2.1. Field sites and sticky trapping

As part of a project to develop an integrated pest management strategy for pollen beetles (Cook et al., 2014), beetle immigration into a total of 178 winter OSR fields, distributed throughout the main UK arable cropping regions, was monitored by sticky trapping during the March–early May periods of 2008–2011. For the purposes of this study, data were analysed from a subset of 41 sites (4, 7, 18 and 12 sites from 2008, 2009, 2010 and 2011, respectively); (see Figure S1 and Table S1 in Supplementary material for further details). With the exception of sites at Rothamsted (Hertfordshire) and Woburn (Bedfordshire), where trapping was conducted by Rothamsted staff, trapping at all other sites was done by volunteer farmers or crop consultants. Some of the volunteers ran traps on more than one field of a farm, either in the same or different years. Sites were selected for analysis based on several criteria. The volunteers were required to have followed experimental protocols accurately, and provided good information on the locations of oilseed rape crops within a 1000 m radius of the traps for the year of sampling and the previous year. In order to be sure that the immigration period had been identified correctly, the number of beetles trapped per day was required to show a peak followed by a decline; sites with no clear peak were excluded.

At each site, either two or four yellow sticky traps (standard 'wetstick', 10 cm x 20 cm) (Oecos, Kimpton, Hertfordshire, UK) were placed on different sides of the field. At all sites, one trap was placed upwind and another downwind along the plane of an assumed WSW prevailing wind direction, and on 15 of the sites, two additional traps were placed in 'cross-wind' positions, at right angles to this plane (Figure 1a). The sticky traps were clipped onto a plastic mount and placed on top of an extendable metal pole so that they could be maintained just above crop canopy height. The traps were angled at 45° to the vertical as this orientation has been shown to be effective for trapping pollen beetles (Blight and Smart, 1999). Traps were placed 3 m into the crop from the field edge and were orientated to face outwards, in order to capture incoming beetles.

Trapping commenced on or shortly after 1st March each year and was continued until the crop reached early flowering (BBCH growth stage code 61, Lancashire et al., 1991). Where possible, traps were changed every 3–4 days, but in practice intervals depended on volunteer time availability, and were often longer (full range 2–21 days but with c. 80% of intervals < 7 days). The collected traps were covered with a polyethylene sheet and returned to Rothamsted for the pollen beetles to be counted. For each site, the sample representing peak beetle count was identified (from the total count from upwind and downwind traps only), and all samples up to and including that date were used for analyses, as representing the period of immigration into the crop. A total of 309 sticky trap samples were used in the analyses.

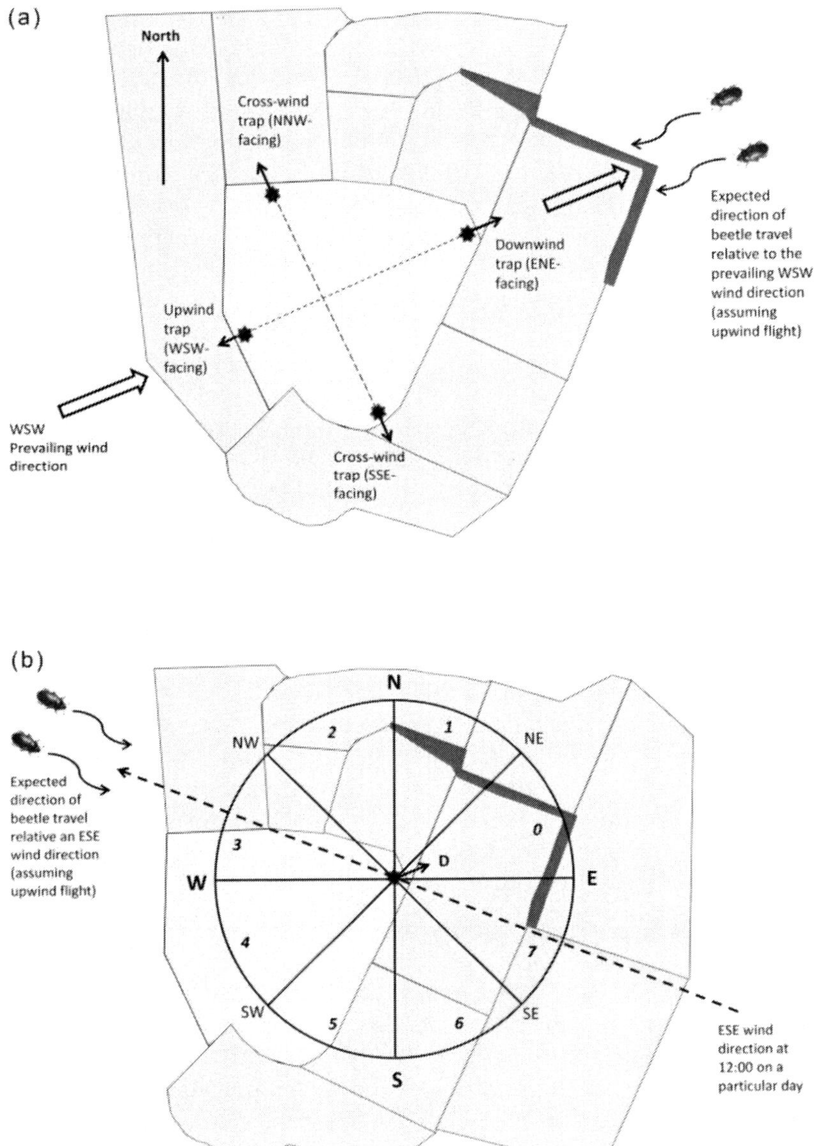

Figure 1. (a) Typical site layout showing directional sticky trap placement in oilseed rape crops relative to an assumed WSW prevailing wind direction (hollow arrows) and the boundaries of a sampled field (textured shading). Upwind (WSW-facing) and downwind (ENE-facing) traps (black stars with arrows to indicate facing direction) were placed 3 m into the crop on opposite sides of the field, along the plane of the WSW-to-ENE prevailing wind direction (short-dashed line). Fifteen of the sites had additional NNW-

and SSE-facing cross-wind traps, placed on a NNW–SSE plane (long-dashed line), perpendicular to that of the prevailing wind. When the wind is from the prevailing WSW direction, pollen beetles flying upwind are expected to enter the crop from the ENE. (b) Landscape mapping zones and assignment of wind and trap directions to specific directional segments or 'octants'. The circular area represents the 1000 m-radius zone within which landscape features were mapped in relation to a downwind trap (D; ENE-facing), and is divided into directional octants (labelled *0–7*). Traps were assigned to the octant corresponding to the direction that they were facing (in the case of the downwind trap shown, octant *0*). If the wind direction was from the ESE at 12:00 on a particular day (dashed straight arrow), ingress of upwind-flying beetles would be expected from a WNW (downwind) direction, and hence wind direction in this situation would be assigned to octant *3*. The difference between wind direction and trap direction can then be calculated in terms of number of octants (in this case there was a 3-octant difference). However, if the wind direction was from the WSW, as in (a), then both the downwind trap and wind direction would be assigned to octant *0* (trap and wind direction aligned, a difference of 0 octants). Note that for modelling at the individual trap scale, only landscape features from the octant facing the trap and its two immediate neighbours (in this case octants *0*, *1* and 7) were used.

2.2. Landscape data

Landscape features were digitally mapped within a 1000 m-radius around each trap using Google Earth (Google Inc., Mountain View, California, USA). The 1000 m scale was chosen as it lies within the range of scales over which previous studies had shown an influence of landscape features on pollen beetle abundance or herbivory (e.g. Thies *et al.*, 2003; Zaller *et al.*, 2008b ; Moser *et al.*, 2009). The features mapped were included on the basis of their potential to provide overwintering habitats for the beetles (woodlands) or act as potential barriers to their movement (hedgerows and lines of trees), or because of their importance as sites for feeding and reproduction (OSR crops – both in the year of trapping and previous year, and residential gardens). Permanent features not already shown on Ordnance Survey base maps (accessed via the UK Digimap service; http://digimap.edina.ac.uk) were identified from Google Earth satellite imagery, while information on the location of OSR crops in the vicinity was provided by the data contributors. ArcGIS 9.2 (ESRI, Redlands, California, USA) was used to extract areas of landscape features (or lengths of linear hedgerows and treelines) from within each of eight 45° sectors, or 'octants', that were created to subdivide the 1000 m-radius circular mapped zone into directional components (Figure 1b). As few fields in our study were particularly large, there was some overlap between landscape features present in sectors around upwind and downwind traps. However, the landscape variables were used as explanatory variables in our model and as the method does not require statistical independence within the explanatory variables, this was not considered a problem from a statistical point of view. A summary of the landscape data is given in Table S2.

2.3. Meteorological data

Meteorological data were obtained from the nearest UK Meteorological Office recording station to each field site, with the exception of Rothamsted and Woburn, where local on-farm data were available (see Table S1 for details of the meteorological stations and their approximate distances from the trap sites.) Daily summaries of the minimum and maximum temperature (°C), rainfall (mm), and wind speed (m/s) and direction were obtained for the duration of sampling at each site.

As maximum temperature is unlikely to be a reliable guide to the temperature over the day, an accumulated temperature measure was calculated. This used a saw-tooth approximation which assumed that the minimum and maximum temperatures occurred at 05:00 and 15:00 respectively, and used linear interpolation between these points. No beetles were found on traps when the maximum daily temperature was less than 10 °C, and so 10 °C was used as the baseline temperature. The accumulated temperature (degree-half-days; dhd) was calculated as the integral of the saw-tooth function during the period 06:00–18:00 (assumed daylight hours) when the interpolated temperature exceeded the baseline 10 °C (Figure S2). For days when the temperature did not reach 10 °C, the accumulated temperature was assigned as zero.

Although wind speed and direction data were available on four occasions throughout the day, (at 00:00, 06:00, 12:00 and 18:00), the reading at 12:00 was chosen as being most relevant to beetle activity as it occurred during the warmer part of the day most likely to coincide with immigration. For each day that a trap was running, wind direction at 12:00 was assigned to one of the octants used for landscape data. This octant corresponded to the expected direction of beetle ingress into the crop if beetles flew upwind, and therefore represented the direction that the wind was going *to*, (i.e. the 'downwind direction'), as opposed to the direction that it was coming *from* (i.e. the 'upwind direction' used in the standard definition of wind direction). For example, if the wind direction was from ESE then it was assigned to octant 3 (the WNW-facing octant opposite; Figure 1b). Each trap was assigned to the octant corresponding to the direction it was facing (out of the crop). For example, a downwind trap facing ENE would be assigned to octant 0. The difference between the actual downwind direction (variable) and the direction of the trap (fixed) was calculated in terms of number of octants (Figure 1b), taking the shortest route around the circle. For example, a trap facing ENE (octant 0) would have a deviation of 3 octants for either wind direction ESE (octant 3) or NNE (octant 5). A trap facing ENE (octant 0) with wind direction WSW (also octant 0) would have a deviation of 0 octants (trap and wind direction aligned).

Summaries of the weather variables are given in Figs. S3 (temperature), S4 (wind speed and direction), and S5 (rainfall).

2.4. Modelling

All analyses and modelling were done using GenStat (17th edition, VSN International, Hemel Hempstead, UK). Before building a model in terms of the explanatory variables, a simple exploratory model was used to ascertain the different sizes and sources of variability and the distribution of the data (details in Section S1 of Supplementary material). The residuals from this analysis and the mean squared residuals were inspected for indications of variance heterogeneity, in particular for any pattern of variance increasing as a quadratic function of the mean, which would be more consistent with a negative binomial distribution than with a Poisson distribution (ver Hoef and Boveng, 2007). The negative binomial distribution was parameterised with mean μ, aggregation parameter k and a variance function of the form:

$$V(\mu) = \mu\left(1 + \frac{\mu}{k}\right)$$

2.4.1. Modelling weather and landscape variables

To explore the effect of weather and landscape on trap catches, a generalised linear mixed model (GLMM) with a composite link function and a negative binomial distribution was used. Since there was a discrepancy between the scale of the weather data (daily) and the sample period for each trap (2–21 days; see Section 2.1), the modelling process required a mechanism to model daily beetle counts as a function of daily weather and then accumulate these counts over the sample period to predict total trap count; this was achieved via the composite link function. As the weather and landscape variables apply to days within fields and traps within fields, respectively, random terms corresponding to fields, days within fields and traps within fields were required to avoid pseudo-replication and ensure that fixed terms were tested at the correct level within the data structure;

this was achieved by use of a GLMM. The composite link method of Thompson and Baker (1981) was implemented within the framework of a GLMM with penalised quasi-likelihood (PQL) estimation (Breslow and Claydon, 1993). The composite link allows accumulation of counts over the sampling period. Its implementation in the GLMM framework models allows the inclusion of random terms to reflect the structure of the data with an appropriate distribution for the observations. The presence of explanatory terms in the model is examined using *t*-tests. The model can operate at the scale of the field (accumulating over traps) or the individual trap. Explanatory variables were classed as weather (accumulated temperature, daytime rainfall, wind speed) or landscape (area of woodlands, gardens, OSR crops in the current and previous year, and lengths of treelines and hedges). In addition, an explanatory variable was included in all models to adjust for higher counts from fields on the Rothamsted farm, identified from the initial exploratory model (see Section S1 of the Supplementary material).

2.4.2. Modelling at the field scale

The total count from the upwind and downwind traps in each field was used as the response when modelling at the field scale. Landscape variables were averaged over the upwind and downwind traps, and field was included as the only random term. The three meteorological variables were added to the model first, as linear and quadratic functions. Non-significant terms were then dropped from this interim model using backwards selection. The landscape variables were then each added individually into the model. Any landscape variables found to be significant were then added as a group to the model containing meteorological variables, and backwards selection repeated to drop any non-significant terms. Predictions for each variable selected were made at the scale of the daily total field trap catch, using fixed specified values of other variables. Confidence intervals for predictions were formed on the log-scale and then back-transformed.

2.4.3. Modelling at the trap scale

Beetle counts from individual traps (upwind, downwind and cross-wind) from each field were used as the response when modelling at the trap scale. Relative wind direction (as defined in Section 2.3 and Figure 1b) was added to the set of explanatory variables used for modelling at the field scale. Based on the assumption that the beetles take a reasonably direct route to the field, landscape data from only the three octants facing each trap (i.e. the octant assigned to the trap plus one either side; Figure 1b) were used to calculate the landscape explanatory variables. Field, day within field and trap within field were included as random terms in the model.

The modelling procedure was the same as that used for field-scale modelling (section 2.4.2), but included the additional meteorological variable, relative wind direction, parameterised as a factor with 5 levels (direction deviation of 0, 1, 2, 3, or 4 octants).

3. RESULTS

The exploratory model indicated that a negative binomial distribution with aggregation parameter $k = 5$ gave acceptable residuals (Section S1 of Supplementary material) so this distribution was adopted for the models.

3.1. Modelling at the field scale

The modelling procedure identified the presence of a field site on the Rothamsted estate, wind speed, accumulated temperature, and the area of OSR crops in the previous year as significant explanatory terms for total field trap counts (Table 1). Details of the fitted model are given in Section S2.1 of the Supplementary material. The estimated values of fixed effects in the final model are shown in Table 1.

Table 1. Estimated parameters from the field-scale model. The t-ratio is calculated as parameter estimate/SE. F probability is the observed significance level from an approximate F-test on sequentially adding each term into the fixed model.

Term	Parameter	Estimate	SE	t-ratio	F prob
Constant (1)	α	0.6967	0.4819	1.445	0.154
Presence on Rothamsted estate (0/1) (R)	β_1	2.1920	0.3364	6.516	<0.001
Linear wind speed (w, m/s)	β_2	−0.4803	0.0404	−11.885	<0.001
Linear accumulated temperature (a, dhd)	β_3	3.0094	0.3098	9.715	<0.001
Quadratic accumulated temperature (a^2)	β_4	−0.3277	0.0727	−4.510	<0.001
Linear area of OSR crops in previous year (p, ha)	β_5	0.013789	0.006957	1.982	0.049

There was a positive effect on daily total field catches of being located on the Rothamsted estate, giving a predicted ratio of beetle numbers at Rothamsted relative to non-Rothamsted fields of 8.95, ie. 8.95 times more beetles at Rothamsted sites if all other variables were equal, with 95% confidence interval (4.54, 17.67). Weather variables were more important (i.e. with larger t-ratios) as explanatory variables than landscape features (Table 1). Wind speed (m/s) showed a statistically significant negative linear relationship with log daily field catch (Table 1) shown in Figure 2a

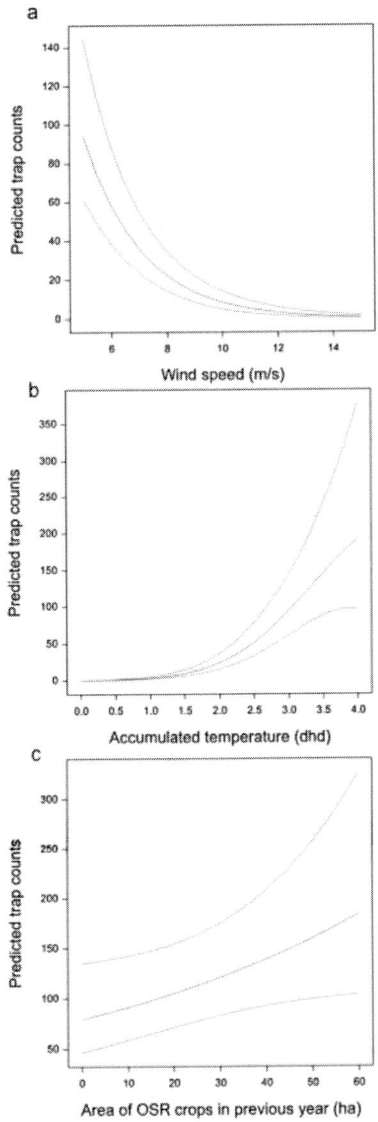

Figure 2. Predicted daily field catch of pollen beetles in traps placed in oilseed rape (OSR) crops (total of upwind and downwind traps) with 95% confidence limits for (a) wind speed (m/s) with 3 dhd accumulated temperature and 12 ha OSR in landscape in previous year, (b) accumulated temperature (dhd) with wind speed of 5 m/s and 12 ha OSR in landscape in previous year, (c) the area of OSR crops within a 1000 m radius with wind speed of 5 m/s and 3 dhd accumulated temperature.

with 95% confidence limits; field catches are expected to increase greatly at lower wind speeds. Accumulated temperature (dhd) showed a statistically significant positive linear relationship with log daily field catch, with a negative quadratic component (Table 1) shown in Figure 2b. Field catches are expected to increase as accumulated temperature increases up to about 4.5 dhd, at which point they are predicted to plateau and then decline. Note that 4.5 dhd corresponds to a constant daytime temperature of 14.5 °C, and is in the top 0.5% of the observed sample so these predictions are based on little data. Rainfall was not a statistically significant variable and the only statistically significant landscape variable (using a P = 0.05 threshold) was a positive linear relationship on the log scale with area of OSR grown in the previous year (Figure 2c).

3.2. Modelling at the trap scale

The modelling procedure identified the presence of a field site on the Rothamsted estate, wind speed, accumulated temperature, a deviation of 2–4 octants between downwind direction and trap direction, and the area of OSR crops in the previous year within the 3 octants facing the trap as important explanatory terms for trap counts (Table 2). Details of the fitted model are given in Section S2.2 of the Supplementary material. Estimates of the fixed effects from the final model are shown in Table 2.

Table 2. Estimated parameters from the trap-scale model. The t-ratio is calculated as parameter estimate/SE. The F probability is the observed significance level from an approximate F-test on sequentially adding each term into the fixed model.

Term	Parameter	Estimate	SE	t ratio	F prob
Constant (1)	α	−3.843	0.4234	−1.054	0.293
Presence on Rothamsted estate (0/1) (R)	β_1	2.099	0.3210	6.540	<0.001
Linear wind speed (w, m/s)	β_2	−0.531	0.0403	−13.174	<0.001
Linear accumulated temperature (a, dhd)	β_3	3.795	0.3233	11.737	<0.001
Quadratic accumulated temperature (a^2)	β_4	−0.548	0.0857	−6.396	<0.001
Linear area of OSR crops in previous year (p, ha)	β_5	−0.01903	0.007893	−2.411	0.018
Deviation of 1 octant in wind direction ($D = 1$)	δ_1	0.019	0.1568	0.123	0.902
Deviation of 2 octants in wind direction ($D = 2$)	δ_2	−0.387	0.1553	−2.494	0.013
Deviation of 3 octants in wind direction ($D = 3$)	δ_3	−1.166	0.1784	−6.534	<0.001
Deviation of 4 octants in wind direction ($D = 4$)	δ_4	−0.729	0.1488	−4.901	<0.001

There was a positive effect on daily total trap catches of being located on the Rothamsted estate, giving a predicted ratio of beetle numbers at Rothamsted relative to non-Rothamsted fields of 8.16, with 95% confidence interval (4.25, 15.68). Weather variables were again more important (i.e. with larger t-ratios) as explanatory variables than landscape features (Table 2). Wind speed (m/s) again showed a statistically significant

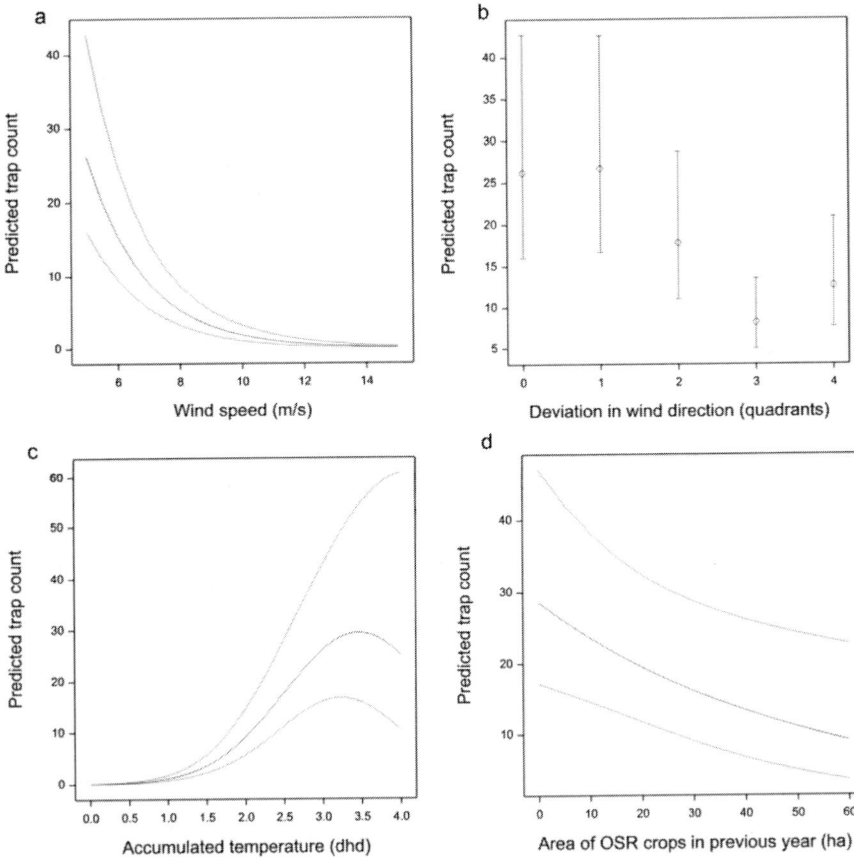

Figure 3. Predicted daily trap catch of pollen beetles with 95% confidence limits for (a) wind speed (m/s) with 3 dhd accumulated temperature, 4.5 ha oilseed rape (OSR) in landscape in previous year and trap aligned with wind direction, (b) the deviation between wind and trap direction in octants with wind speed of 5 m/s, 3 dhd accumulated temperature and 4.5 ha OSR in landscape in the previous year, (c) accumulated temperature (dhd) with windspeed of 5 m/s, 4.5 ha OSR in landscape in the previous year and trap aligned with wind direction, (d) the area of OSR crops within a 1000 m radius with windspeed of 5 m/s, 3 dhd accumulated temperature and trap aligned with wind direction.

negative linear relationship with log daily field catch (Table 2) shown in Figure 3a with 95% confidence limits. Deviations of more than 1 octant between wind and trap direction were associated with lower trap catches (Table 2), although trap catches were slightly higher for 4 octants deviation (trap facing upwind) than for 3 octants deviation, which gave the lowest trap catches (shown in Figure 3b). Accumulated temperature (dhd) showed a statistically significant positive linear relationship with log daily field catch, with a negative quadratic component (Table 2), shown in Figure 3c. Relative field catches are expected to increase as accumulated temperature increases up to about 3.5 dhd (equivalent to a constant daytime temperature of 13.5 °C), at which point it plateaus and then decreases. Again rainfall was not significant and the only landscape variable showing statistical significance was a negative linear relationship on the log scale with area of OSR grown in the previous year (Figure 3d).

4. DISCUSSION

In our study, pollen beetle catches on sticky traps during beetle immigration into winter OSR crops were influenced by a combination of meteorological and landscape factors. At both field and single trap scales, meteorological influences, particularly wind speed and accumulated temperature, were dominant over landscape features as explanatory variables for trap catch, implying that the model could be used to predict periods of high beetle immigration risk, based on accumulated temperature and wind speed alone. Our results represent the most direct evidence to date that pollen beetles use upwind anemotaxis at a landscape scale during immigration into the crop. They also highlight that positioning of monitoring traps relative to field boundaries and wind direction is important in determining trap efficiency, and suggests that the area of OSR grown within the surrounding landscape in the previous season affects the abundance of beetles immigrating into OSR crops.

It is important to note that the models developed in this study are regression-based and are built on correlation between the explanatory variables and responses. Caution is therefore required when attributing causal effects to explanatory variables, as they may simply be representing the causal effect of a third variable that they are correlated with. At both the field and trap scales, larger numbers of pollen beetles were trapped at Rothamsted compared with other sites. These results were not unexpected due to the estate's history of growing OSR crops without insecticide treatment, and both winter- and spring-sown OSR, but the magnitude of the differences (predicted catches at Rothamsted of c. 8–9 times those of other sites) was surprising. Over the last decade Rothamsted has, however, hosted many experiments investigating the trap cropping potential of turnip rape (*Brassica rapa* L.) for pollen beetle control. At the green bud stage, turnip rape is more attractive to pollen beetles than OSR, and the presence of a turnip rape border surrounding the crop can reduce within-crop beetle infestation to below spray threshold levels (Cook *et al.*,

2004; Cook *et al.*, 2006 ; Cook *et al.*, 2007), but the early flowering of winter-sown turnip rape (if left untreated with insecticide) can lead to a large proportion of the pollen beetle larval population escaping parasitisation (Skellern et al. in prep), and a model developed by Vinatier et al. (2012) has suggested that turnip rape trap cropping could increase densities of the pest, especially if untreated. This, together with a very high proportion of gardens in the Rothamsted area relative to other sites could explain the larger populations of beetles found.

The insignificant influence of rainfall on trap catches, once accumulated temperature and wind speed had been accounted for, was unexpected as generally the activity of pollen beetles in the field appears reduced in wet conditions (S.M Cook and M. Skellern, personal observations) and precipitation of 1 mm was selected as a cut off value in previous models (Junk et al., 2016). However, there were relatively few days with >1 mm rain in our dataset (Figure S5) and rainfall may tend to be associated with lower temperatures. In addition, other factors such as relative humidity and barometric pressure can affect insect flight activity (Fournier *et al.*, 2005; Zhang *et al.*, 2008; Tansey *et al.*, 2010 ; Liu *et al.*, 2011). It is possible, for example, that if elevated post-rainfall humidity conditions favour beetle migration (to our knowledge the influence of humidity on pollen beetle flight activity has not yet been characterised), this could have offset any negative influence of rainfall on trap catches.

The observation that pollen beetles were not found on traps until the maximum temperature rose above 10 °C is consistent with recent laboratory findings that the 10% flight temperature threshold for the pest lies within the range 10.9–12.5 °C (Ferguson et al., 2015), and with lower limits for field-observed flight of 10.2 °C (Laska and Kocourek, 1991) but higher than that predicted (8.0° C) by immigration models for beetles in Luxembourg (Junk et al., 2016). The quadratic effect of accumulated temperature indicated an increase in predicted beetle catch as accumulated temperature rose from 0 to 4.5 dhd for the field-scale model then predicted a decline. The trap-scale model showed a similar increase with accumulated temperature up to around 3.5 dhd, but then plateaued and predicted the decline at a slightly lower accumulated temperature. While pollen beetles show a sigmoidal temperature-response curve with 50% flight temperature threshold estimates of 15.5–16.2 °C (Ferguson et al., 2015), our predicted decrease in trap catch with higher temperatures for the trap-scale model should be regarded with caution; there were few observations above 4 dhds, so predictions for these higher temperatures may be unreliable, as indicated by the very wide confidence limits (Figs. 2 b and 3 c). It should also be borne in mind that during spring, higher temperatures are more likely to occur later in the season, towards the end of beetle immigration, when the available pool of beetles yet to migrate would be diminished.

Hence, the observed catch may be lower than it would have been for the same environmental conditions earlier in the season.

Previous studies have indicated the tendency of pollen beetles to locate OSR crops by odour-driven upwind anemotaxis (Evans and Allen-Williams, 1994; Williams *et al.*, 2007 ; Moser *et al.*, 2009), but because we have used directional sticky traps placed at the field edge, we show this for the first time directly at the point of entry into an OSR crop – at a landscape scale. The largest numbers of beetles were caught over periods when traps were aligned with the downwind direction, suggestive of upwind anemotaxis. Accordingly, as the prevailing wind came from a WSW–SSW direction, those traps facing in a downwind direction (ENE) caught more beetles than upwind-facing (WSW) traps. These results indicate that for optimal trapping efficiency, pollen beetle monitoring traps should be placed on the downwind side of OSR crops, facing downwind relative to the prevailing wind direction. The tendency of beetles to fly upwind was clearly not an exclusive pattern, however, as beetles were still caught on traps facing other directions, but their numbers were lower, and tended to decrease as the trap faced away from the downwind direction. Interestingly, beetle numbers from traps where the deviation in trap–downwind direction was 4 octants (i.e. 180°) were higher than those from the cross-wind direction where the deviation was 3 octants (i.e. 135°); this observation is consistent with beetles sometimes travelling with the wind rather than flying upwind.

Previous studies that have investigated landscape effects of surrounding OSR area on pollen beetle abundance or herbivory in the same year have had variable results, but the majority, in agreement with the current study, have shown no influence (Thies and Tscharntke, 1999; Thies *et al.*, 2003; Thies *et al.*, 2008; Gladbach *et al.*, 2011; Scheid *et al.*, 2011 ; Rusch *et al.*, 2013). Others have shown negative relationships, often attributed to dilution effects (Zaller *et al.*, 2008a; Moser *et al.*, 2009 ; Schneider *et al.*, 2015), or positive relationships (Valantin-Morison et al., 2007). These discrepancies could arise from unknown differences among study regions (Rusch et al., 2013) or be attributable to sampling methodology differences, particularly in relation to scale, or to the temporal dynamics of the relationship between the beetles and their host crop. Indeed, Beduschi et al. (2015) showed that the effect of surrounding OSR area on beetle abundance changed with time, from being negative during peak flowering, to positive after flowering, probably reflecting dilution and crowding effects, respectively. In the current study, however, trapping took place mostly during the green–yellow bud stages, before flowering, when inward migrating beetle populations would have been concentrated around crop edges; any potential dilution effects of crop area on beetle abundance may not manifest until the later flowering stages, when the beetles are likely to be more dispersed.

Relatively few studies have considered the landscape-scale effects of OSR area in the previous season on the current season's pollen beetle abundance. Schneider et al. (2015) showed that an inter-annual increase in landscape proportions of OSR resulted in beetle dilution effects, while Thies et al. (2008) observed no influence of this variable on pollen beetle herbivory. Beduschi et al. (2015) found that landscape proportions of OSR negatively influenced beetle abundance one or two years later, and that this effect was mediated through changes in parasitism. Although the present study shows landscape-scale effects of the previous season's OSR area on the abundance of beetles immigrating into OSR crops, there were inconsistencies between the trap- and field scale models in the direction of these effects. While the field-scale model showed a positive influence of this variable, for the trap-scale model the relationship was negative. The landscape data used for the two models differed in that only landscape features facing the trap (in the same or neighbouring octants) were used in the trap-scale model whereas all surrounding landscape was used in the field-scale model, so we might not expect complete agreement. However, it is possible that these differences are the result of spurious correlation in one or both models as the distributions of previous season's OSR crop areas and of the trap counts were both positively skewed, with many more occurring low than high values. If there was no relationship, there would likely be many fewer high trap counts for the few large areas simply because of the smaller number of observations. Exclusion of the top 10% of the distribution of previous season's OSR crop areas from the analysis (analysis not shown) meant that the negative trap-scale relationship became non-significant, which supports this notion. However, excluding the top 10% of this distribution for the field-scale analysis increased the significance of the positive relationship, suggesting that, in contrast with the results of previous studies, this positive relationship may be real. Factors such as differences in parasitism rates and the extent to which regional scale landscape structure necessitates long-distance migration (Rusch et al., 2013) may explain these differences. For example, in landscapes that are reasonably balanced in terms of breeding and overwintering sites (limited migration necessary), it might be expected that, particularly where parasitism is low, beetle abundance may reflect surrounding OSR area in the previous year. By contrast, in landscapes that promote long distance migration because they are more compartmentalised at the regional scale (i.e. containing some areas with large open fields and other distinct areas with more complex landscapes), relationships between beetle abundance and the previous season's OSR area are unlikely to be found, particularly at relatively fine (e.g. 1000 m) sampling scales.

Increasing proportions of woodland in the landscape are often associated with higher pollen beetle densities or damage (Valantin-Morison *et al.*, 2007; Zaller *et al.*, 2008b; Zaller *et al.*, 2009; Rusch *et al.*, 2012b ; Rusch *et al.*, 2013), probably due to their role as overwintering sites which later become the springtime source of emerging beetles. In the current study, however, the area of woodland in the surrounding landscape was not found to be an important determinant of trap catch, for either model. The reasons for this are unclear. As observed by Zaller et al. (2008b), the differing effects of landscape-related features among studies may relate to differences in methodological approach. The scale over which landscape features were mapped is important; if the scale was too large then the inclusion of irrelevant information (noise) may obscure relationships, or if the scale was too small, then relevant information would be missing. The 1000 m scale chosen is unlikely to have been inappropriate, however, as it was not dissimilar to those used by other studies that have shown effects of landscape features on pollen beetle abundance or damage (e.g. Thies *et al.*, 2003; Zaller *et al.*, 2008a; Moser *et al.*, 2009; Beduschi *et al.*, 2015; Schneider *et al.*, 2015). The factors affecting beetle immigration may also be subtly different to those affecting beetle abundance on the crop, or crop damage, particularly when factors such as natural enemies and crop management are considered. Grassland habitats, which were not considered in the present study, can be important as pollen beetle overwintering sites (Rusch et al., 2012a) and as landscape-scale determinants of beetle abundance (Rusch et al., 2013), but Rusch et al. (2012a) also found that emerging beetles were more associated with local habitat characteristics such as low soil moisture and a thick litter layer than with habitat type *per se*. It is possible, particularly considering the wide geographical range of our study, that a range of habitats were providing conditions suitable for beetle overwintering; woodlands may have been important as overwintering sites in some regions, but in others different habitat types may have provided more suitable conditions, leading to a diminished effect of woodland area overall.

In conclusion, our study has shown that wind speed and accumulated temperature were more important than landscape variables in predicting the abundance of pollen beetles immigrating into OSR fields. The efficiency of pollen beetle monitoring traps could be optimised by placing them on the downwind side of a crop, facing downwind relative to the prevailing wind direction. The area of OSR crops grown in the surrounding landscape during the previous season was positively related to trap catch at the field scale and could potentially contribute to assessment of potential pest pressure for individual OSR crops. While proPlant.expert DSS already incorporates wind speed and accumulated temperature into its underlying model and has been shown to accurately predict periods of pollen beetle migration risk (Ferguson et al., 2016), optimal placement of monitoring traps could complement the DSS in terms of further reducing monitoring effort and costs. Inspection of the traps when the DSS has indicated a period of high migration risk would determine whether immigration into a specific crop has actually begun, and whether inspection of the crop is warranted, particularly as Ferguson et al. (2016), using the same data pool as the present study, found that at >91% of sites beetles were caught on monitoring traps before they were observed in the crop. The proPlant DSS is now freely available to UK growers and widespread DSS uptake by OSR growers is promising (Ferguson and Cook, 2014). This, in combination with the use of optimally-placed monitoring traps could help to reduce prophylactic insecticide applications, thus reducing impacts on non-target organisms and ameliorating the risk of further resistance development.

ACKNOWLEDGEMENTS

This work was supported under the UK Department for Environment, Food and Rural Affairs[1] Sustainable Arable LINK programme (Project LK09108) by funding from the UK Health and Safety Executive (Chemicals Regulation Directorate)[2] and the UK Agriculture and Horticulture Development Board[3] (Project RD-2007-3394), and received financial and/or in kind support the following Project Partners: Association of Independent Crop Consultants[4]; Bayer CropScience[5]; Elsoms Seeds Ltd[6], KWS UK[7], Limagrain UK[8], Monsanto-UK[9], Oecos Ltd[10]; proPlant GmbH[11]; Saaten- Union UK[12], Syngenta Crop Protection UK[13], Syngenta Seeds[14]; VSN International[15]. Academic partners included Queen Mary University of London[16] and Rothamsted Research[17], which receives strategic funding from the UK Biotechnology and Biological Sciences Research Council (BBSRC). SMC was part-funded by BBSRC Project BB/J004286/1. We are grateful to Rod Blackshaw (Plymouth University) who acted as the LINK Committee Project Officer and to all Project Partners (affiliations indicated by superscript) for their generously given expertise and time, and for excellent open and collaborative discussions during project meetings that ensured the success of the project: Eileen Bardsley[5], Jemillah Bailey[1,3] Jo Bowman[8], Sean Burns[14], Matthew Clarke[9], Jackie Davies[10], Thomas Döring[16], Andrew Ferguson[17], Carl Gibbard[7], Andreas Johnen[11], Richard Jennaway[12], Richard Meredith[5], Darren Murray[15], Mark Nightingale[6], Nigel Padbury[14], Colin Patrick[12], Julia-Sophie von Richthofen[11], George Rothschild[2], Lesley Smart[17], Peter Taylor[4], Michael Tait[13] and Peter Werner[7]. Special thanks to all the volunteers who gave generously of their time to run monitoring traps, in particular: Andy Barr, Tim Elwess, Will Green, John Hawthorne, Robert Hobill, Allan Lock, Ian Mennell, Heledd Parry, Colin Patrick, Doug Pickup, Joe Scott, Duncan Wakelin, Neil Watson, and for Bayer CropScience: Steve Ashberry, Roger Bradbury, Ben Giles, James Lewis and Laurence Oades. We also thank the two anonymous reviewers whose comments helped to improve this manuscript.

REFERENCES

1. AHDB-HGCA, 2013. Monitoring and Control of Pollen Beetle in Oilseed Rape. Information Sheet 18/Spring. , pp. 2013.
2. Alford, D.V., Nilsson, C., Ulber, B., 2003. Insect pests of oilseed rape crops. In: Alford, D.V. (Ed.), Biocontrol of Oilseed Rape Pests. Blackwell Science, Oxford, pp. 9–42.
3. Beduschi, T., Tscharntke, T., Scherber, C., 2015. Using multi-level generalized path analysis to understand herbivore and parasitoid dynamics in changing landscapes. Landscape Ecol. 30, 1975–1986.
4. Blight, M.M., Smart, L.E., 1999. Influence of visual cues and isothiocyanate lures on capture of the pollen beetle, Meligethes aeneus in field traps. J. Chem. Ecol. 25, 1501–1516.
5. Breslow, N.E., Claydon, D.G., 1993. Approximate inference in generalized linear mixed models. J. Am. Stat. Assoc. 88, 9–25.
6. Cook, S.M., Watts, N.P., Hunter, F., Smart, L.E., Williams, I.H., 2004. Effects of a turnip rape trap crop on the spatial distribution of Meligethes aeneus and Ceutorhynchus assimilis in oilseed rape. IOBC/WPRS Bull. 27, 199–206.
7. Cook, S.M., Smart, L.E., Martin, J.L., Murray, D.A., Watts, N.P., Williams, I.H., 2006. Exploitation of host plant preferences in pest management strategies for oilseed rape (Brassica napus). Entomol. Exp. Appl. 119, 221–229.
8. Cook, S.M., Rasmussen, H.B., Birkett, M.A., Murray, D.A., Pye, B.J., Watts, N.P., Williams, I.H., 2007. Behavioural and chemical ecology underlying the success of turnip rape (Brassica rapa) trap crops in protecting oilseed rape (Brassica napus) from the pollen beetle (Meligethes aeneus). Arthropod-Plant Interact. 1, 57–67.
9. Cook, S.M., Skellern, M.P., Döring, T.F., Pickett, J.A., 2013. The potential for manipulation of petal colour in control strategies for the pollen beetle (Meligethes aeneus). Arthropod Plant Interact. 7, 249–258.
10. Cook, S.M., Ferguson, A.W., Skellern, M.P., Watts, N.P., Martin, J.L., Smart, L.E., Woodcock, C.M., Pickett, J.A., 2014. Developing an integrated pest management strategy for pollen beetles in winter oilseed rape: results from the defra SA-LINK project LK09108. IOBC/WPRS Bull. 104, 96.
11. Ellis, S., Berry, P., 2012. Re-evaluating Thresholds for Pollen Beetle in Oilseed Rape. HGCA publication PR495. AHDB-HGCA. Eurostat, 2016a. Agriculture Database. http://ec.europa.eu/eurostat/web/ agriculture/data/database. (last accessed October 2016). Eurostat, 2016b. Eurostat Yearbook: Agricultural Products. http://ec.europa.eu/ eurostat/statistics-explained/index.php/Agricultural_ products. (last accessed October 2016).
12. Evans, K.A., Allen-Williams, L.J., 1994. Laboratory and field response of the pollen beetle Meligethes aeneus to the odour of oilseed rape. Physiol. Entomol.19, 285–290.
13. Ferguson, A.W., Cook, S.M., 2014. Results of a small survey amongst farmers and advisers on their evaluation of the proPlant pollen beetle migration tool and its influence on their practice. IOBC/WPRS Bull. 104, 97–104.
14. Ferguson, A.W., Klukowski, Z., Walczak, B., Clark, S.J., Mugglestone, M.A., Perry, J.N., Williams, I.H., 2003. Spatial distribution of insect pests in oilseed rape: implications for integrated pest management. Agric. Ecosyst. Environ. 95, 509– 521.
15. Ferguson, A.W., Nevard, L.M., Clark, S.J., Cook, S.M., 2015. Activity-temperature relationships in Meligethes aeneus: Implications for pest management. Pest Manag. Sci. 71, 459–466.
16. Ferguson, A.W., Skellern, M.P., Johnen, A., von Richthofen, J.-S., Watts, N.P., Bardsley, E., Murray, D.A., Cook, S.M., 2016. The potential of decision support systems to improve risk assessment for pollen beetle management in oilseed rape. Pest Manag. Sci. 72, 609–617.

17. Fournier, F., Pelletier, D., Vigneault, C., Goyette, B., Boivin, G., 2005. Effect of barometric pressure on flight initiation by Trichogramma pretiosum and Trichogramma evanescens (Hymenoptera: Trichogrammatidae). Environ. Entomol. 34, 1534–1540.
18. Free, J.B., Williams, I.H., 1978. Responses of the pollen beetle Meligethes aeneus, and the seed weevil Ceuthorhynchus assimilis, to oilseed rape Brassica napus, and other plants. J. Appl. Ecol. 15, 761–774.
19. Garthwaite, D.G., Thomas, M.R., Parrish, G., Smith, L., Chippindale, C., Pietraville, S., 2011. Pesticide Usage Survey Report 235 Arable Crops in the United Kingdom 2010. Defra, London.
20. Gladbach, D.J., Holzschuh, A., Scherber, C., Thies, C., Dormann, C.F., Tscharntke, T., 2011. Crop-noncrop spillover: arable fields affect trophic interactions on wild plants in surrounding habitats. Oecologia 166, 433–441.
21. Gotlin Culjak, T., Pernar, R., Juran, I., Anci 9 ı c, M., Bažok, R., 2016. Impact of oilseed rape crop management systems on the spatial distribution of Brassicogethes aeneus (Fabricius 1775): Implications for integrated pest management. Crop Prot. 89, 129–138.
22. Goulson, D., 2013. An overview of the environmental risks posed by neonicotinoid insecticides. J. Appl. Ecol. 50, 977–987.
23. Hanson, H.I., Smith, H.G., Hedlund, K., 2015. Agricultural management reduces emergence of pollen beetle parasitoids. Agric. Ecosyst. Environ. 205, 9–14.
24. Heckel, D.G., 2012. Insecticide resistance after silent spring. Science 337,1612–1614.
25. Heimbach, U., Mueller, A., 2013. Incidence of pyrethroid-resistant oilseed rape pests in Germany. Pest Manag. Sci. 69, 209–216.
26. Johnen, A., von Richthofen, J.-S., 2013. The decision-support system proPlant expert: a computer-based tool for integrated pest management in Europe. IOBC/WPRS Bull. 96, 99–105.
27. Junk, J., Jonas, M., Eickermann, M., 2016. Assessing meteorological key factors influencing crop invasion by pollen beetle (Meligethes aeneus F.) past observations and future perspectives. Meteorol. Z. 25, 357–364.
28. Lancashire, P.D., Bleiholder, H., Vandenboom, T., Langeluddeke, P., Stauss, R., Weber, E., Witzenberger, A., 1991. A uniform decimal code for growth-stages of crops and weeds. Ann. Appl. Biol. 119, 561–601.
29. Laska, P., Kocourek, F., 1991. Monitoring of flight activity in some crucifer-feeding pests by means of yellow water traps. Acta Entomol. Bohemos. 88, 25–32.
30. Liu, Z.F., McNeil, J.N., Wu, K.M., 2011. Flight mill performance of the lacewing Chrysoperla sinica (Neuroptera: Chrysopidae) as a function of age, temperature, and relative humidity. J. Econ. Entomol. 104, 94–100.
31. Mänd, M., Williams, I.H., Viik, E., Karise, R., 2010. Oilseed rape, bees and integrated pest management. In: Williams, I.H. (Ed.), Biocontrol-based Integrated Managemant of Oilseed Rape Pests. Springer-Verlag, London, pp. 357–379.
32. Moser, D., Drapela, T., Zaller, J.G., Frank, T., 2009. Interacting effects of wind direction and resource distribution on insect pest densities. Basic Appl. Ecol. 10, 208–215.
33. Nauen, R., Zimmer, C.T., Andrews, M., Slater, R., Bass, C., Ekbom, B., Gustafsson, G., Hansen, L.M., Kristensen, M., Zebitz, C.P.W., Williamson, M.S., 2012. Target-site resistance to pyrethroids in European populations of pollen beetle Meligethes aeneus. F. Pestic. Biochem. Phys. 103, 173–180.
34. Nilsson, C., 1987. Yield losses in summer rape caused by pollen beetles (Meligethes spp.). Swed. J. Agric. Res. 17, 105–111.
35. Nilsson, C., 1988. The pollen beetle (Meligethes aeneus) in winter and spring rape at Alnarp 1976–1978: I. Migration and sex ratio. Vaxtskyddsnotiser 52, 134–139.
36. Ouvrard, P., Hicks, D.M., Mouland, M., Nicholls, J.A., Baldock, K.C., Goddard, M.A., Kunin, W.E., Potts, S.G., Thieme, T., Veromann, E., Stone, G.N., 2016. Molecular taxonomic analysis of the plant associations of adult pollen beetles (Nitidulidae

Meligethinae), and the population structure of Brassicogethes aeneus. Genome 59, 1101–1116.

37. Robinson, R.A., Sutherland, W.J., 2002. Post-war changes in arable farming and biodiversity in Great Britain. J. Appl. Ecol. 39, 157–176.
38. Roubos, C.R., Rodriguez-Saona, C., Isaacs, R., 2014. Mitigating the effects of insecticides on arthropod biological control at field and landscape scales. Biol. Control. 75, 28–38.
39. Rusch, A., Valantin-Morison, M., Roger-Estrade, J., Sarthou, J.P., 2012a. Local and landscape determinants of pollen beetle abundance in overwintering sites. Agric. For. Entomol. 14, 37–47.
40. Rusch, A., Valantin-Morison, M., Roger-Estrade, J., Sarthou, J.P., 2012b. Using landscape indicators to predict high pest infestations and successful natural pest control at the regional scale. Landscape Urban Plan. 105, 62–73.
41. Rusch, A., Valantin-Morison, M., Sarthou, J.P., Roger-Estrade, J., 2013. Effect of crop management and landscape context on insect pest populations and crop damage. Agric. Ecosyst. Environ. 166, 118–125.
42. Scheid, B.E., Thies, C., Tscharntke, T., 2011. Enhancing rape pollen beetle parasitism within sown flower fields along a landscape complexity gradient. Agric. For. Entomol. 13, 173–179.
43. Schneider, G., Krauss, J., Riedinger, V., Holzschuh, A., Steffan-Dewenter, I., 2015. Biological pest control and yields depend on spatial and temporal crop cover dynamics. J. Appl. Ecol. 52, 1283–1292.
44. Skellern, M.P., Clark, S.J., Ferguson, A.W., Watts, N.P., Cook S.M. Banker plant bonuses? The benefits and risks of including plants in field margins to promote conservation biocontrol of specialist pests, (In prep.). Stoate, C., Boatman, N.D., Borralho, R.J., Rio Carvalho, C., de Snoo, G.R., Eden, P., 2001. Ecological impacts of arable intensification in Europe. J. Environ. Manage. 63, 337–365.
45. Tansey, J.A., Dosdall, L.M., Keddie, A., Olfert, O., 2010. Flight activity and dispersal of the cabbage seedpod weevil (Coleoptera: Curculionidae) are related to atmospheric conditions. Environ. Entomol. 39, 1092–1100.
46. Thieme, T., Heimbach, U., Muller, A., 2010. Chemical control of insect pests and insecticide resistance in oilseed rape. In: Williams, I.H. (Ed.), Biocontrol-based Integrated Management of Oilseed Rape Pests. Springer-Verlag, London, pp. 131–335.
47. Thies, C., Tscharntke, T., 1999. Landscape structure and biological control in agroecosystems. Science 285, 893–895.
48. Thies, C., Steffan-Dewenter, I., Tscharntke, T., 2003. Effects of landscape context on herbivory and parasitism at different spatial scales. Oikos 101, 18–25.
49. Thies, C., Steffan-Dewenter, I., Tscharntke, T., 2008. Interannual landscape changes influence plant-herbivore-parasitoid interactions. Agric. Ecosyst. Environ. 125, 266–268.
50. Thompson, R., Baker, R., 1981. Composite link functions in generalized linear models. J. Roy. Stat. C App. 30, 125–131.
51. Valantin-Morison, M., Meynard, J.-M., Dore', T., 2007. Effects of crop management and surrounding field environment on insect incidence in organic winter oilseed rape (Brassica napus L.). Crop. Prot. 26, 1108–1120.
52. Vanbergen, A.J., 2013. Threats to an ecosystem service: pressures on pollinators. Front. Ecol. Environ. 11, 251–259.
53. ver Hoef, J.M., Boveng, P.L., 2007. Quasi-poisson vs. negative binomial regression: how should we model overdispersed count data? Ecology 88, 2766–2772.
54. Vinatier, F., Gosme, M., Valantin-Morison, M., 2012. A tool for testing integrated pest management strategies on a tritrophic system involving pollen beetle, its parasitoid and oilseed rape at the landscape scale. Landscape Ecol. 27, 1421– 1433.
55. The major insect pests of oilseed rape in Europe and their management: an overview. In: Williams, I.H. (Ed.), Biocontrol-based Integrated Management of Oilseed Rape Pests. Springer-Verlag, London, pp. 1–43.

56. Williams, I.H., Cook, S.M., 2010. Crop location by oilseed rape pests and host location by their parasitoids. In: Williams, I.H. (Ed.), Biocontrol-based Integrated Management of Oilseed Rape Pests. Springer-Verlag, London, pp. 215–244.

57. Williams, I.H., Frearson, D., Barari, H., McCartney, A., 2007. Migration to and dispersal from oilseed rape by the pollen beetle Meligethes aeneus, in relation to wind direction. Agr. Forest. Entomol. 9, 279–286.

58. Zaller, J.G., Moser, D., Drapela, T., Schmoeger, C., Frank, T., 2008a. Effect of within-field and landscape factors on insect damage in winter oilseed rape. Agr. Ecosyst. Environ. 123, 233–238.

59. Zaller, J.G., Moser, D., Drapela, T., Schmöger, C., Frank, T., 2008b. Insect pests in winter oilseed rape affected by field and landscape characteristics. Basic Appl. Ecol. 9, 682–690.

60. Zaller, J., Moser, D., Drapela, T., Schmoger, C., Frank, T., 2009. Parasitism of stem weevils and pollen beetles in winter oilseed rape is differentially affected by crop management and landscape characteristics. Biocontrol 54, 505–514.

61. Zhang, Y., Wang, L.M., Wu, K.M., Wyckhuys, K.A.G., Heimpel, G.E., 2008. Flight performance of the soybean aphid, Aphis glycines (Hemiptera: aphididae) under different temperature and humidity regimens. Environ. Entomol. 37, 301–306.

62. Zimmer, C.T., Nauen, R., 2011. Pyrethroid resistance and thiacloprid baseline susceptibility of European populations of Meligethes aeneus (Coleoptera: nitidulidae) collected in winter oilseed rape. Pest. Manag. Sci. 67, 599–608. M

CHAPTER 17

Agriculture and Forest Transition: Understanding of Land Use Change in a Cultural Landscape

José Eduardo dos Santos[1], Angela Terumi Fushita[1], Imyra Maíra Martins de Souza[2], Lia Martucci Amorim[2], Valéria Aparecida David Andrade[2]

[1]*Department of Hydrobiology, Federal University of São Carlos, São Carlos, Brazil*

[2]*Postgraduate Program in Ecology and Natural Resources, Federal Universityof São Carlos, São Carlos, Brazil*

ABSTRACT

Land use trajectory analysis provides both the time and kind of land cover changes to monitor land use dynamics. Land cover change trajectories for three different dates (1965, 1989 and 2014) extracted from satellite images by vectorization of the thematic information were studied This study was carried out on a cultural landscape (São Carlos municipality, SP, Brazil) with three major land use-cover classes (forest, agriculture and other uses). The results showed the spatio- temporal variability of landscape pattern and forest and agriculture change trajectories. Analysis based on these landscape trajectories demonstrates that agriculture and forest cover changes have been caused by human activities. The results reflect the conflicting interactions between environmental and human systems in the study area. A key question is that the pressure exerted on forest land use-cover depends on the incentives that move society from a conflicting relation with a municipal territory. A more sustainable landscape transition and trajectory for São Carlos municipality are extremely dependent on the regulatory role of the government through strategies related to the implementation of specific categories of legally protected areas (Legal Reserves and Areas of Permanent Preservation).

Keywords: Land Use-Cover, Change Trajectories, Landscape Sustainability, Legally Protected Areas, Remote Sensing

1. INTRODUCTION

Recent studies have revealed the environmental impacts of land use from global to local scales. These range from changes in regional climates, changes in carbon and hydrologic cycles, declines in biodiversity through the loss and fragmentation of habitats, degradation of soil and water and overexploitation of native species [1] [2] .

Land-use practices are essential for human well-being because they provide natural resources and ecosystem services. On the other hand, these practices can also degrading the ecosystems and services upon which humans depend [3] . This process takes different forms and depends on circumstances, national, regional and local contexts and government policies.

Forest transitions assert that stocks of forest change as societies undergo development, industrialization and urbanization. Changes occur at various scales and may characterize an entire country, a province within a country or a county within a geographical region [4] . The impact of a forest transition on biodiversity varies from place to place and begins during a period of deforestation. Initially, natural areas decline in extent as forested lands are converted into agricultural areas to meet human population demands [5] .

Socio-economic activities have been one of the most important factors for land cover change trajectories. In place of temporal series dates of change in satellite images, researchers are now more focused on temporal land cover change trajectories [6] .

Trajectory analysis is a new approach for land cover change research based on each pixel in a time series [7] and is developed for long term mapping. Temporal trajectory is using to identify land use trends over time from the relationships among the factors that shape the human-environment relation and their influence within a particular region. The analysis of a land cover trajectory provides the timing and nature of land cover changes and can also provide new information about what kind of land cover change occurred for a landscape [6] .

This paper presents land cover change trajectory analysis for forest and agriculture areas and other areas for five decades (1965, 1989 and 2015) throughout São Carlos municipality, São Paulo, Brazil, highlighted the importance of understanding landscape dynamics for sustainability and conservation purposes. This research seeks to (1) determine the spatio-temporal variability of landscape trajectories in the municipal territory and (2) provide evidence as to whether factors revealed by a land cover change trajectory can ensure long-term environmental sustainability by reducing the negative environmental impacts of land use while maintaining ecosystem goods and services.

2. MATERIAL AND METHODS

2.1. Study Area

The municipality of São Carlos covers an area of 1136.907 km^2 and is located in the northeast region of São Paulo state, Brazil, between 47°30'W and 48°30'W and 21°30'S and 22°30'S. It was selected for study by presenting scenery related to land use fast changes, typical for municipalities located in the northeast region of São Paulo state, Brazil (Figure 1).

According to a demographic census, São Carlos municipality had a total population of 238,958 inhabitants in [9] . Its primary vegetation cover is characterized by semideciduous and riparian forests and various savanna physiognomies with arboreal savanna, sparsely arboreal savanna, and short-shrub savanna. Its maximum altitude is about 950 m with a minimum of 520 m. The climate of the region is classified between CWA and AW in Köppen's system, representing a transition from warm tropical with dry winter climate to tropical with dry summer and humid winter. The average annual rainfall is 1468 mm, with rain predominating between November and February [8] .

2.2. Methods

The trajectories of land use changes for the São Carlos (SP) municipality were analyzed over a period of 50 years (1965-2014) based on information from topographic maps for the year 1965 [10] and Landsat imagery for

1989 and 2014. The land use typologies for 1965 were obtained by vectorization of the thematic information from topographic maps of the study area [10] . The land use typologies for 1989 and 2014 were obtained through Landsat images with a pixel size of 30 × 30 m, collected on October 24, 1989 (TM-5 sensor) and on June 07, 2014 (OLI sensor). The spectral bands 3, 4 and 5 were processed using ENVI 4.7 software, resulting in an R5G4B3 color composition for Landsat imagery. The land use typologies were discriminated by visual interpretation method through on screen digitizing based on tone, texture and context [11] [12] .

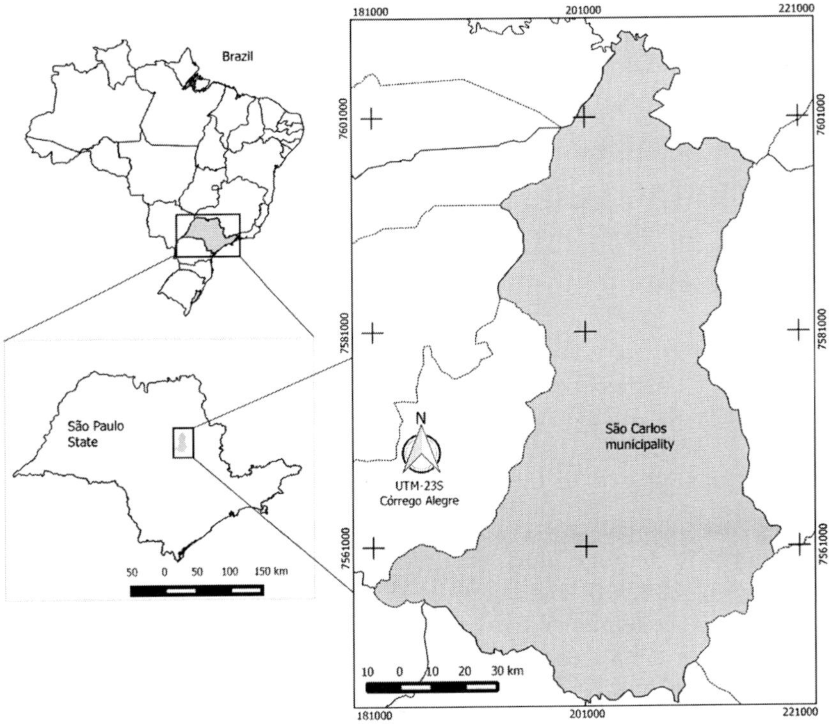

Figure 1. Location of São Carlos municipality, SP, Brazil.

To simplify the spatial and temporal dynamics of landscape trajectories for illustration purposes, three land use classes were defined: Forest (secondary forest, savanna and short-shrub savanna), Agriculture (perennial and annual crops, bare soil, pasture and forestry) and Others (settlements, road networks, industrial complexes, mining areas and water bodies). The spatial representation of trajectories of land use changes for the periods 1965, 1989 and 2014 were obtained based on the IDRISI Selva software command CROSSTAB.

The survey of conservative (Legal Reserves and Areas of Permanent Preservation) and punitive (environmental damages process) strategies to protect natural vegetation on private farmland in São Carlos municipality were based on the Brazilian Forest Code. Legal Reserves (LR) information (spatial distribution, area and establishment period) between 1983 and 2014 was obtained through a geodatabase from the Coordination of Biodiversity and Natural Resources-Environmental Secretary of the State of São Paulo. The Areas of Permanent Preservation (APPs) information (spatial distribution and area) was obtained through topographic maps from the Geographic and Cartographic Institute [10] . Areas of Permanent Preservation thematic maps were overlaid with a forest coverage map, resulting in quantitative data to estimate the total APPs percentage area.

The environmental damages processes from 2003 to 2014 were obtained through documents registered in the Coordination of Environmental Inspection-Environmental Secretary of the State of São Paulo.

3. RESULTS AND DISCUSSION

The recent release of forest cover change estimates for Brazil provides an opportunity to assess variable forest transitions and explore implications in terms of promoting sustainability. The trajectory of forest loss in Brazil between 2000 and 2010 (236,600 km^2) was mostly due to the expansion of agricultural activities, representing 65% of total deforestation in the period. In the corresponding period, the expansion of pasture areas accounted for the other 35% of deforestation, totaling 127,200 km^2 of areas of the Amazon or the Atlantic Forest. Particularly, between 2010 and 2012, there was an increase in the processes of land use and cover in Brazil, where changes in the landscape reached 3.5% of the country. The expansion of the agricultural frontier has intensified, with the deforestation of an area of 77,520 km^2 (68% of total deforestation), while the expansion of planted pastures accounted for 28% of deforestation, or 32,120 km^2 [13] .

Natural landscape conversion for anthropogenic use or changing management practices has also transformed a large proportion of São Paulo province's land surface. By clearing Atlantic forest, practicing subsistence agriculture, intensifying sugar-cane production and expanding urban centers, human actions are changing the province's landscape in significant ways [14] . In a similar way, patches of tropical savanna (cerrado vegetation) are continuously under threat due edge effects and also to the following land use types around these remnants: pasture, annual crop areas, perennial crop areas, sugar cane cultivation areas, forestry areas, settlements, road networks and industrial complexes [15] .

Landscape change through human influence for São Carlos municipal territory shows that natural vegetation has been replaced by intensifying sugar-cane production and expanding urban area. A legal proceedings survey for São Carlos municipal territory only involving deforestation activities from 1991 to 2000 identified the loss of 1121.62 ha of natural vegetation, roughly 100 ha/year. This total of natural and semi natural vegetation area was estimated at 14.1% of the municipal territory's total area [16] . The natural vegetation occupied 28% (31,776.4 ha) of the São Carlos municipality total area in 2012 [17] .

Multi-temporal information regarding the processes and patterns of land use change is particularly useful in cultural landscapes where changes have been rapid. Furthermore, this information provides the opportunity to understand how historical land use changes affect environmental and socio-economic sustainability. The spatial and temporal aspects of land use and cover change trajectories for five decades (1965, 1989 and 2015) of the study area are illustrated in Figure 2. The time series of remote sensing data revealed that land use and cover changes do not take place in a progressive and gradual way in São Carlos municipality. A period of rapid and abrupt change (1965-1989) was followed by recovery of forest areas (1989-2014) for a similar quantitative condition existing as of 1989.

For agricultural land use and cover, pasture areas, perennial and annual crop areas, bare soil and forestry areas makes up the largest percent with 54.17%, 66.60% and 63.26% respectively for 1965, 1989 and 2014. Forest areas, including secondary forest, savanna and short-shrub savanna, make up the next largest land use and occur predominantly in areas with greater relief. Forest area decreased during the first period of study from 41.75% to 26.91% (1965-1989) but then reached a subsequent recovery in the proportion of forested area (26.39%) during the second period (1989-2014). The class other land uses (including settlements, road networks, industrial complexes, mining areas and water bodies) increased during all period (4.08%, 6.49% and 10.35% respectively for 1965, 1984 and 2014) (Figure 2). The expansion of the urban area in the 50-year period reflects rapid and significant municipal population growth, with a population around 70,000 in 1965 and a total population of 238,958 inhabitants in 2014 and the potential to reach 241,389 in 2015 (http//www.cidades.ibge.gov.br/).

Change trajectories between the years 1965, 1989 and 2014 were compared on a pixel-by-pixel basis to examine possible land use change. Land use classification trajectories for 1965-1989-2014 are shown in Figure 3, where "F" refers to Forest, "A" to Agriculture, and "O" to Others.

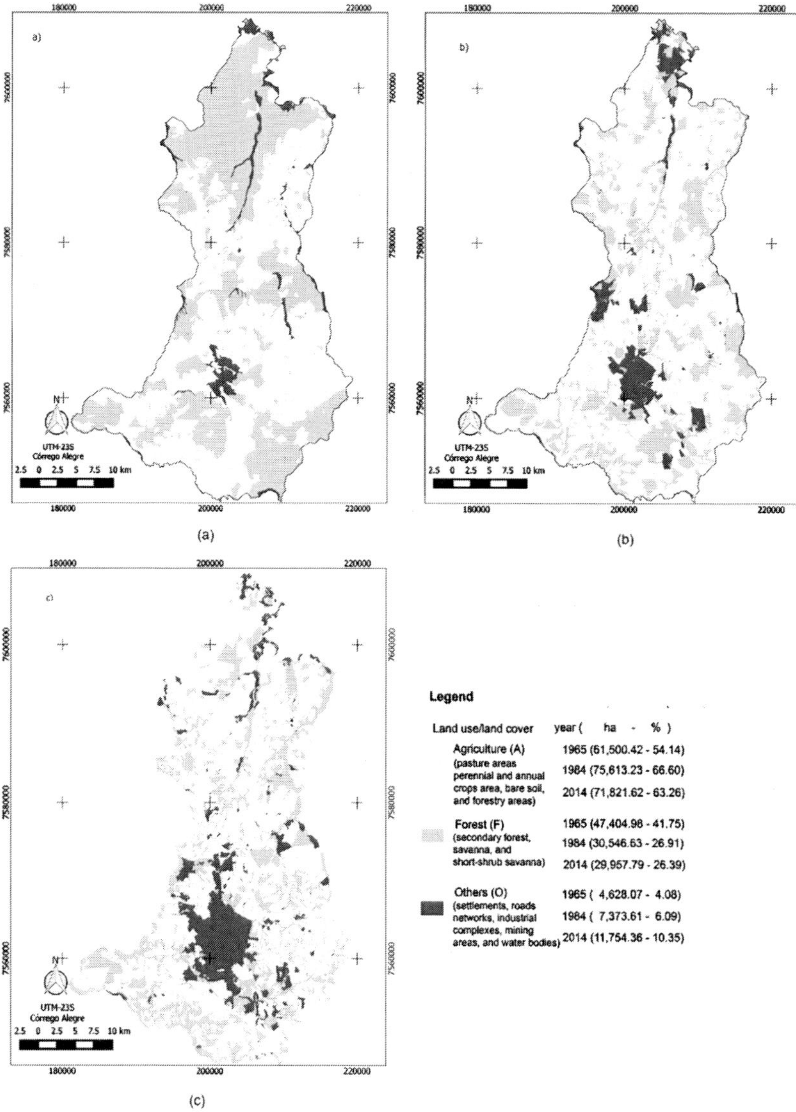

Figure 2. Kind of land use and cover (Agriculture, Forest and Others) changes in São Carlos municipality for (a) 1965, (b) 1989 and (c) 2014.

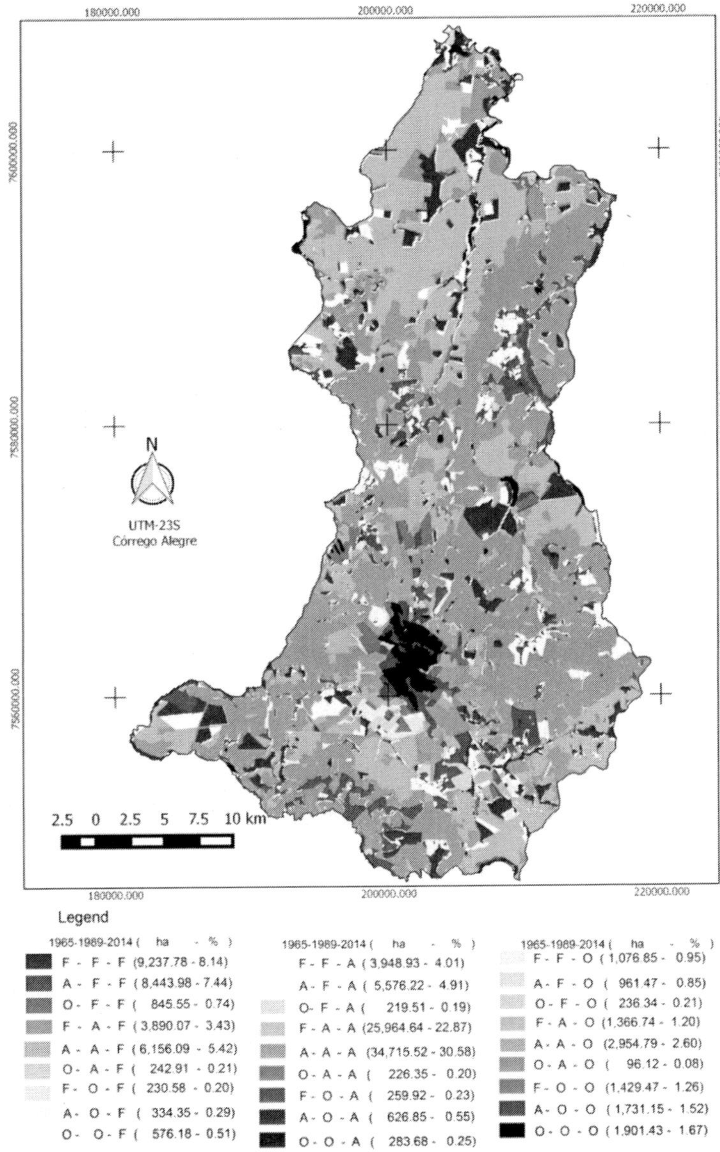

Figure 3. Land use classification trajectories for 1965-1989-2014 at São Carlos municipality, with "F" refers to forest, "A" to agriculture, and "O" to other use classes (settlements, road networks, industrial complexes, mining areas and water bodies).

Around forty percent of the landscape remained in the same land use class from 1965 to 2014, with 9237.78 ha (8.14%) of stable forest cover (F-F-F), 34,715.52 ha (30.58%) of stable agriculture use (A-A-A) and 1901.43 ha

(1.67%) of stable other areas class (O-O-O) in the last five decades. For the total forest area existing in 2014 (26.39%), which remained in the same type of use (9237.78 ha/8.14%/F-F-F), the remaining area (18.25%) refers to forest area converted to other uses and which returned to a forest condition (3.63%/F-A-F and F-O-F), the conversion of agricultural areas to forest area (13%:15%/A-F-F; A-A-F and A-O-F) and the conversion of other types of land use to forest area (1%:46%/O-F-F; O-A-F and O-O-F) (Figure 3).

The total agricultural area existing in 2014 (63.26%), around 34,715.52 ha, (30.58%/A-A-A) remained in the same type of use. Agriculture took the place of 30,173.49 ha (27.11%) of forest from 1965-2014 (1.4%/F-F-A; 22.87%/F-A-A and 0.23%/F-O-A). Agriculture took the place of 729.54 ha of other land uses from 1965-2014 (O-F-A/0.19%; O-A-A/0.20% and O-O-A/0.25%) (Figure 3).

The other land uses class in 2014 replaced 3.41% of areas occupied by forest in 1965 (F-F-O/0.95%; F-A-O/1.20% and F-O-O/1.26%). Other land uses took the place of 5647.41 ha of agriculture land use from 1965-2014 (A-F-O/0.85%; A-A-O/2.60% and A-O-O/1.52%) (Figure 3).

During the first period (1965-1989), the main trajectories were dominated by deforestation transitions that led to the increase of agricultural land as a result of logging practices. In this first period, nearly 15% of São Carlos municipality was replaced by agriculture or other uses (Figure 2). It seems that these changes were induced by organized human activities that coincided with the local practical situation. Deforestation was caused by either permanent or sudden forces that took place in the municipal territory.

In the municipal territory, the period of random change, coincidentally also of a large amount of deforestation, coincides with the Brazilian National Alcohol Program (ProAlcohol), characterized by trade policies, which opened up the economy to international trade in favor of alcohol production. This program was created in 1975 to concentrate all national efforts to increase alcohol production and use. During phase one (1975-1979), government policies searched to facilitate both distillery expansion and higher conversion rates in the gasoline-al- cohol mixture. In phase two, from 1980, the Brazilian government authorized and subsidized a vast expansion of sugar cane production capacity and industrial investment in mills and in distilleries [18] .

The study area suffered serious forest losses as a result of agricultural expansion in the 50-year period, which brought great damage to local biodiversity and to local residents. Agricultural land use in 2014 occupies nearly 64% of the total study area. Around 30.58% of the study area remained in the same type of land use (A-A-A) since 1965. About 5% of agricultural land use of the study area in 1965 has undergone a transition for forest land use use in 1989, returning again for agricultural land use in 2014. Agriculture replace of 27.11% of forest land use (F-F-A/4.01%; F-A-A/22.87%, and F-O-A/0.23%), and 0.64% of other land uses (O-F-A/0.19%; O-A-A/0.20%, and O-O-A/0.25%) during 1965-2014 (Figure 4(a)).

From Figure 4(b), it can be seen that most of the contiguous forest land use in the north and southeast region in 1965 was replaced (F-F-A/3.48% and F-A-A/22.87%) by agriculture. Only 8.14% of São Carlos municipality was occupied by forest area that had remained for five decades.

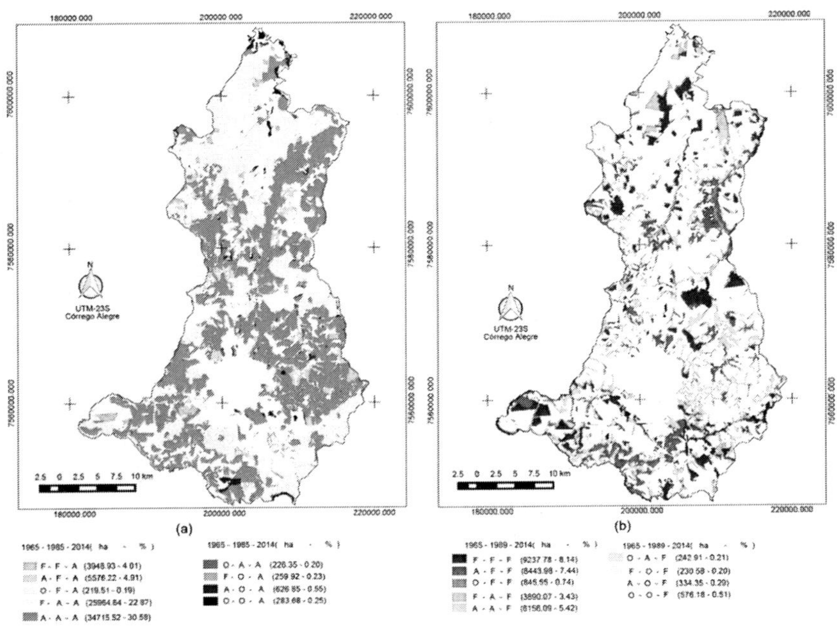

Figure 4. Agriculture (a) and (b) forest land use trajectory during 1965 to 2014 at São Carlos municipality, with “F” refers to forest, “A” to agriculture, and “O” to other use classes.

In 2014, the vegetation in legally protected areas in São Carlos municipality occupied 23,085.91 ha (20.6%) with 20,968.24 ha (18.50%) of Legal Reserves and 2387.67 ha (2.10%) of Areas of Permanent Preservation with vegetation (Figure 5).

The amount of remaining vegetation (6871.58 ha/5.78%) of São Carlos municipality not legally protected and was more susceptible to anthropic occupation. This unprotected natural vegetation has come become under increased conversion pressure from agriculture to compensate for the lost production associated with the restrictions that apply to Areas of Permanent Preservation and Legal Reserves [19] .

Governance and law enforcement were important tools for re-conversion of current agricultural land to forest land in São Carlos municipality. The increase in vegetation area was associated with the establishment of Legal Reserves between 1999 and 2014 (Figure 6). The absence of established Legal Reserves from 2010-2014 may be explained by the reduction of legal requirements for conservation in Brazil's Forest Code [19] .

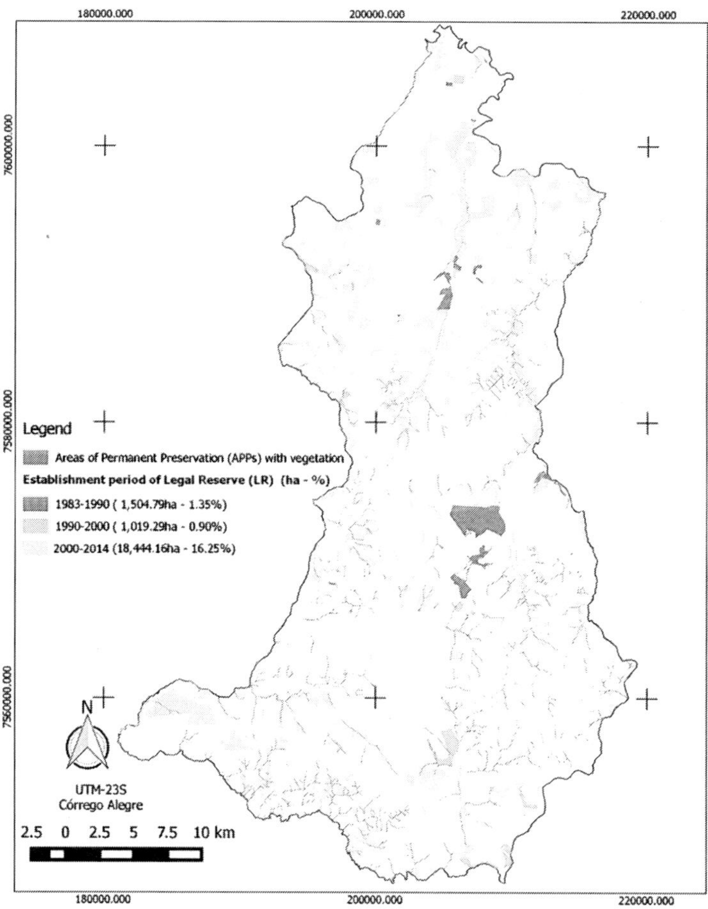

Figure 5. Specific categories of legally protected areas (ha and %) in São Carlos municipality: Legal Reserves (LR) and Areas of Permanent Preservation (APPs). Establishment period (1983-2014) of Legal Reserve.

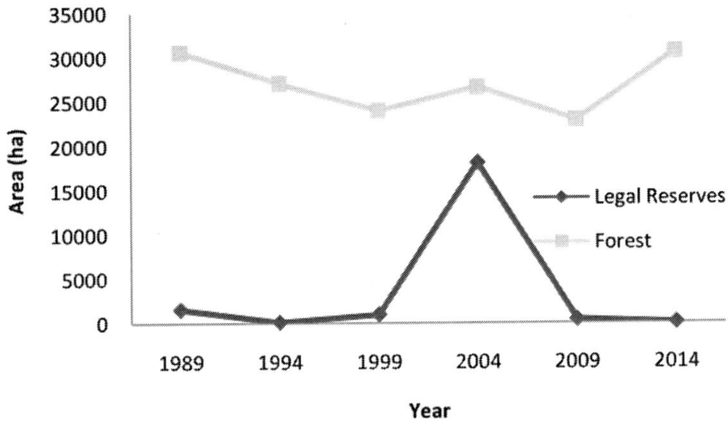

Figure 6. Establishment period of Legal Reserves and Forest trajectory areas in São Carlos municipality between 1989 and 2014.

The Brazilian Forest Code is the central piece of legislation regulating land use and management on private property. The substitutive Forest Code (Law N° 12.651, 25th May 2012) reduced requirements for protection of natural vegetation on private farmland and allowed protection of natural vegetation outside the farm [20] .

The establishment of Legal Reserves and Areas of Permanent Preservation can be considered a preventive environmental practice that requires not only defense against imminent damages, but that also encourages the protection and regular use of natural resources. Furthermore, the Brazilian legislation has punitive mechanisms related to the denouncement of environmental damage.

In São Carlos municipality, the forest area increase between 2010 and 2014 (Figure 6) can be associated with the actions of environmental damage repair. There were 126 occurrences in this period (2010-2014), compared to a total of 194 occurrences between 2003 and 2014.

The punitive approach legally penalizes the infringer and requires repair of effects of environmental damage, for example, when damage related deforestation is identified, an official document (Environmental Recovery Commitment Agreement/TCRA) is signed by the infringer and environmental agencies to formalize measures to be implemented for native vegetation restoration.

Accurate land use change trajectories analysis was useful for devising future changes for landscape sustainability and planning strategies for the territorial municipality. A very important process to estimate is the rate, pattern and type of land cover change trajectories in order to predict changes for sustainable development.

Such changes often caused by the interaction of climatic and land use factors and have an important impact on ecosystem processes. However, the land cover changes for the São Carlos municipality data-deforestation, changes in the extent of cultivated lands and urbanization-are processes of conversion that are not strongly affected by interannual climatic variability.

4. CONCLUSIONS

The present level of local biodiversity in São Carlos municipality is related to 29,957.49 ha (26.39%) of forested area, set by a change in landscape structure with a reduction in size, an increase in the number and in the isolation of fragments and a loss of landscape connectivity (Figure 4(b)).

The transition from forest for agricultural and urban land use is a recognized process of global occurrence. However, at a regional and local level that it is possible to know and act on the balance of this transition. The cultural landscape related to the municipal territory of the State of São Paulo, as well as other cities of Brazil are experiencing a transition and increasing loss of natural habitats resulting from the interaction between nature, development, and society. This scenario makes it essential to know the amount of the remaining natural capital of a cultural landscape, as well as the implementation of strategies to control or minimize the loss of habitat and local biodiversity.

The question is what amount of forest area should be remained under current trends and policies given trade-offs with economic development and agriculture to ensure environmental sustainability of cultural landscape.

According a review based on species of temperate areas, the amount can be considered a threshold of around 30% of remaining habitat, above which the effects of biodiversity loss would be due to habitat loss [21] [22] . Below this threshold, there may be a drastic effect on the spatial distribution of the habitat. This threshold has no empirical support, as shown by results obtained in tropical regions reporting fragmentation effects on the habitat loss process [23] . However, there is evidence that landscapes with less than 30% of natural habitat tend to have only small fragments and be isolated, thus supporting impoverished communities and different taxonomic groups [24] [25] .

Thus, the threshold of 30% could be considered as the lower limit of a forest that has a landscape managed by man to balance economic use and biodiversity conservation [26] . However, even considering the area values of Legal Reserve and Permanent Protection Areas (Figure 5), the municipality of São Carlos does not have a forest cover area above this threshold.

São Carlos municipality has achieved a reduction in landscape deforestation. However, this profit is not yet assured. Our study suggests that the Forest Code will allow additional deforestation due to local development actions. Fortunately, government initiatives are aligning to assist conservation efforts in expanding legally protected areas in São Carlos municipality. These implemented regional strategies can support the complete set of ecosystem services and improve managed landscape sustainability.

ACKNOWLEDGEMENTS

Financial support was provided by the Coordination for the Improvement of Higher Education Personnel (CAPES).

CITE THIS PAPER

José Eduardo dos Santos, Angela Terumi Fushita, Imyra Maíra Martins de Souza, Lia Martucci Amorim, Valéria Aparecida David Andrade (2015) Agriculture and Forest Transition: Understanding of Land Use Change in a Cultural Landscape. *Open Journal of Applied Sciences*,**05**,797-807.

REFERENCES

1. Lambin, E.F., Geist, H.J. and Lepers, E. (2003) Dynamics of Land-Use and Land-Cover Change in Tropical Regions. Annual Review of Environment and Resources, 28, 205-241.
2. Foley, J.A., Defries, R., Asner, G.P., Bonan, G., Carpenter, S.R., Chapin, F.S., Coe, M.T., Daily, G.C., Gibbs, H.K., Helkowski, J.H., Holloway, T., Howard, E.A., Kucharik, C.J., Monfreda, C., Patz, J.A., Prentice, I.C., Ramankutty, N. and Snyder, P.K. (2005) Global Consequences of Land Use. Science, 309, 570-574.
3. DeFries, R.S., Foley, J.A., and Asner, G.P. (2004) Land-Use Choices: Balancing Human Needs and Ecosystem Function. Frontiers in Ecology and the Environment, 2, 249-257.
4. Mather, A.S. and Needle, C.L. (1998) The Forest Transition: A Theoretical Basis. Area, 30, 117-124.

5. Rudel, T.K., Coomes, O.T., Moran, E., Achard, F., Angelsen, A., Xu, J. and Lambin, E. (2005) Forest Transitions: Towards a Global Understanding of Land Use Change. Global Environmental Change, 15, 23-31.
6. Borri, M.S. and Vozenilek, V. (2014) Remote Sensing and Land Use/Land Cover Trajectories. Journal of Geophysics & Remote Sensing, 3, 123.
7. Skalos, J. and Engstova, B. (2010) Methodology for Mapping Non-Forest Wood Elements Using Historic Cadastral Maps and Aerial Photographs as a Basis for Management. Journal of Environmental Management, 91, 831-843.
8. Soares, J.J., Silva, D.W. and Lima, M.I.S. (2003) Current State and Projection of the Probable Original Vegetation of the São Carlos Region of São Paulo State. Brazilian Journal of Biology, 63, 527-536.
9. IBGE (2014) Resolution No. 2 of August 26, 2014. Estimates of the Population for the States and Municipalities with Reference Date on July 1, 2014. Official Federal Gazette, Brazil 28 August, Section 1, 115.
10. IBGE (1971) Topographic Maps: Topographic Mapping of Brazil. Scale 1:50.000. Rio de Janeiro.
11. Crosta, A.P. (1992) Remote Sensing Digital Image Processing. IG/UNICAMP, Campinas.
12. Moreira, M.A. (2011) Remote Sensing Fundamentals and Application Methods. 4th Edition, UFV, Viçosa, MG.
13. IDS (2015) Indicators of Sustainable Development: Brazil. Coordination of Natural Resources and Environmental Studies. IBGE, Rio de Janeiro.
14. Ribeiro, M.C., Mettzger, J.P., Martensen, A.C., Ponzoni, F.J. and Hirota, M.M. (2009) The Brazilian Atlantic Forest: How Much Is Left, and How Is the Remaining Forest Distributed? Implications for Conservation. Biological Conservation, 142, 1141-1153.
15. Durigan, G., Siqueira, M.F. and Franco, G.A.D.C. (2007) Threats to the Cerrado Remnants of the State of São Paulo, Brazil. Scientia Agricola (Piracicaba. Braz.), 64, 355-63.
16. Cintra, R.H.S., Santos, J.E., Moschini, L.E., Pires, J.S.R. and Henke-Oliveira C. (2006) Qualitative and Quantitative Analysis of Environmental Damages through Instauration and Registers of Lawful Documents. Brazilian Archives of Biology and Technology, 49, 989-999.
17. Bertini, M.A., Fushita, A.T. and Lima, M.I.S. (2015) Vegetation Coverage in Hydrographic Basins in the Central Region of the State of São Paulo, Brazil. Brazilian Journal of Biology, 75, 709-717.
18. Oliveira, J.A. (2002) The Policymaking Process for Creating Competitive Assets for the Use of Biomass Energy: The Brazilian Alcohol Programme. Renewable and Sustainable Energy Reviews, 6, 129-140.
19. Sparovek, G., Berndes, G., Barreto, A.G.O.P., Leoname, I. and Klug, F. (2012) The revision of the Brazilian Forest Act: Increased Deforestation or a Historic Step towards Balancing Agricultural Development and Nature Conservation? Environmental Science & Policy, 16, 65-72.
20. Soares-Filho, B., Rajão, R., Macedo, M., Carneiro, A., Costa, W., Coe, M., Rodrigues, H. and Alencar, A. (2014) Cracking Brazil's Forest Code. Science, 344, 363-364.
21. Andrén, H. (1994) Effects of Habitat Fragmentation on Birds and Mammals in Landscape with Different Proportions of Suitable Habitat: A Review. Oikos, 71, 355-366.
22. Fahrig, L. (2003) Effects of Habitat Fragmentation on Biodiversity. Annual Review of Ecology, Evolution and Systematic, 34, 487-515.
23. Develey, P.F. and Metzger, J.P. (2006) Emerging Threats to Birds in Brazilian Atlantic Forests: The Roles of Forest loss and Configuration in a Severely Fragmented Ecosystem. In: Laurance, W.F. and Peres, C.A., Eds,, Emerging threats to Tropical Forests, University of Chicago Press, Chicago, 269-290.

24. Martensen, A.C., Pimentel, R.G. and Metzger, J.P. (2008) Relative Effects of Fragment Size and Connectivity on Bird Community in the Atlantic Rain Forest: Implications for Conservation. Biological Conservation, 141, 2184-2192.
25. Metzger, J.P., Martensen, A.C., Dixo, M., Bernacci, L.C., Ribeiro, M.C., et al. (2009) Time-Lag in Biological Responses to Landscape Changes in a Highly Dynamic Atlantic Forest Region. Biological Conservation, 142, 1166-1177.
26. Metzger, J.P. (2010) The Forest Code has Scientific Basis? Brazilian Journal of Nature Conservation, 8, 92-99.

CHAPTER 18

Climate Change Effects on Evapotranspiration in Mexico

Martín Mundo-Molina

Investigation Centre, Faculty of Engineer, Autonomous University of Chiapas, Tuxtla Gutiérrez, Mexico

ABSTRACT

In Mexico's case, the scarcity of water in the north of the country is worrying, even more from the agricultural point of view because according to the results of general circulation models, a major impact due to global warming is expected in this region, and it will have important repercussions on the rural sector in the north of Mexico. According to [1] the results of the sensitivity analysis made in this study indicate that the most vulnerable zone is the north of Mexico, wherein the increase of evapotraspiration (ET) is greater in comparison to the rest of the country; up to 8% annual average for a +3°C growth in mean annual temperature. Due to some limitations in this preliminary investigation (e.g., global temperature data was used without regionalizing it), it was decided to make more detailed studies to estimate the climate change effects on ET on a regional scale, using the downscaling method to adjust temperature data. In this study a new methodology to estimate the ET before climate change scenarios is introduced, which includes the selection of the Hargreaves-Samani method (HS), calibrated and compared against the Penman-Monteith ASCE method in various irrigation districts in the northern part of the country, obtaining ET estimations with a 93% precision. This procedure was applied to nine states in north Mexico: Baja California, Baja California Sur, Chihuahua, Sinaloa, Sonora, Tamaulipas, Nuevo León, Coahuila y Durango. The principal results are enunciated as follows: the ET variations between the contemporary scenario and the 2030 scenario are quite significant, according to the data of 160 meteorological stations; for temperature variations between 0.1°C to 0.45°C the corresponding ET fluctuation goes from 2% in the current scenario to 7% in the 2030 scenario. These obtained percentages are greater than the ones expected to happen for the precision of the method. It is important to note that a 7% rise of ET (related to a regional temperature increase of approximately one degree) would represent in practice having more millions of m3 of water in dams to satisfy the water demand of crops.

Keywords: Climatic Change, Reference Evapotranspiration, Irrigation Agriculture

1. INTRODUCTION

The observation networks of the World Meteorological Organization (WMO) have registered a raise in worldwide mean temperature from 3°C to 6°C during the last hundred years because of increment of greenhouse gases in the atmosphere during the last 2000 years.

According to stored data and physical evidence, the increase of greenhouse gases has originated Earth's over- heating. In 1998, the United Nations Environment Programme (UNEP) and WMO created the Intergovernmental Panel on Climate Change (IPCC). The objective of this group is to analyze thoroughly and objectively scientific, technical and socioeconomic information to understand the risks and impact of climatic change, as well as the possible adaptation and reduction methods of it. The IPCC consists mainly of three work groups and a special team (which makes inventories about greenhouse gases) and has published up until now four scientific reports on climate aspects, mitigation and vulnerability (the last two reports focus on climate change impact in natural and socioeconomic systems). IPCC work team I (scientific group) published in Paris, on February 2007, the fourth evaluation report about global warming. It includes the results of the studies made on the last six years, and future perspectives about their possible effects.

In Mexico's case, the scarcity of water in the north of the country is worrying, even more from the agricultural point of view. According to the results of general circulation models, a major impact due to global warming is expected in this region, and it will have important repercussions on the rural sector in the north of Mexico. This area has been subject to dry seasons for long periods of time and this situation will increase the problems of water availability for crop irrigation. The major agricultural districts are at the north of the country and they consume 80 of every 100 liters of water used in Mexico, from which 50% of water resource is wasted. Therefore, considering the potential risk and vulnerability of the area, it is important to estimate the effects of climatic change in evapotranspiration (ET) in the principal irrigation districts of the country to calculate the possible growth on water demand, for it would mean to make use of a greater water volume from the dams in a region where this resource is limited.

However, it is not easy to quantify climate change effects on agriculture for several scientific reasons; the most important of them are three: first, the estimation of the quantitative effects of global warming over future climate involves the use of numeric models on global scale that do not contain all variables influencing the phenomenon (e.g. ocean's CO_2 absorption). It also uses inconsistent data; for example the rate of greenhouse gases emission depends on the anthropogenic effect, especially in the consumption and emission of fossil fuels, choice of technology, economy evolution and demographic growth of each country. In second place the results of global models that estimate future climate (increase or decrease of temperature or precipitation) generate data that cannot be directly applied to specific regions, so it's necessary to regionalize it in a process that, if not executed, directly affects the accuracy of the calculations. In third place, empiric (temperature and radiation models) and semi-empiric (Penman-Monteith models) evapotranspiration models were not designed to estimate ETo under climatic change conditions. In its current state, this set of equations is limited to model climate change and its effects on ETo, and can be used to perform sensitivity analysis at most.

Due to the complexity of global warming effects on temperature estimations, IPCC established different scenarios of climate change that take on account expected variability of conditions instead of a single calculation of greenhouse gas emission, and its relationship with the raise of global temperature. Each exponential shows answers based on different actions to mitigate climate change impact; for example: The higher tendency (upper exponential) is constructed under a vision of total failure in the efforts to diminish the effects of global climate change. The two center exponentials are associated with the hypothesis of the proper function of carbon market, i.e., the state and private actions are effective to diminish the greenhouse gas emission which allows a moderately lesser impact of climate change. Finally, the lower exponential of the graph assumes that all involved countries sign the Kyoto Protocol to mitigate climate change impact.

2. BACKGROUND

Based in these scenarios, efforts have been made in order to estimate climatic change effect on irrigated agriculture and dryland farming in Mexico. In dryland farming case, sensibility studies were elaborated with two specific models, the CERES-MAIZE and GFDL-R30 models. The CERES-MAIZE model simulated the performance of different corn crop productions on non-irrigated areas, with varied weather, seed and soil conditions. On the other hand, the GFDL-R30 model created theoretical agro climatic maps (i.e.), maps of potential suitable regions for certain crop production. Both models were applied in the following municipalities: Atlacomulco (state of Mexico), Izúcar and Ixcamilpa (Puebla), Tuxpan and Coatepec (Veracruz) and Magdalena and La Huerta (Jalisco) (Figure 1). These models were used considering arbitrary temperature (+2°C and +4°C) and pre-cipitation (±10% and ±20%) increases, and even combinations of both. The CERES-MAIZE model simulation results indicate a reduction in corn crop performance in all studied regions (except Atlacomulco), which means a negative climate change effect. Moreover the GFDL-R30 model results show that the simulated regions and their agro climatic areas are unfit for crop production under potential climatic change scenarios. For irrigated agriculture, the first studies elaborated in Mexico about climate change effects in evapotranspiration of crops was made in the Mexican Institute of Water Technology (IMTA in its Spanish acronym), about a decade ago. This preliminary study consisted in the application of a FAO radiation model to estimate the effects of climate change on ET annually in Mexico. Figure 2 shows the obtained results with the radiation model, which consist of a set of ET isolines on a national scale, calculated under a climate change scenario with actual temperature conditions. Figure 3 depicts the ET annual estimates under a 3°C increase of global temperature scenario [1] .

The results of the sensitivity analysis made in this study indicate that the most vulnerable zone is the north of Mexico, wherein the increase of ET is greater in comparison to the rest of the country; up to 8% annual average for a +3°C growth in mean annual temperature. Although the results of this study are indicative, they allow an idea of the increment of water for crop irrigation. Even without considering a climate change scenario, an important rise in irrigation water demand is expected by 2025 in México, as explained in Table 1 (this increase is estimated by 16.5% between 1995 and 2025).

Figure 1. Location where were applied the The CERES-MAIZE models: Atlacomulco (state of Mexico), Izúcar and Ixcamilpa (Puebla), Tuxpan and Coatepec (Veracruz) and Magdalena and La Huerta (Jalisco).

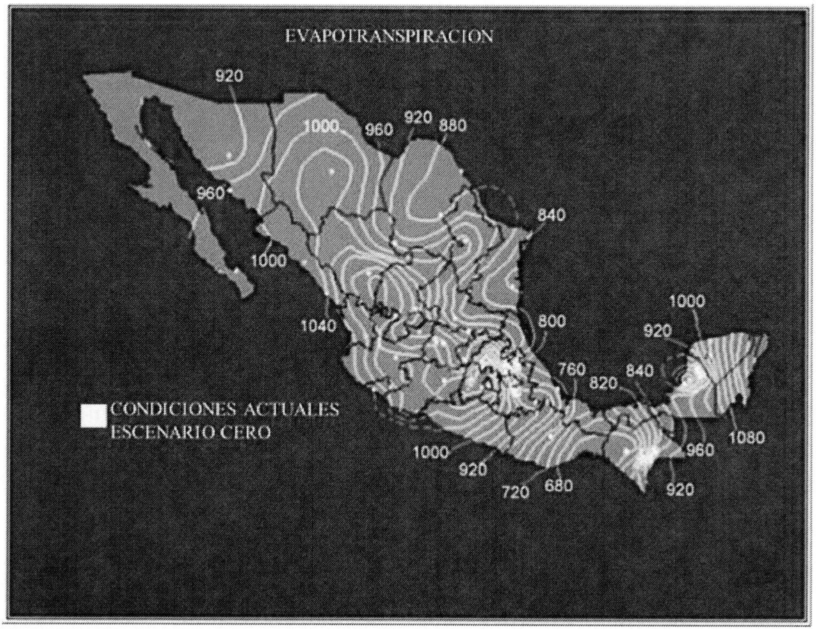

Figure 2. ET isolines under current temperature conditions [1] .

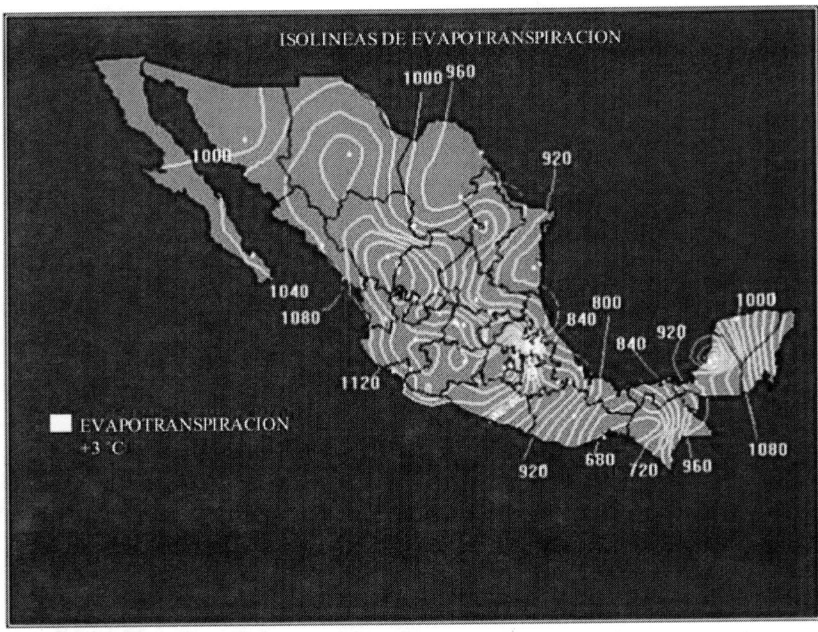

Figure 3. ET isolines with a 3°C increase on temperature [1] .

3. ET ESTIMATIONS MADE WITH HARGREAVES-SAMANI METHOD AND REGIONAL DATA

The indicative preliminary study on climate change effects in ET that used the radiation method mentioned above has several additional limitations to the ones indicated in the introduction. These are the most important: a) The IPCC scenarios were used without data on regional temperature; the process of deriving information from large scale models (global models) to regional scale models was not done; b) The same temperature increments were used for different seasons (not regionalized); c) Equal temperature increments were used for all months of the year.

A research was conducted to correct these limitations. It was made under a cooperative agreement between IMTA and the Autonomous University of Chiapas (UNACH in its Spanish acronym). The following methodology was used:

a) A method for calculating ET on the northern region of Mexico was selected. This method was to meet certain requirements, including: the usage of few variables to estimate ET, for the lack of meteorological data prevented the use of the Penman-Monteith (PM) equation; accuracy in the ET estimate results, or close to parallel the precision of the PM equation and to serve as reference for different researchers as a precise method for ET assessment.

b) The site to calibrate the method was selected. Irrigation district 041 in Valle del Yaqui, state of Sonora was chosen, due to Sonora being one of the most important states in the northern region of the country and having a network of automated meteorological stations for the calibration of the selected method.

c) The method was calibrated to estimate ET. The daily and monthly results were compared against the estimated results obtained from the witness method (Penman-Monteith ASCE).

d) The northern states of Mexico which include major irrigation districts were selected. Nine states from the north were designated, as well as 160 meteorological stations that hold records with more than 30 years worth of data.

e) Climate change scenarios were selected. Once the method for evapotranspiration estimation was calibrated, two types of scenarios were picked, the A1B and the A2. The results obtained with the A1B scenario are presented in this work.

f) Data of average anomalies in temperature obtained from IPCC models on A1B scenarios was used. The results processed from 16 climate models that participated in IPCC for the fourth evaluation report were analyzed by the hydro-meteorology IMTA team using the downscaling technique. The data was later used in the research.

Table 1. Water consumption for irrigation in Mexico, real and predicted [2]

Population ad water data	Historical data							Predictec
	1950	1960	1970	1980	1990	1995	2010	2025
Population (M)	27.49	36.71	50.4	67.05	84.5	93.97	118	136.72
Irrigation area (M ha)	1.6	3	3.58	4.98	5.6	6.2	6.8	7.1

g) The ET data in the "zero scenario" (current conditions), and the average temperature anomalies calculated for year 2030 were compared. The ET was evaluated with the current scenario anomalies (present temperature conditions) and the estimated stipulation for year 2030; after that, a comparison was made between both sets.

Therefore, for a precise ET calculus under climate change scenarios (sensibility analysis) it would be convenient to use the Penman-Monteith ASCE equation for all the northern part of the country, as the one stated below [3] :

$$\lambda Et_o = \left(\frac{\Delta}{\Delta+\gamma^*}\right)(Rn-G)+\left(\frac{\gamma}{\Delta+\gamma^*}\right)\frac{k_1(0.622)\rho\lambda}{P_a}\frac{1}{r_a}\left(e_z^o-e_z\right) \quad (1)$$

where:

ET_o = Evapotranspiration of referenced crop (mmd^{-1})

Rn = Net solar radiation on the surface of the crop ($MJ\cdot m^{-2}\cdot d^{-1}$)

G = Sensitive heat flux from soil ($MJ\cdot m^{-2}\cdot d^{-1}$)

Δ = Vapor pressure curve inclination (kPa)

γ = Psychometric constant (kPa°C^{-1})

γ^* = Modified psychometric constant (kPa°C^{-1})

ρ = Density of the fluid at a constant pressure ($kg\cdot m^{-3}$)

λ = Latent vaporization heat ($MJ^{-1}\cdot$°C^{-1})

P_a = Atmospheric pressure (kPa)

e_z^o = Saturated vapor pressure of air at z height (kPa)

e_z = Vapor pressure of air at z height (kPa)

r_c = superficial resistance to vapor transfer (sm^{-1}), or simply surface resistance.

r_a = Aerodynamic resistance to sensible heat and vapor transfer (sm^{-1}), or aerodynamic resistance.

However this equation needs variables and data that are not available in all studied places. For this reason an analysis was made on over a dozen of empiric methods that would allow calculating with precision the ET. Thus the Hargreaves-Samani (HS) method was selected to estimates these variables. The HS method originates from [4] and [5] though various adjustments were made to the initial equation introduced on 1975. This HS equation estimates the evapotranspiration of a harvest using grass as reference. The HS equation is [3] :

$$ET_o = 0.0023R_a(TD)^{1/2}(t_m + 17.8) \tag{2}$$

where:

TD = Difference between maximum and minimum temperature (°C)

R_a = Extraterrestrial radiation (mmd^{-1}), (R_a Variable is obtained from tables or with a equation)

Rs = Solar radiation (mmd^{-1})

This method is recommended by several investigators, [3] [6] - [9] for its precision. It can even, if required, be used to estimate ET for periods of five days and on daily scale [10] and [11] . It is a simple method that only needs extraterrestrial radiation and minimum and maximum temperature data to calculate ET. For its calibration the daily and monthly information collected from 10 stations in irrigation district 041, Valle del Yaqui (Figure 4) was used, Stations have daily measurements of radiation, relative humidity, wind speed, temperature and precipitation; some with over ten years of data.

In Figure 5 and Figure 6 are shown the results of the comparison between the HS and the Penman-Monteith ASCE methods for the CIANO station, located in block 910, lot 3, in the following coordinates of Yaqui Valley: 27°22'14 North latitudes, 109°55'4 West longitudes.

In both cases the linear regression coefficient is higher than 0.92, with a standard deviation from the model of 1.63 mm, a quadratic error of 0.44 in the first case and 0.6 in the second case, as well as 0.93 model efficiency for the daily scale and 0.97 for the monthly scale. Once the precision of the method was proven, the selection of the IPCC climate change scenarios proceeded. The results presented in this work were obtained under the A1B scenario considerations. The A1 family is characterized by a rapid economic growth, a global population that will reach its peak in 2025 and a society that adopts and uses efficient technologies. This family divides into two: A1F1 (intense use of petroleum fuels) and A1B (balanced use of different energy sources); the latter is considered as the most likely scenario to occur, according to the scientific community.

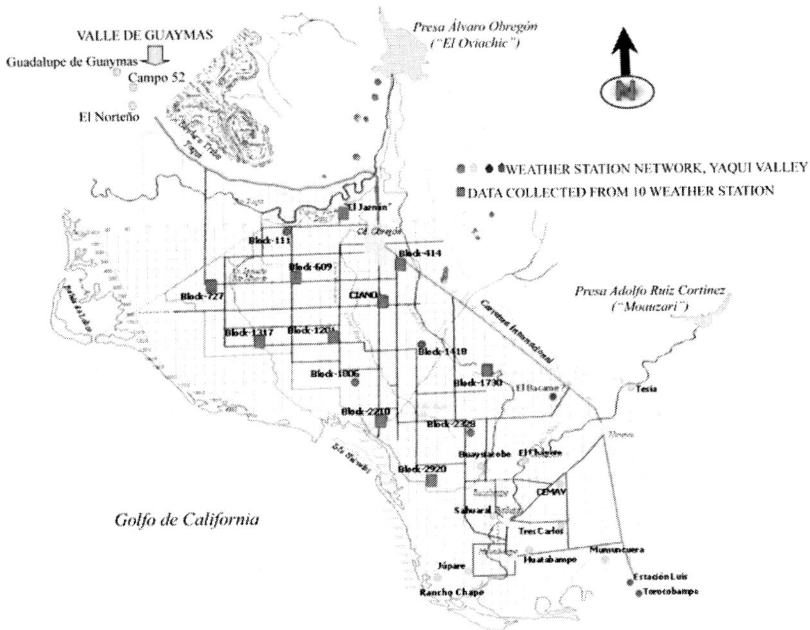

Figure 4. The small circles and small rectangle represents the weather stations network, Yaqui Valley Sonora Mexico.

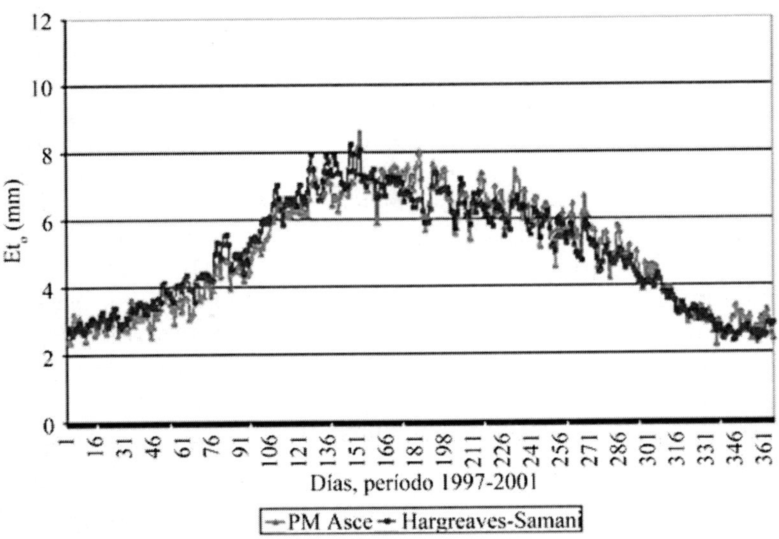

Figure 5. Penman-Monteith (PM) ASCE vs. Hargreaves-Samani comparison on a daily scale.

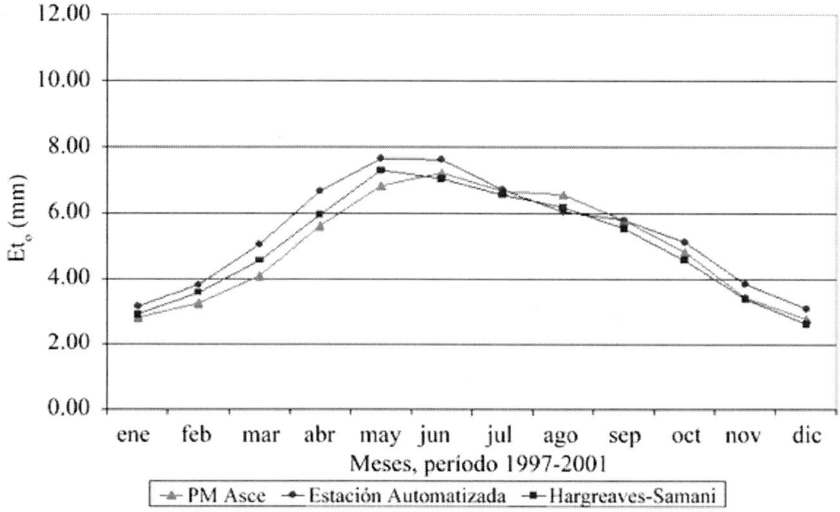

Figure 6. Penman-Monteith (PM) ASCE vs. Hargreaves-Samani comparison on a monthly scale.

When the scenario was selected, an analysis of the obtained information on the global IPCC models followed. Large scale climate models are not able to adequately simulate climate variations at regional level (regional climate behavior is a stochastic process, conditioned by global climate), therefore it is necessary to use downscaling techniques to complete the process. Downscaling techniques allow the derivation of information from global models to regional ones through the statistic inference of the relationship between both scales. Consequently, investigators from the sub-coordination of hydrometeorology of IMTA[2] studied the results of 16 climate models addressed in the Fourth Evaluation Report (2007), from where the monthly average anomalies registered during the 1960-1989 period for the A1B scenario were obtained from the models using statistic downscaling techniques. With the estimated temperature anomalies for year 2030 (in A1B scenarios), simulations were performed to calculate ET in the principal irrigation districts of the following states: Baja California, Baja California Sur, Chihuahua, Sinaloa, Sonora, Tamaulipas, Nuevo León, Coahuila y Durango (Figure 7). Figure 8 shows, in °C, the difference between the actual and the 2030 scenarios for more than 160 stations in this states (the variability observed in Figure 8 is due to regionalization of temperature through downscaling techniques).

The result of the sensitivity analysis with the HS method is shown in Figure 9, where the differences in percentage of ET between the actual and the 2030 scenarios can be observed.

Figure 10 shows the ET isolines for July in the current scenario while Figure 11 illustrates ET isolines for the 2030 scenario.

When both figures are compared, an increment by 7% in the ET is clear between the actual scenario and the 2030 scenario, as well as a major density of isolines in 2030 that indicates greater temperature gradients.

4. CONCLUSIONS

Global climate change is a theme that has gained relevance in the last decade, and one that is studied from different perspectives: socioeconomic, technical and scientific. There is worldwide concern about the potential repercussions of climate change on natural resources; in Mexico's case, the climate change effects are worrisome, especially in the agriculture sector in the north of the country for it is a wide region with scarce water resources. One of the first studies in Mexico about these potentials risks was made about a decade ago in the Mexican Institute of Water Technology. This preliminary study consisted in the application of the FAO radiation model to estimate the climate change effects on evapotranspiration, and the results obtained were a set of ET isolines under diverse scenarios that indicated the most vulnerable zone in the country, the northern region of Mexico. In this region the annual increment of ET is 8% greater than the rest of the country for a raise of 3°C in mean temperature. Due to some limitations in this preliminary investigation (e.g., global temperature data was used without regionalizing it), it was decided to make more detailed studies to estimate the climate change effects on ET on a regional scale, using the downscaling method to adjust temperature data. In this study a new methodology to estimate the ET before climate change scenarios is introduced, which includes the selection of the Hargreaves-Samani method (HS), calibrated and compared against the Penman-Monteith ASCE method in various irrigation districts in the northern part of the country, obtaining ET estimations with a 93% precision. This procedure was applied to nine states in north Mexico: Baja California, Baja California Sur, Chihuahua, Sinaloa, Sonora, Tamaulipas, Nuevo León, Coahuila y Durango. The principal results are enunciated as follows: the ET variations between the contemporary scenario and the 2030 scenario are quite significant, according to the data of 160 meteorological stations; for temperature variations between 0.1°C to 0.45°C the corresponding ET fluctuation goes from 2% in the current scenario to 7% in the 2030 scenario. These obtained percentages are greater than the ones expected to happen for the precision of the method.

Figure 7. Principal irrigationdistricts in thenorth of Mexico: Baja California, Baja California Sur, Chihuahua, Sinaloa, Sonora, Tamaulipas, Nuevo León, Coahuila y Durango.

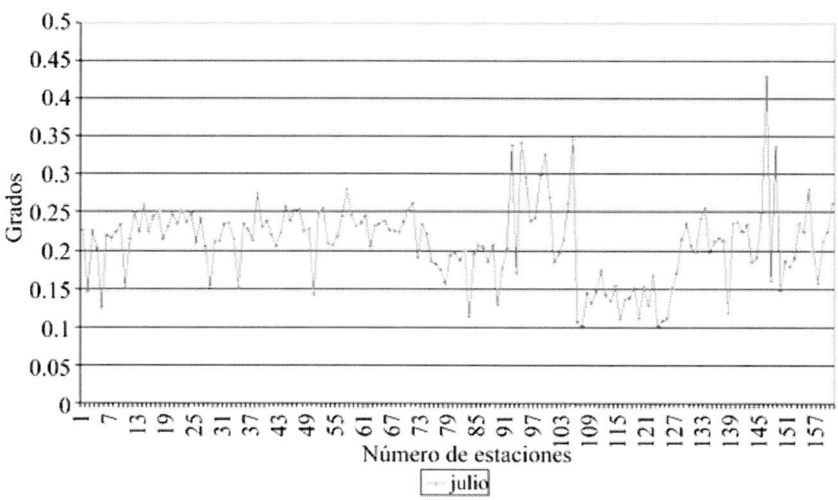

Figure 8. ET Difference between the current and the 2030 scenarios for the month of July (in °C).

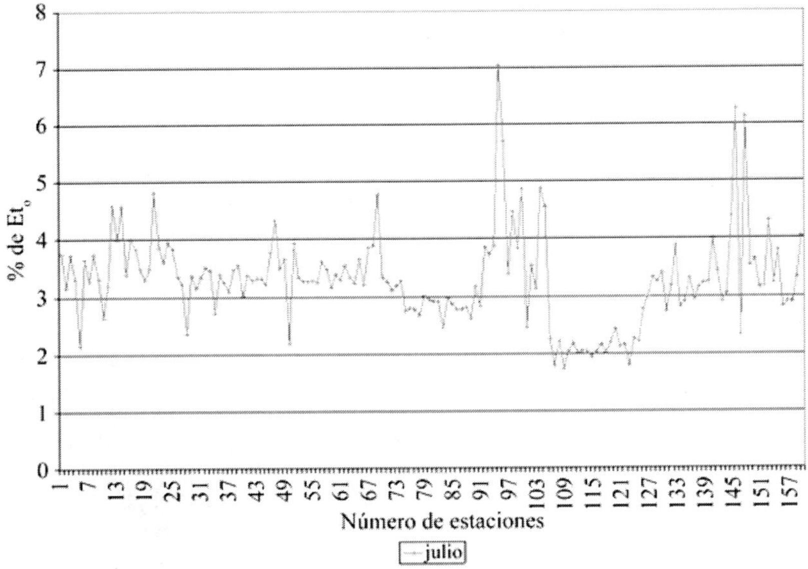

Figure 9. Percentage difference of ET for the month of July between the current and 2030 scenarios.

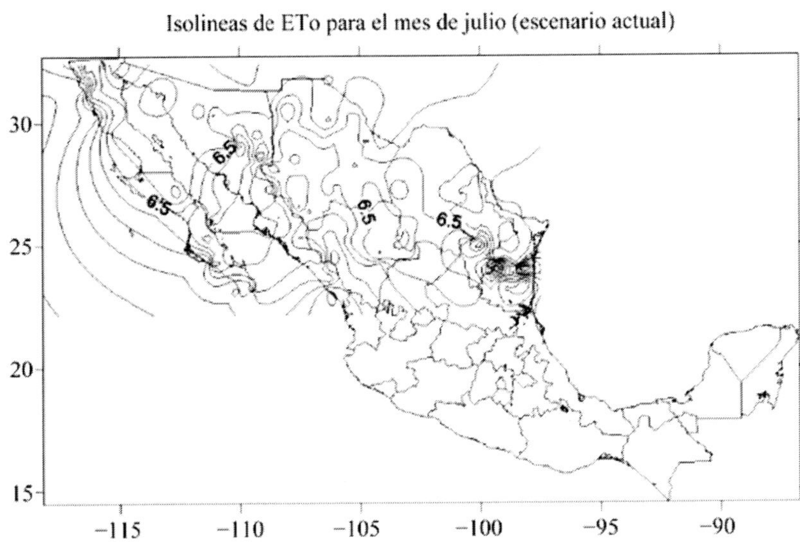

Figure 10. ET isolines in July for the current scenario.

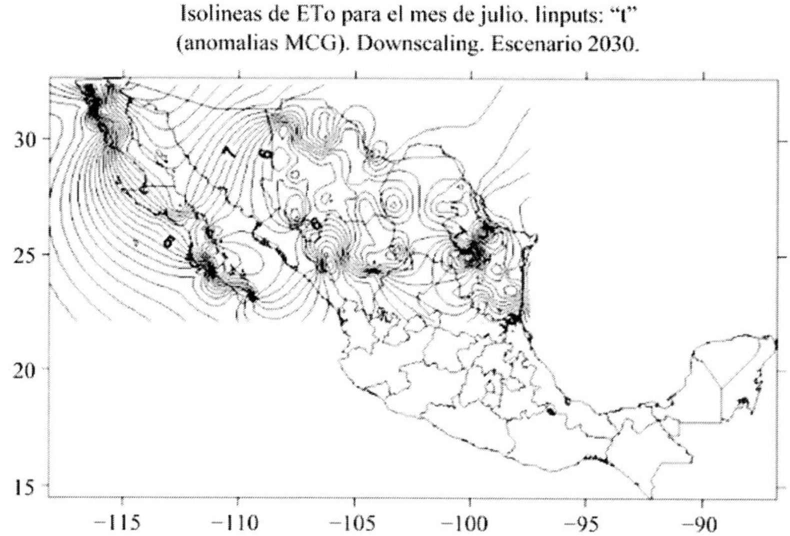

Figure 11. ET isolines in July for the 2030 scenario.

It is important to note that a 7% rise of ET (related to a regional temperature increase of approximately one degree) would represent in practice having more millions of m^3 of water in dams to satisfy the water demand of crops. With the intention to proportionate an idea of this additional water volume, data from the irrigation water volumes used in Sonora in 2004 was taken. That year, a surface of 631 200 hectares was irrigated with a total of 6315 Mm^3 of water for agricultural purposes only (superficial and underground water).

Considering that the average efficiency in water consumption for irrigation is 48%, 3283 Mm^3 of water were lost, and only 3031.20 Mm^3 were employed. The 7% of the used water volume is 211.58 Mm^3, which is almost a third of the Adolfo Ruiz Cortinez dam capacity. Therefore 211.58 Mm^3 would be the additional water demand for a 0.5°C temperature increase in Sonora. It is prudent to point out that these results are only indicative because of the limitations the evapotranspiration methods have to do ET estimations under climate change scenarios and the complexity of the theme, however, it is also important to highlight that the results allow to form an idea of the potential effect of mean temperature increase in irrigation zones. Because of the importance of the topic and the need to calculate more accurately the effects of climatic variability in irrigation and seasonal areas, the investigation must continue as well as the developing of new calculation methods that include not only the climate change effect, the needs and availability of water for crops but also the consequence of these changes on crop performance. It's important to develop and adapt methodologies that allow the estimating of crop functionality by using biological simulations, also ones that consider the potential effect of carbon dioxide and water availability. It is also important to obtain the anomaly values of "t" in a lower scale (downscaling), e.g. on a watershed scale or an irrigation district scale to obtain more accurate ET estimations under climate change scenarios.

ACKNOWLEDGEMENTS

Martín Mundo-Molina (A1B scene) and Laura Montserrat Mundo-Navarro for reviewing this paper.

REFERENCES

1. Martínez, A.P., Hernández, L. and Mundo-Molina, M. (1998) Global Warming Effects on the Water Balance in México. 7th International Conference Computer in Agriculture, Florida, 26-30 October 1998. ASAE, The Society for Engineering and Agricultural, Food and Biological System.
2. Shiklomanov, I.A. (1998) World Water Resources at the Beginning of the 21st Century. Monograph Prepared and Submitted to UNESCO, Division of Water Sciences Hydrological Institute, International Hydrological Programme (IHP), UNESCO.
3. Jensen, M.E., Burman, R.D. and Allen, R.G., Eds. (1990) Evapotranspiration and Irrigation Water Requirements. ASCE Manuals and Reports on Engineering Practices No. 70, New York, 332 pp.
4. Hargreaves, G.H. and Samani, Z.A. (1982) Estimating Potential Evapotranspiration. Journal of Irrigation and Drainage Engineering—American Society of Civil Engineers (ASCE), 108, 225-230.
5. Hargreaves, G., Hargreaves, G. and Riley, J. (1985) Irrigation Water Requirements for Senegal River Basin. Journal of Irrigation and Drainage Engineering, 111, 265-275.
6. Choisnel, E., de Villete, O. and Lacroze, F. (1992) Une approche uniformisee du calcul de Lévapotranspiration potentielle pour Lénsemble des PAYS de la communaute europeenne. Unsysteme D'Information agronomique pour La Communaute Europeenne, Centre Commun de Recherche, Commission des Communaute Europeennes.
7. Hargreaves, H.G. (1994) Defining and Using Reference Evapotranspiration. Journal of Irrigation and Drainage Engineering, ASCE, 120, 1132-1139.
8. Allen, R.G. (1995) Evaluation of Procedures for Estimating Grass Reference Evapotranspiration Using Air Temperature Data Only. Rep. Prepared for the United Nation Food and Agricultural Organization. Rome, Italy. In Jensen, D.T., Hargreaves, H.G., Temesgen, B. and Allen, R.G. (1997) Computation of Eto under Non Ideal Conditions. Journal of Irrigation and Drainage Engineering, ASCE, 123, 394-400.
9. Amatya, D.M., Skaggs, R.W. and Gregory, J.D. (1995) Comparison of Methods for Estimating REF-ET. Journal of Irrigation and Drainage Engineering, ASCE, 121, 427-435.

10. Snyder, R.L. (2000) Daily Reference Evapotranspiration (Eto) Calculator. Department of Land, Air and Water Resources, University of California. Davis, California.
11. Snyder, R.L. (2002) Epistolary Communication.

CHAPTER 19

Farmers, Food and Climate Change: Ensuring Community-Based Adaptation is Mainstreamed into Agricultural Programmes

Helena Wright[a*], Sonja Vermeulen[b], Gernot Laganda[c], Max Olupot[d], Edidah Ampaire[e] and M.L. Jat[f]

[a]Centre for Environmental Policy, Imperial College London, London, UK; [b]CGIAR Research Program on Climate Change, Agriculture and Food Security (CCAFS), Coordinating Unit, University of Copenhagen, Copenhagen, Denmark; [c]International Fund for Agricultural Development, Rome, Italy; [d]African Forum for Agricultural Advisory Services, Kampala, Uganda; [e]International Institute of Tropical Agriculture (IITA), Kampala, Uganda; [f]International Maize and Wheat Improvement Centre (CIMMYT), New Delhi, India

ABESTRACT

Climate change creates widespread risks for food production. As climate impacts are often locally specific, it is imperative that large-scale initiatives to support smallholder farmers consider local priorities and integrate lessons from successful autonomous adaptation efforts. This article explores how large-scale programmes for smallholder adaptation to climate change might link effectively with community-led adaptation initiatives. Drawing on experiences in Bangladesh, Mozambique, Uganda and India, this article identifies key success factors and barriers for considering local priorities, capacities and lessons in large-scale adaptation programmes. It highlights the key roles of extension services and farmers' organizations as mechanisms for linking between national-level and community-level adaptation, and a range of other success factors which include participative and locally driven vulnerability assessments, tailoring of adaptation technologies to local contexts, mapping local institutions and working in partnership across institutions. Barriers include weak governance, gaps in the regulatory and policy environment, high opportunity costs, low literacy and underdeveloped markets. The article concludes that mainstreaming climate adaptation into large-scale agricultural initiatives requires not only integration of lessons from community-based adaptation, but also the building of inclusive governance to ensure smallholders can engage with those policies and processes affecting their vulnerability.

Keywords: climate change; community-based adaptation; mainstreaming; agriculture; adaptation

INTRODUCTION

Climate-related risks and opportunities play a prominent role in agricultural development, but are not always recog-nized in sector programming and investment planning. Climate change is now affecting crop productivity and the ability of farmers to harvest and process agricultural produce, with direct impacts on the nearly 70% of people in developing countries living in rural areas where agricul-ture is the main livelihood (Vermeulen, Campbell, & Ingram, 2012). In many rural areas, episodes of extreme weather interrupt access to markets, while restricted liveli-hood options and insufficiently diversified energy systems perpetuate the degradation of those ecosystems which are needed more than ever as natural buffers against floods, landslides and soil erosion. Climate-related disasters can disrupt social networks and wipe out years of financial savings, rolling back decades of development progress (Carter, Little, Mogues, & Negatu, 2007). Climate change materializes predominantly as a threat multiplier for poor rural households, adding new dimensions to the portfolio of risks, opportunities and longer term trends facing people whose livelihoods depend on agriculture.

Adaptation responses to these risks and trends have been distinguished in the literature between 'autonomous adaptation' at the individual, household or farm level and 'planned adaptation' usually at the level of national govern-ment. In reality these levels are functionally linked (Adger, Huq, Brown, Conway, & Hulme, 2003; Eriksen et al., 2011), but nonetheless the distinction between them pro-vides a useful framework for distinguishing two contrasting mechanisms for mainstreaming community-based adap-tation (CBA): either by 'bottom-up' scaling up or out from local community-driven, autonomous adaptations, or alternatively by being built into the design of 'top-down' planned adaptation programmes. Both of these mechanisms offer opportunities to integrate community-based approaches to climate change adaptation with larger agricultural planning and investment processes, thereby benefitting a high number of farmers at large geo-graphic scales. As the impacts of climate change are felt

locally, there is a growing body of research arguing that community-identified and -led activities are integral to effective adaptation for smallholder farmers (Heltberg, Siegel, & Jorgensen, 2009; Kansiime, 2012; Reid et al., 2009). However, as we elaborate below, the transfer of knowledge and practice from local adaptation experience appears to be rare, despite its potential value.

At present it is unclear to what extent CBA has been mainstreamed into agricultural programmes and funding agencies across different levels. The purpose of this article is therefore to use insights from four case studies to explore the barriers to and opportunities for mainstream-ing CBA in agriculture at broader scales. The next section defines and explores the implications of key terms such as CBA, mainstreaming and scaling up, as well as summar-izing progress at national and international levels. Section 3 summarizes four cases: two experiences of mainstream-ing adaptation into agricultural planning in Mozambique and Bangladesh, and two local-level examples from Uganda and India in which community-based agricultural adaptation endeavours have made efforts to influence policy at higher levels (to scale up). We analyse the oppor-tunities and barriers for community-based approaches to provide effective inputs to higher level policy or larger scale programmes, and for large-scale programmes to respond effectively. Section 4 considers the key role of extension as a conduit for two-way learning between the national and community levels, and concludes with some emerging suggestions for more effective exchange and co-learning.

2. MAINSTREAMED VERSUS COMMUNITY-BASED ADAPTATION: FROM GAPS TO MEETING POINTS

Climate change mainstreaming is now widely promoted as a more effective approach than stand-alone interventions on climate change adaptation or mitigation. In its broadest sense, climate change mainstreaming entails incorporation of climate change considerations into public policy and practice, at all planning levels, across all sectors and invol-ving public, private and civil society actors. Substantial practical guidance now exists for policy-makers and prac-titioners to mainstream climate adaptation into develop-ment policies and programmes, including specifically in the agriculture sector (CARE, 2009; FAO, 2012a; UNDP-UNEP, 2011).

Most of this mainstreaming guidance emphasizes the importance of engaging stakeholders from the start, includ-ing local stakeholders in affected communities, and being responsive to their expressed priorities and needs. Yet most guidance fails to clarify whether the mainstreaming process is envisaged as a bottom-up process through which successful CBA actions are replicated and scaled-up to reach new target groups and geographic areas, or as a top-down process in which large-scale government action plans deliver pre-determined adaptation benefits. This question is especially relevant as it helps to determine the ambition and institutional layout of a climate main-streaming process: such a process could entail the compara-tively straightforward dissemination of new technical information and know-how through established channels (such as extension services), or the more complex strength-ening of innovative processes and institutional mechanisms through which relevant adaptation knowledge can be gen-erated or internalized.

In principle, agricultural programmes that mainstream adaptation into their planning do not preclude commu-nity-based approaches, but there may be considerable chal-lenges in reaching scale while also assuring local 'ownership' (control over decisions and resources) and accommodating the diversity that comes with differing local priorities. CBA can be defined as 'a community-led process, based on communities' priorities, needs, knowl-edge, and capacities, which should empower people to plan for and cope with the impacts of climate change' (Reid et al., 2009, p. 13). CBA builds on a long history of community-driven approaches to development, which rose to global prominence in the 1990s, following percep-tions that large-scale centralized development programmes were performing poorly, and that poor people could and should be the central decision-makers in their own develop-ment (Mansuri & Rao, 2004). It has been suggested using farmers as the agents of change is more likely to be sustain-able, while recognizing pitfalls relating to maladaptation, that is, potential negative impacts of adaptation across spatial or temporal scales (Vincent et al., 2013). While pol-icies proliferate at the national level, which can lead to duplication, households tend to adapt to multiple stresses in an integrated way (Stringer, Mkwambisi, Dougill, & Dyer, 2010). Community-driven development, in which decisions are made and budgets are allocated locally, can deliver sustainable outcomes at lower cost than centrally managed programmes, while also conferring better govern-ance in terms of local accountability, transparency and empowerment (Binswanger-Mkhize, De Regt, & Spector, 2010). Yet participation does not automatically guarantee success, which depends additionally on sustained facili-tation and careful design (Blaikie, 2007; Mansuri & Rao, 2004) as well as an enabling policy environment.

Most CBA endeavours do not arise from large-scale programmes but rather from local innovation, either driven purely by communities or else prompted and facili-tated by a non-governmental organization (NGO), develop-ment agency or research organization operating at the local scale. For these endeavours to reach scale presents a differ-ent set of challenges than for large-scale mainstreamed pro-grammes to be responsive to community-level priorities. Some barriers identified from global experiences in scaling up community-driven development initiatives included hostile institutional settings, barriers to accessing finance, lack of compatible incentives, stakeholders with differing values, geographical or socio-political differences and logistical challenges (Binswanger-Mkhize et al., 2010). However, it must not be assumed participative processes are generating an effective and consensual output: the term 'participation' is often used to describe 'very rudimen-tary levels of consultation between professionals and com-munity members' (Taylor, 2003, p. 122), in contrast to active engagement involving two-way information flows (Reed et al., 2009).

Arguably, a supportive enabling environment with par-ticipation of local communities is required to integrate adaptation into development (Sietz, Boschutzz, & Klein, 2011). Scaling up means more than just physical scaling up (mass replication); but also social scaling up (increasing social inclusiveness) and conceptual scaling up in terms of moving beyond participation to embedding empowerment in the entire development process (Binswanger-Mkhize et al., 2010). This links to the concept of 'procedural justice' in adaptation (Thomas & Twyman, 2005), leading us to seek not only mainstreaming of climate change adaptation into agriculture, but also inclusive gov-ernance whereby farming communities can engage with policies and processes affecting their vulnerability. Adapt-ing our conceptual framework from Linn (2012), we argue mainstreaming CBA requires an 'enabling environment' with institutional, political, fiscal, market, resource, cultural and learning space for CBA to occur as a process. Main-streaming CBA is more than scaling up of specific adap-tation practices or knowledge, it is about mainstreaming institutional and organizational approaches that allow this knowledge to be generated.

In policy and practice, there has been some progress in mainstreaming adaptation in agriculture. Under the United Nations Framework Convention on Climate Change (UNFCCC), many least developed countries identified agriculture as a vulnerable sector in their National Adap-tation

Programmes of Actions (NAPAs). Countries are now beginning to mainstream adaptation into agricultural policies and programmes through National Adaptation Plans (NAPs), which should enable the adoption of a 'par-ticipatory and fully transparent approach, taking into con-sideration vulnerable groups, communities and ecosystems' (UNFCCC, 2011, p. 80). Multi-lateral funds and financial agencies are in the early stages of integrating mechanisms for managing climate risks and trends into their programming. The World Bank has guidelines for mainstreaming of adaptation into agriculture and natural resources management projects (World Bank, 2010), while the International Fund for Agricultural Development (IFAD) recently established the world's single largest fund for adaptation in smallholder agriculture. IFAD's new Adaptation for Smallholder Agriculture Programme (ASAP) provides a new source of grant co-financing to scale up and integrate adaptation across IFAD's approximately US$1 billion annual new investments and introduces a systematic appraisal of climate-related risks and vulnerabilities into agricultural investment planning (IFAD, 2012a). ASAP investments under development at the country level include a range of CBA components, including participatory mapping and vulnerability assess-ment, delegation of priority-setting for spending of adap-tation funds to community groups, and mechanisms for community-to-community learning across administrative and geographic boundaries.

3. MAINSTREAMING CBA IN AGRICULTURE: INSIGHTS FROM CASE STUDIES

Case studies were selected that cover a range of different geographical regions and climate change vulnerabilities across four countries in Asia and Africa. Experiences are drawn from national-level experiences in mainstreaming adaptation into agricultural planning in Mozambique and Bangladesh using secondary sources, and from local (sub-national) level experiences in Uganda and India using primary data. Each case study introduces the relevant institutions, mechanisms and project activities (whether governmental or provided by NGOs), drawing lessons from these in terms of the success factors and barriers encountered in mainstreaming CBA into agricultural pro-gramming, and exploring the policy implications.

3.1. Mainstreaming of CBA in agriculture in Bangladesh

In Bangladesh, initial steps have been taken to mainstream processes and lessons from CBA, but there are various institutional and communication barriers. Experiences with mainstreaming climate change adaptation into agricul-tural planning under the 'Livelihood Adaptation to Climate Change' (LACC) project were reviewed using secondary literature and reports. Located in the low-lying Ganges– Brahmaputra delta, Bangladesh is at risk of increasing flooding, more intense cyclones and sea level rise in a warmer climate (Huq, Rahman, Konate, Sokona, & Reid, 2003). The LACC project under the Comprehensive Disas-ter Management Programme (CDMP) promoted livelihood adaptation among vulnerable communities, implemented jointly by the Department of Agricultural Extension (DAE) and Food and Agriculture Organisation of the UN (FAO) (Baas & Ramasamy, 2008). Project outputs included learning lessons from CBA. The project assessed existing locally specific risk-coping strategies and technologies, monitored local agro-meteorological data and downscaled climate scenarios (Baas & Ramasamy, 2008), intending to create an overlap between local and scientific knowledge (Torres, 2009). Due to the lack of reliable downscaled climate data, pilot projects focused on 'no regrets' options for field testing, such as drought-tolerant crops in the North-West (Baas & Ramasamy, 2008).

In the first pilot phase (2005–2007), mainstreaming and scaling up were not effectively addressed (Baas & Rama-samy, 2008), but in later stages, lessons were learnt and the broad-based reach of DAE's 12,000 agricultural exten-sion workers were tapped (FAO, 2010a). An independent CDMP evaluation found that LACC was successful and proposed further embedding climate-related knowledge in forthcoming projects, but found that gender issues require further attention (Russell, Mahbub, Khan, & Islam, 2009). Political turmoil and staff continuity were also chal-lenges (Luxbacher, 2011). A programme review recognized information and communication gaps, such as illiteracy and the bias towards production of printed materials, the under-utilized role of information and communication technol-ogies (ICT), and the absence of a communications plan (FAO, 2010a). To overcome these pitfalls, a deliberate effort was proposed to mainstream adaptation within national policy and development planning, including advo-cacy and policy briefs for law-makers and local officials, and better inclusion of communication activities at every level (FAO, 2010a). Collaboration with the Agricultural Information Service (AIS) and inclusion of climate change into education curricula were also recommended (FAO, 2010a). A continuing barrier was the reach of exten-sion workers, with the ratio of extension workers to farmers at 1:12,000 (FAO, 2010a). General lessons were that inte-gration of disaster risk reduction (DRR) and adaptation into operational local-level frameworks are crucial to initiate long-term processes, and there is no need to set up separate institutional structures within sectoral line agencies (FAO, 2010b). Adaptation was highlighted as a social learning process, and inclusive and participatory mechanisms can contribute to this learning (FAO, 2011).

Drawing on lessons from the previous projects, the Dis-aster and Climate Risk Management in Agriculture (DCRMA) project aims to mainstream disaster and climate risk management in the DAE and strengthen its capacity. There is now collaboration with AIS in dissemi-nating success stories from the grassroots level. In a current project 'Agricultural Adaptation in Climate Risk Prone Areas of Bangladesh', many lessons from the LACC are being built upon, including collaboration with farmer field schools, as well as improving community-based early warning systems and rural communication ser-vices (BCCRF, 2013). The project seeks to focus on com-munity-based and field-level adaptive research and participatory extension approaches, as well as commu-nity-based seed and grain storage infrastructure, water har-vesting and small-scale irrigation, drawing on local knowledge to develop regional agro-ecological databases and community-based DRR plans (DAE, 2013). Replica-tion and scaling up of agricultural adaptation options are intended to occur through farmer clubs and water-management groups (DAE, 2013). It was argued there is no need to create separate 'climate field schools' because farmer field schools will iteratively adjust to climate-related changes like salinity, but communication may be needed to ensure effective innovation (AEC, 2011).

Overall, the case provides nascent evidence of main-streaming of climate change into the activities of national agricultural institutions and programmes in Bangladesh, building upon years of progress in collaboration with FAO. Climate is also being integrated into the research pri-orities of the Bangladesh Agricultural Research Council, particularly the high-priority areas of climatic impacts on fisheries, water resources management, forests and disaster management (Hussain & Iqbal, 2011). However, while CBA approaches are articulated within project documents under the DCRMA, it is too early to explore whether these will be effective in practice. Insufficient capacity and lack of coordination among research scientists, exten-sion workers and farmers remain key challenges for the forthcoming project (Rahman, 2011). Furthermore, biodi-versity loss and inadequate use of indigenous knowledge in food-related contexts are additional barriers to adaptation (Mallick, Amin, & Rahman, 2012). Lack of inclusion of farmers in research and general lack of awareness about climate change are also barriers. Mainstreaming adaptation in agriculture is an on-going process in Bangladesh. Pro-posed ways forward include greater communication efforts, coordination amongst stakeholders and collabor-ation with existing organizations including farmer field schools.

3.2. Adapting to climate change in semi-arid environments in Mozambique

In Mozambique, experiences with mainstreaming CBA show that major challenges include the capacity of farmers' organizations and extension services, farmers' access to markets and coordination across implementing agencies, particularly at local levels. National experiences with mainstreaming adaptation into agriculture were reviewed using secondary reports and literature, with a focus on the three-year (2008–2010) UN Joint Program (UNJP) on Environmental Mainstreaming and Adaptation to Climate Change, which aimed to help Mozambique inte-grate climate change into national policy and set up pilot adaptation projects. The programme was designed to align with government planning and strategies, create synergies and avoid duplication. Pilot projects were implemented in Chicualacuala, Gaza Province (FAO, 2011). The activities arguably contributed to realization of Mozambique's NAPA (FAO, 2012b), which prioritizes early warning systems, increasing producer capacity and management of water resources. For example, to strengthen early warning systems, the UNJP assisted the government by rehabilitating and re-equipping a weather station and expanding the reach of the Chicualacuala community radio station. In an area where livestock is crucial to livelihoods, a network of trained community animal health workers (CAHW) was established. Taking actions at the community level meant problems could be identified more accurately and locally appropriate preventive measures could be taken.

An independent evaluation found that the establishment of community groups was highly effective, but the project's relevance was reduced by not fully responding to commu-nities' market-related challenges (Eucker & Reichel, 2012). It was recommended that adaptation strategies ought to be flexible, with a greater focus on 'how' results are achieved (process) rather than 'what' is achieved (Eucker & Reichel, 2012). A further study in Gaza showed communities have multiple viable strategies for reducing climate risks, includ-ing livestock management and livelihood diversification, which could be expanded and strengthened through a greater service provision by the government, notably weather forecasts and climate information services (Sacra-mento, Matavel, Basílio, & Bila, 2012). Expert interviews revealed various institutional barriers to mainstreaming adaptation, including lack of human resources, insufficient data, lack of inter-institutional coordination and communi-cation and scarce financial resources (Sietz et al., 2011). In Mozambique, underlying structural issues such as weak markets for agricultural commodities, poor infrastructure and limited access to micro-finance exacerbate difficulties for smallholder farmers (Osbahr, Twyman, Adger, & Thomas, 2008).

Communities involved in the programme identified that the human resources most important for their livelihoods were health, education, farming skills and extension ser-vices (farm and veterinary) (FAO, 2012b). Farmers' organ-izations also provided social capital for adaptation (FAO, 2012b). The National Directorate of Agrarian Extension (DNEA) is the main institution responsible for agricultural extension. It does not have any climate-specific pro-grammes, demonstrating that adaptation is insufficiently mainstreamed at present. However, Mozambique's Third Poverty Reduction Strategy (PARPA III) identified some climate-related activities, including water-management and improved seed varieties. One challenge is that Mozam-bique's extension services are relatively new, formed only in 1987 amidst a challenging political environment (Gemo, Eicher, & Eclemariam, 2005). It is difficult to finance extension services in subsistence and semi-subsis-tence economies without taxable agricultural exports (Eicher, 2004). The majority of farmers face challenges accessing extension services, with some 2000 extension workers covering a rural population of over 14 million (FAOSTAT, 2012). This demonstrates a need to scale up existing extension services to increase farmers' food secur-ity and resilience. Adaptation is now being integrated in the ASAP-supported value chain development project in Mozambique, with extension services recognized as a barrier, as well as gaps in financial services and smallholder market access (IFAD, 2012b). Building the capacity of farmers' organizations is another key priority (IFAD, 2012b). Overall, although Mozambique has a supportive national-level legislative environment and awareness among donors is high, there is still limited institutional capacity for mainstreaming initiatives at provincial and dis-trict levels (Sietz et al., 2011). In spite of the decentraliza-tion process, lack of communication, coordination, funding and poor information dissemination impede mainstreaming adaptation at local levels.

3.3. Climate-smart adaptation research in Rakai District, Uganda

In Uganda, the main finding is that non-functional policies and regulations at national or sub-national levels inhibit mainstreaming of CBA. Participatory research has been undertaken by the International Institute of Tropical Agri-culture (IITA) with producers in assessment of vulner-ability and evaluation of adaptation options, as part of the Climate Change Agriculture and Food Security (CCAFS) research programme. Climate change threatens to decrease yields, reduce farm revenues, worsen food insecurity and deepen rural poverty (Nabikolo, Bashaasha, Mangheni, & Majaliwa, 2012; UNDP, 2013; Waithaka, Nelson, Thomas, & Kyotalimye, 2013). Participatory vulnerability assessments were conducted in Rakai

District to capture interactions among biophysical, social, political–institutional, socio-cultural, economic and environmental vari-ables. Twenty focus-groups were conducted in 10 different zones, 2 per zone, separating men and women to capture gendered differences in perceptions. Participatory discussions were held on climatic and environmental changes that have occurred in the last 2–3 decades, changes to farming practices, climate constraints experi-enced and adaptation practices farmers use to cope with climate challenges. In addition, in-depth key informant interviews were conducted with selected smallholder farmers, political leaders, public extension entities, NGOs and businesses (agro-produce marketers and agro-input dealers). Furthermore, a formal survey that utilized struc-tured questionnaires was administered to individual farmers selected randomly from sites, to complement the analysis and interpretation of findings (Kyazze & Kristjan-son, 2011). The research aimed to analyse collected infor-mation, and generate and present different climate-smart scenarios to male and female producers to help them evaluate their applicability and sustainability. Smallholder farmers took centre stage in developing 'climate-smart' options on the premise that effective participation of vul-nerable communities is likely to enhance design, adoption and ownership of adaptation.

Although what works in a pilot might not necessarily work elsewhere, lessons so far learnt from the project provide useful insights by highlighting success and con-straining factors that could be applied by large-scale programmes that intend to scale up CBA. The participation of a broad range of stakeholders enabled shared learning and fostered commitment to undertake actions. Since the climate-smart options agreed upon measure up to realities on the ground, their adoption has relatively higher chances of sustainability. Sustainability strategies have been weaved in right at the start to avoid over-dependence, including identifying the right stakeholders for institutional support, building relevant capacities of different stakeholders, specifying roles for different actors and securing commitment from them to deliver on roles.

Various factors may constrain adoption of climate-smart options if they are not dealt with. In focus group dis-cussions, it emerged that local policies exacerbated farmer vulnerability; for instance in the past farmers had access to communal grazing lands which were utilized during periods of fodder scarcity. Similarly, farmers used to produce crops in wetlands during droughts and return to their upland plots during the rainy season (Turyahabwe, Kakuru, Tweheyo, & Tumusiime, 2013). Yet communal grazing lands and wetlands were leased out by the district land board to a few well-off farmers, who have either fenced them off or used them to establish commercial euca-lyptus woodlots (Ampaire, 2013). As a result, poor small-holder farmers no longer have access to these resources. In addition, planting of eucalyptus in the wetlands resulted in lowering of the water table and drying of community wells (Ampaire, 2013). Women and children were particu-larly affected as they have to travel up to four kilometres to fetch water during the dry season. There was little incentive for adopting 'climate-smart' practices such as agroforestry, as the land is limited and farmers cannot accommodate the time lag on the return on investment.

Based on these insights, a more detailed study was con-ducted to understand policy formulation and implemen-tation processes and constraints to CBA. A range of national-level policies and regulations currently exist on the paper to guide access to and use of natural resources, but these are almost non-functional at local levels (Rwaka-kamba, 2009). Examples include the Uganda Forestry Policy (2001), National Environment Act (1998) and National Wetlands Policy (1995). Findings affirm that CBA is constrained by lack of policy implementation, which is brought about by multiple factors, including exclusion of implementers in the formulation process, inadequate knowledge about policies (Glass, 2007), poor coordination among actors, lack of clarity in roles, limited resources and political interference, coupled with corruption. In addition, land tenure is insecure and the Uganda National Land Policy was only recently approved in 2013, following prior allegations of land grabbing from land owners, media and civil society. At present, smallholders feel helpless as there are no laws implemented to assure their access to land and other natural resources. Climate-smart solutions will only be effective if there is political will among national and local leaders to address constraints jointly. IITA and partners are planning policy engagement actions at national and sub-national levels, and sharing evidence both on technical issues, such as 'no regret' climate-smart technologies in coffee–banana systems, and on institutional issues, such as gender inclusion and resource access. While it is too early to explore whether specific measures will be mainstreamed at national levels, the aim is to enable more inclusive implementation of natural resource use policies.

3.4. Participatory research and mobilization of young farmers in Karnal District, India

The key lesson from India is that CBA can mobilize young farmers and provide a platform for scaling out adaptation technologies. The International Maize and Wheat Improve-ment Centre (CIMMYT), under the CCAFS research pro-gramme, is conducting participatory action research with farmers in Karnal, an agriculturally vibrant region of Haryana state

(Aggarwal et al., 2010). This case study draws on primary data including household surveys (Singh, 2013) and participatory technology evaluation trials undertaken in the climate-smart villages, in which data were collected by CIMMYT–CCAFS (Table 1). Since the mid-1960s, increases in agricultural productivity, rapid industrial growth and expansion of the non-formal rural economy have quadrupled per capita Gross Domestic Product (GDP), and markedly reduced poverty. However, securing these gains is becoming a challenge in the context of soaring food and fuel prices, volatile markets, global economic downturn, diversion of human capital from agriculture, soil degradation, shrinking farm sizes, depletion of water resources and overarching effects of climate change (Ambast, Tyagi, & Paul, 2006; Humphreys et al., 2010; Jat et al., 2012). Climate change is projected to lead to uncertain onset of monsoons and more frequent extremes of weather (Aggarwal, Joshi, Ingram, & Gupta, 2004). Also, other competitive sectors and schemes such as 'MNREGA' (Mahatma Gandhi National Rural Employment Guarantee Act) have diverted farm labour. Conven-tional agricultural technologies, farming practices and linear out-scaling approaches (Swanson, 2008) under emer-ging climatic risks further exacerbate the challenges and make farming unattractive to farmers in general, and youth and women in particular (GWA, 2012).

Conservation agriculture (CA) management technol-ogies offer some solutions to the emerging challenges of climate change across the Ganges River basin, including Haryana, by maintaining soil fertility and water-holding capacity in conditions of unpredictable monsoons (Erenstein, Farooq, Malik, & Sharif, 2008; Gathala et al., 2011; Jat et al., 2009). Out-scaling (replication) of these relatively knowledge-intensive technologies and practices is more difficult than Green Revolution technologies (new seeds, fertilizers and irrigation). Significant efforts are being made on development and dissemination of new technologies through various institutions, but adoption remains slow. Major bottlenecks include the increasing average age of farmers, traditional mindsets, youth moving out of farming, individualistic and linear technol-ogy development, adaptation and dissemination.

Discussing with communities the ways to break the impasse, CIMMYT decided to undertake technology devel-opment with young farmers in the belief that engaging young farmers in CBA will facilitate adaptation and adop-tion of new technologies. CIMMYT also recognized the advantage of bringing young farmers together to influence policy-makers to support technology promotion, targeting not only adaptation and mitigation but also improving farm profitability and generating alternate employment for rural youth through technology-led business opportu-nities. The other perceived benefit was to develop suitable institutional mechanisms for buying and sharing assets such as expensive farm machinery, and for using resources more precisely at community-level. CIMMYT interacted with a group of young farmers from Taraori village, Karnal District. The response was overwhelming. During the interactions, farmer groups showed keen interest in new-generation technologies to resolve problems of seeding rice with less labour, precision in levelling to save irrigation water, residue management towards improv-ing soil fertility and water-holding capacity, eliminating tillage to save on fuel, energy and water, and improving nutrient-use efficiency. All of these actions enhance rural livelihoods by increasing farmers' incomes, thus reducing vulnerability (Aggarwal et al., 2004), and enabling adap-tation benefits such as ability to respond rapidly and cost effectively to delayed and unpredictable start of the growing season, or minimizing crop losses during dry periods.

Table 1. Climate-smart technologies adopted and disseminated, and their monetary advantages.

	Technologies adapted and disseminated	Climate-smart category	Yield gains over local farmers' practices (kg ha^{-1})	Monetary gains over local farmers' practices (US$ ha^{-1})	Number of farmers who benefited
1	Laser land levelling	Water smart	480	144	250
2	No-till wheat with residue retention (turbo seeder)	Carbon, energy and water smart	600	174	60
3	Direct dry seeded rice	Water and energy smart	00	180	60
4	Site-specific nutrient management, nutrient expert decision support tool rice–wheat system	Nutrient smart	550	127	82
5	GreenSeeker sensor guided nitrogen application	Nutrient smart	275	72	10
6	*Diversification/intensification*				
A	Relay mungbean in wheat	Carbon smart	855	217	5
B	Dual purpose wheat		1230	313	5
C	Introduction of maize replacing rice	Water smart	[a]	315	2

[a]Two different crops, hence comparisons of yield are not made.

Enthusiasm was so high that a group of 20 young farmers from the village took the initiative to form a society registered as 'Society for Conservation of Natural Resources and Empowering Rural Youth'. Since the incep-tion of this society, policy-makers have visited and inter-acted with these farmers to learn more about resource-efficient, climate-smart and profitable technologies. Also, as farmers' participation in technology development and adaptation is critical, the CCAFS research programme established a participatory strategic research platform at the village level to serve as a capacity-building aware-ness-creation platform for different stakeholders. As sum-marized in Table 1, adaptive technologies were demonstrated and disseminated to a large number of farmers in the local area, and more widely across Haryana. Input and output data were collected from selected farmers using a simple checklist, and subsequently market prices for input costs and crop production were used to calculate net returns.

Farmers of the society have been active in publicizing the technologies through print and electronic media, includ-ing local and national newspapers, on television, and par-ticipation in various state, national and international level meetings. Recognizing the innovative contributions of these young farmers, the State-level Innovative Farmer Award was presented to them by the Chief Minister of Haryana in December 2012. The Chief Minister also announced state-wide incentives for community-based climate-smart and resource-efficient technologies, primar-ily CA and resource-efficient mechanization, providing evi-dence that CBA is being mainstreamed at the sub-national level. The new Haryana State Agriculture Policy, adopted in 2014, recognizes emerging threats due to climate change, and emphasizes adaptive measures to minimize these consequences (ADH, 2014). The fact adaptation actions were community-based captured national media and political attention. This village became a role model for rural youth in five more such young farmer coopera-tives. CBA built the capacity of young farmers so they not only adopted new technologies but also provided ser-vices to other farmers to earn money. An important element of this was the participatory approach and non-linear flow of information. In traditional extension systems, different organizations work in isolation and often deliver conflicting messages to farmers, but in this approach, farmers formed a common platform to debate and reach consensus on the new technologies.

DISCUSSION

There are emerging evidence adaptation programmes and strategies for agriculture and are more likely to be effective if they directly involve communities that are innovating and implementing CBA at local levels. In India, the enthu-siasm of young farmers in community-based organizations enabled the adoption, piloting and subsequent wider disse-mination of adaptation technologies, overcoming social barriers to adaptation. It is perhaps unsurprising this suc-cessful case study has targeted younger farmers, who are most likely to be interested in the long-term future of farming. In Uganda, a participatory approach to assess-ment of climate-related risks and vulnerabilities and devel-opment of profitable climate-smart options captured the priorities and preferences of different categories of local stakeholders and created local ownership. These insights are particularly relevant when we consider the locally specific impacts of climate change and the uncertainty about how impacts may manifest themselves. Local per-ceptions by farmers about climate can also be matched up with meteorological data at national weather stations, as has been done in Uganda (Osbahr, Dorward, Stern, & Cooper, 2011). Recognizing adaptation as a process of social learning (Collins & Ison, 2009), policy-makers will do well to recognize the value of locally specific knowledge from CBA, through farmer field schools and other means.

The major opportunity to bring social learning on adap-tation to the national level is via existing advisory services rather than creation of new networks and institutions. Extension services provide a vital social learning role (Eicher, 2004) and are a medium through which farmers access climate-relevant information, including market information and technologies, and so through which CBA could potentially be scaled out. But in many cases, includ-ing Bangladesh and Mozambique the current reach of extension services is limited. For example, in Mozambique, extension services are available only to a minority of farmers (from 4% in Inhambane to 7% in Maputo province; IFAD, 2012b). Furthermore, extension services usually target relatively wealthier households and thus may not reach the most vulnerable (Cunguara & Moder, 2011), pre-senting a challenge for scaling up CBA.

Mainstreaming CBA in agriculture faces particular institutional, social, policy, market and financial barriers. In Mozambique, there are barriers to national-level main-streaming related to non-alignment of policies, strategies and plans, poor institutional coordination, and limited human and financial resources. In Bangladesh, key pro-blems identified were lack of inter-agency coordination, communication barriers and literacy gaps. These barriers to CBA, including communication and literacy gaps, are highlighted in other studies (Spires, Shackleton, & Cundill, 2014). Proposed solutions that may be cost-effec-tive as well as institutionally feasible include: farmers' organizations and low-cost platforms for shared learning on adaptation, particularly across local and national levels; better use of non-print communications media to overcome literacy gaps, in particular verbal communi-cations by radio, mobile phones and face-to-face exchange; as well as tackling market barriers through trade reform, improved transport and storage facilities.

Furthermore, CBA activities implemented in these case studies were not necessarily responding to climate change but also to other challenges, such as conserving fuel and water in India. Climate factors may have less influence than other socio-economic stresses in shaping agricultural livelihoods (Mertz, Mbow, Reenberg, & Diouf, 2009; Zier-vogel, Bharwani, & Downing, 2006). Successful adap-tation policy in agriculture will thus need to create synergies with agricultural development to enhance adap-tive capacity, while recognizing that 'modern farming' can have both positive and negative impacts on adaptive capacity (Dixon, Stringer, & Challinor, 2014).

In the Ugandan case, gaps in the implementation of existing land and forestry legislation made farmers more vulnerable to climate risks. Overall, CBA can be con-strained by both lack of policy implementation, and by policy implementation. Existing institutions may not be inclusive or community-focused, limiting the extent to which local-level CBA can mainstreamed (scaled up) into national policies. These findings support the view that scaling up requires an enabling policy, political and insti-tutional space, as well as financial and market space for an initiative to grow (Linn, 2012). Mainstreaming CBA needs to move beyond identifying and promoting best prac-tices, towards tackling drivers of vulnerability and institu-tionalizing an enabling environment for CBA to occur as a process.

CONCLUSIONS AND IMPLICATIONS

This article draws upon large-scale and local-level cases in Bangladesh, Mozambique, Uganda and India to appraise the opportunities and barriers for community-based approaches to provide effective inputs to higher level policy or larger scale programmes, and for large-scale pro-grammes to respond effectively. Extension services and farmers' organizations are highlighted as mechanisms for linking between national-level and community-level adap-tation, while success factors include participative and locally driven vulnerability assessments and tailoring of adaptation technologies to local contexts, mapping local institutions and working in partnership across institutions. Barriers include weak governance, gaps in the regulatory and policy environment, high opportunity costs, low lit-eracy and underdeveloped markets.

Mainstreaming CBA in agriculture raises issues of what constitutes 'additionality' in adaptation. Is it sufficient to mainstream treatment of climate risks into existing agricul-tural development programmes and extension services, or is there need to extend such services in climate-vulnerable areas? The evidence from Bangladesh and Mozambique suggests mainstreaming adaptation into existing services may not be sufficient to reduce vulnerability, and additional investment is required to scale up support. The new ASAP-supported IFAD programme in Mozambique recognizes that access to technology, extension and infrastructure alone does not demonstrably increase household incomes if these efforts are isolated from access to value chains (IFAD, 2012b), and is thus investing in the fundamentals of rural development (infrastructure, market access and information). Implicit in this is an understanding that the key to building adaptive capacity is to address the existing 'development deficit' (Parry et al., 2009). Crucially, main-streaming must not become a 'ploy by developed countries' to avoid providing additional adaptation finance (Klein, 2010, p. 45).

Since CBA encounters barriers at both national and sub-national levels, approaches or policies may be needed to overcome these gaps at different scales. Further research is needed on policies and reforms to strengthen adaptation as a social learning process, recognizing that it may be necessary to address barriers at broader scales (institutions, regulations or markets) in order to overcome local resource constraints. Creating space for both development partners and farmers to convey their adaptation priorities to policy-makers is a novel and potentially important modality for adaptation mainstreaming in agriculture, again a social learning process (Kristjanson, Harvey, Van Epp, & Thornton, 2014). It may be prudent to consider a broad interpretation of mainstreaming, to include improve-ment of structural and legal frameworks towards the objec-tive of reducing the underlying vulnerability of all farmers. The experiences reported in this article demonstrate that mainstreaming adaptation in agriculture needs to go beyond 'climate-proofing' agricultural development, towards tackling the underlying drivers of poverty that exacerbate vulnerability and constrain adaptation.

REFERENCES

1. Adger, N., Huq, S., Brown, K., Conway, D., & Hulme, M. (2003). Adaptation to climate change in the developing world. Progress in Development Studies, 3(3), 179–195.
2. ADH. (2014). Haryana state agriculture policy. Haryana: Agriculture Department. Retrieved June 16, 2013, from http://agriharyana.nic.in/ Agriculture% 20Policy/English% 20Haryana_State_Agriculture_Policy_Draft.pdf
3. AEC. (2011). Farmer field schools in the agricultural extension component (AEC). Retrieved September 22, 2013, from http://bangladesh.ipm-info.org/library/ documents/aec_ffs_ process_documentation.pdf
4. Aggarwal, P.K., Ayyappan, S., Chand, R., Gupta, H.S., Gupta, R., Jat, M.L., ... Virmani, S.M. (2010). State of Indian agriculture The Indo-Gangetic Plain. National Academy of Agricultural Sciences, 56, Retrieved from http://naasindia.org/documents/ Annual%20Report%202010-11.pdf
5. Aggarwal, P.K., Joshi, P.K., Ingram, J.S.I., & Gupta, R.K. (2004). Adapting food systems of the Indo-Gangetic Plains to global environmental change: Key information needs to improve policy formulation. Environmental Science & Policy, 7(6), 487–498.
6. Ambast, S.K., Tyagi, N.K., & Paul, S.K. (2006). Management of declining groundwater in the trans-Indo-Gangetic Plain (India): Some options. Agricultural Water Management, 82 (3), 279–296.
7. Ampaire, E. (2013). Unpublished IITA report, internal document. Baas, S., & Ramasamy, S. (2008). Community-based adaptation in action. Environment and

Natural Resources Management Series 14. Rome: FAO. Retrieved from http://www. preventionweb.net/english/professional/publications/v.php?id=8311

8. BCCRF. (2013). Retrieved from http://bccrf-bd.org Binswanger-Mkhize, H.P., De Regt, J.P., & Spector, S. (2010).
9. Local and community driven development: Moving to scale in theory and practice. New Frontiers of Social Policy, Washington, DC: World Bank.
10. Blaikie, P. (2007). Is small really beautiful? Community-based natural resource management in Malawi and Botswana. World Development, 34(11), 1942–1957.
11. CARE. (2009). Mainstreaming climate change adaptation. Retrieved September 22, 2013, from http://www.careclimatechange.org/ files/adaptation/CARE_VN_Mainstreaming_Handbook.pdf
12. Carter, M.R., Little, P.D., Mogues, T., & Negatu, W. (2007). Poverty traps and natural disasters in Ethiopia and Honduras. World Development, 35(5), 835–856.
13. Collins, K., & Ison, R. (2009). Living with environmental change: Adaptation as social learning. Environmental Policy and Governance, 19(6), 351–357.
14. Cunguara, B., & Moder, K. (2011). Is agricultural extension helping the poor? Evidence from rural Mozambique.Journal of African Economics, 20(4), 562–595.
15. DAE. (2013). Draft project outline documents. Unpublished manuscript. Department of Agricultural Extension (DAE).
16. Dixon, J.L., Stringer, L.C., & Challinor, A.J. (2014). Farming system evolution and adaptive capacity: Insights for adaptation support. Resources, 3(1), 182–214.
17. Eicher, K.C. (2004). Mozambique: Building African models of agricultural extension. In W. Rivera & G. Alex (Eds.),
18. Extension reform for rural development (pp. 12–19). Agriculture and Rural Development Discussion Paper 12, Washington, DC: World Bank.
19. Erenstein, O., Farooq, U., Malik, R.K., & Sharif, M. (2008). On-farm impacts of zero tillage wheat in South Asia's rice–wheat systems. Field Crop Research, 105, 240–252.
20. Eriksen, S., Aldunce, P., Bahinipati, C.S., Martins, R.D., Molefe, J.I., Nhemachena, C., ... Ulsrud, K. (2011). When not every response to climate change is a good one: Identifying prin-ciples for sustainable adaptation. Climate and Development, 3(1), 7–20.
21. Eucker, D., & Reichel, B. (2012). Final evaluation: Environment mainstreaming and adaptation to climate change. MDG Fund. Retrieved from http://www.mdgfund.org/sites/default/ files/Mozambique%20-%20Environment%20-%20Final%20Evaluation%20Report.pdf
22. FAO. (2010a). Communication assessment and action plan for the LACC Project. CSDI Technical Paper, FAO, Rome. Retrieved from http://www.africa-adapt.net/media/resources/467/CSDI. LACC%20Project.pdf
23. FAO. (2010b). General lessons from livelihood adaption to climate change project. FAO, Rome. Retrieved from http:// www.fao.org/climatechange/24647–05e72d0b0d6 6b38b187f 892ca94bdaa3b.pdf
24. FAO. (2011). Lessons from the field: Experiences from FAO climate change projects. FAO, Rome. Retrieved from http:// www.fao.org/docrep/014/i2207e/i2207e.pdf
25. FAO. (2012a). Incorporating climate change considerations into agricultural investment programmes: A guidance document. Rome: FAO Investment Centre. Retrieved from http://www. fao.org/docrep/016/i2778e/i2778e.pdf
26. FAO. (2012b). Adaptation to climate change in semi-arid environ-ment: Experience and lessons from Mozambique.
27. Environment and Natural Resource Management, 19, Retrieved from http://www.fao.org/docrep/015/i2581e/ i2581e00.pdf
28. FAOSTAT. (2012). Retrieved from http://www.faostat.fao.org Gathala, M., Ladha, J.K., Balyan, V., Saharawat, Y.S., Kumar, V.,

29. & Sharma, P.K. (2011). Effect of tillage and crop establish-ment methods on physical properties of a medium-textured soil under 7-year rice–wheat rotation. Soil Science Society of America Journal, 75, 1–12.
30. GCWA. (2012). Proceedings of first Global Conference on Women in Agriculture (GCWA), 13–15 March, 2012, New Delhi, India. Indian Council of Agricultural Research and Asia Pacific Association of Agricultural Research Institutions, p. 82.
31. Gemo, H., Eicher, C., & Eclemariam, S. (2005). Mozambique's experience in building a national extension system. Michigan: Michigan State University Press.
32. Glass, S. (2007). Implementing Uganda's national wetlands policy: A case study of Kabale district. Independent Study Project (ISP) Collection. Paper 101. Retrieved from http:// digitalcollections.sit.edu/isp_collection/101
33. Heltberg, R., Siegel, P., & Jorgensen, S. (2009). Addressing human vulnerability to climate change: Toward a 'no regrets' approach. Global Environmental Change, 19(1), 89–99.
34. Humphreys, E., Kukal, S.S., Christen, E.W., Hira, G.S., Balwinder-Singh, Sudhir-Yadav, & Sharma, R.K. (2010). Halting the ground water decline in north-west India – Which crop technologies will be winner. Advances in Agronomy, 109, 155–217.
35. Huq, S., Rahman, Q., Konate, M., Sokona, Y., & Reid, H. (2003).Mainstreaming adaptation to climate change in least devel-oped countries (LDCs). London: IIED.
36. Hussain, S., & Iqbal, A. (2011). Research priorities in Bangladesh agriculture. Dhaka: Bangladesh Agricultural Research Council.
37. IFAD. (2012a). Adaptation for smallholder agriculture pro-gramme (ASAP). Rome, Italy. Retrieved from http://www. ifad.org/climate/asap/asap.pdf
38. IFAD. (2012b). Pro-poor value chain development in the Maputo and Limpopo corridors (PROSUL). Retrieved from http://www.ifad.org/ operations/projects/ design/106/ mozambique.pdf
39. Jat, M.L., Gathala, M.K., Ladha, J.K., Saharawat, Y.S., Jat, A.S., Kumar, V., ... Gupta, R. (2009). Evaluation of precision land leveling and double zero-till systems in the rice–wheat rotation: Water use productivity, profitability and soil physical properties. Soil Tillage Research, 105, 112–121.
40. Jat, M.L., Malik, R.K., Saharawat, Y.S., Gupta, R., Bhag, M., & Paroda, R. (Eds.). (2012). Regional dialogue on conservation agricultural in South Asia. Proceedings and recommen-dations, New Delhi, India, 1–2 November 2011, p. 34.
41. Kansiime, M. (2012). Community-based adaptation for improved rural livelihoods: A case in eastern Uganda. Climate and Development, 4(4), 275–287.
42. Klein, R.J.T. (2010). Mainstreaming climate adaptation into development: A policy dilemma. In A. Ansohn & B. Pleskovic (Eds.), Climate governance and development, Berlin Workshop Series 2010 (pp. 35–52). Washington, DC: World Bank.
43. Kristjanson, P., Harvey, B., Van Epp, M., & Thornton, P.K. (2014). Social learning and sustainable development. Nature Climate Change, 4, 5–7.
44. Kyazze, F.B., & Kristjanson, P. (2011). Summary of baseline household survey results: Rakai district, south central Uganda. Copenhagen, Denmark: CGIAR Research Program on Climate Change, Agriculture and Food Security (CCAFS).
45. Linn, J. (Ed.). (2012). Scaling up in agricultural, rural develop-ment and nutrition. IFPRI. Retrieved from http://www.ifpri. org/publication/scaling-agriculture-rural-development-and-nutrition
46. Luxbacher, K. (2011). Bangladesh' comprehensive disaster man-agement programme. Inside Stories: CDKN.
47. Mallick, D., Amin, A., & Rahman, A. (2012). Case study on climate compatible development in agriculture for food secur-ity in Bangladesh. Bangladesh Centre for Advanced Studies (BCAS), Dhaka, September 2012. Retrieved from https:// germanwatch.org/en/download/8347.pdf
48. Mansuri, G., & Rao, V. (2004). Community-based and -driven development: A critical review. The World Bank Research Observer, 19(1), 1–39.

49. Mertz, O., Mbow, C., Reenberg, A., & Diouf, A. (2009). Farmers perceptions of climate change and agricultural adaptation strategies in rural Sahel. Environmental Management, 43, 804–816.
50. Nabikolo, D., Bashaasha, B., Mangheni, M., & Majaliwa, J.G.M. (2012). Determinants of climate change adaptation among male and female headed farm households in eastern Uganda. African Crop Science Journal, 20(2), 203–212.
51. Osbahr, H., Dorward, P., Stern, R., & Cooper, S. (2011). Supporting agricultural innovation in Uganda to respond to climate risk: Linking climate change and variability with farmer perceptions. Experimental Agriculture, 47, 293–316.
52. Osbahr, H., Twyman, C., Adger, W.N., & Thomas, D.S.G. (2008). Effective livelihood adaptation to climate change disturbance: Scale dimensions of practice in Mozambique. Geoforum, 39 (6), 1951–1964.
53. Parry, M., Arnell, N., Berry, P., Dodman, D., Fankhauser, S., Hope, C., ... Wheeler, T. (2009). Assessing the costs of adap-tation to climate change. London: IIED.
54. Rahman, M.M. (2011). Country report: Bangladesh, ADBI-APO workshop on climate change and its impact on agriculture, Seoul, Republic of Korea, 13–16 December 2011.
55. Reed, M.S., Graves, A., Dandy, N., Posthumus, H., Hubacek, K., Morris, J., ... Stringer, L.C. (2009). Who's in and why? A typology of stakeholder analysis methods for natural resource management. Journal of Environmental Management, 90, 1933–1949.
56. Reid, H., Alam, M., Berger, R., Cannon, T., Huq, S., & Milligan, A. (2009). Community-based adaptation to climate change: An overview. In Community-based adaptation to climate change, Participatory Learning and Action (PLA) 60 (pp. 11–33). London: IIED. Retreived from http://pubs.iied.org/ pdfs/14573IIED.pdf
57. Russell, N., Mahbub, A. Q. M., Khan, M.H., & Islam, N. (2009). Bangladesh comprehensive disaster management pro-gramme: Terminal evaluation. Retrieved from https://www. climate-eval.org/sites/default/files/evaluations/371%20Comprehensive%20Disaster%20Management%20Program.pdf
58. Rwakakamba, T. (2009). How effective are Uganda's environ-mental policies? Mountain Research and Development, 29 (2), 121–127.
59. Sacramento, A., Matavel, A., Basílio, M., & Bila, S. (2012). Climate change impacts and coping strategies in Chicualacuala district. Gaza Province: UNEP. Retrieved from http://www.unep.org/climatechange/adaptation/Portals/133/ documents/Chicualacuala_Report-Climate_Change_Impacts_ n_Coping_Strategies.pdf
60. Sietz, D., Boschutzz, M., & Klein, R. J. T. (2011). Mainstreaming climate adaptation into development assistance: Rationale, institutional barriers and opportunities in Mozambique.
61. Environmental Science & Policy, 14(4), 493–502.
62. Singh, R.K.P. (2013). Summary of baseline household survey results. May 2013, CCAFS. Retrieved from http://ccafs.cgiar. org/publications/summary-baseline-household-survey-results-karnal-harayana-state-india
63. Spires, M., Shackleton, S., & Cundill, G. (2014). Barriers to implementing planned community-based adaptation in devel-oping countries: A systematic literature review. Climate and Development, 6(3), 277–287.
64. Stringer, L.C., Mkwambisi, D.D., Dougill, A.J., & Dyer, J.C. (2010). Adaptation to climate change and desertification: Perspectives from national policy and autonomous practice in Malawi. Climate and Development, 2, 145–160.
65. Swanson, B. (2008). Global review of good agricultural extension and advisory service practices. Rome: FAO.
66. Taylor, M. (2003). Public policy in the community. Hampshire: Palgrave Macmillan.
67. Thomas, D., & Twyman, C. (2005). Equity and justice in climate change adaptation amongst natural-resource-dependent societies. Global Environmental Change, 15(2), 115–124.

68. Torres, C. (2009). Advancing adaptation through communication for development. Rome: FAO. Retrieved from http://www. fao.org/docrep/012/i1553e/i1553e00.pdf
69. Turyahabwe, N., Kakuru, W., Tweheyo, M., & Tumusiime, D.M. (2013). Contribution of wetland resources to household food security in Uganda. Agriculture & Food Security, 2(1), 5.
70. UNDP. (2013). Climate risk management for sustainable crop production in Uganda: Rakai and Kapchorwa districts. New York, NY: United Nations Development Programme (UNDP), Bureau for Crisis Prevention and Recovery (BCPR).
71. UNDP-UNEP. (2011). Mainstreaming climate change adaptation into development planning: A guide for practitioners. Retrieved from http://www.unep.org/pdf/ mainstreaming-cc-adaptation-web.pdf
72. UNFCCC. (2011). Decision CP.17. FCCC/CP/2011/9/Add.1. United Nations Framework Convention on Climate Change. Retrieved from http://unfccc.int /files/adaptation/cancun_ adaptation_framework/national_adaptation_ plans/applicat ion/pdf/decision_5_cp_17.pdf
73. Vermeulen, S.J., Campbell, B.M., & Ingram, J.S.I. (2012). Climate change and food systems. Annual Review of Environment and Resources, 37, 195–222.
74. Vincent, K., Cull, T., Chanika, D., Hamazakaza, P., Joubert, A., Macome, E., & Mutonhodza-Davies, C. (2013). Farmers' responses to climate variability and change in southern Africa– Is it coping or adaptation? Climate and Development, 5(3), 194–205.
75. Waithaka, M., Nelson, G.C., Thomas, T.S., & Kyotalimye, M. (2013). East African agriculture and climate change. Washington, DC: IFPRI.
76. World Bank. (2010). Mainstreaming adaptation to climate change in agriculture and natural resources management projects. Washington, DC: World Bank. Retrieved from http:// preventionweb.net/go/17066
77. Ziervogel, G., Bharwani, S., & Downing, T.E. (2006). Adapting to climate variability: Pumpkins, people and policy. Natural Resources Forum, 30, 294–305.

Index